JASMONATES AND PLANT DEFENSE

An Eco-Friendly and Sustainable Approach for Stress Tolerance

JASMONATES AND PLANT DEFENSE

An Eco-Friendly and Sustainable Approach for Stress Tolerance

Edited by

Uzma Azeem, PhD

Khalid Rehman Hakeem, PhD

APPLE ACADEMIC PRESS

First edition published 2025

Apple Academic Press Inc.
1265 Goldenrod Circle, NE,
Palm Bay, FL 32905 USA

760 Laurentian Drive, Unit 19,
Burlington, ON L7N 0A4, CANADA

CRC Press
2385 NW Executive Center Drive,
Suite 320, Boca Raton FL 33431

4 Park Square, Milton Park,
Abingdon, Oxon, OX14 4RN UK

Library and Archives Canada Cataloguing in Publication

CIP data on file with Canada Library and Archives

Library of Congress Cataloging-in-Publication Data

CIP data on file with US Library of Congress

ISBN: 978-1-77491-616-2 (hbk)
ISBN: 978-1-77491-617-9 (pbk)
ISBN: 978-1-00345-424-3 (ebk)

Dedication

Dedicated to COVID-19 frontline workers

About the Editors

Uzma Azeem, PhD
Department of Botany, Sanmati Government College of Science Education and Research, Ludhiana, Jagraon, Punjab, India

Uzma Azeem, PhD, has received her postgraduate, MPhil, and doctorate degrees from Punjabi University, Patiala, Punjab, India. She spent more than a decade there from 2006 to 2018. She has done her PhD in Botany with specialization in Mycology and Plant Pathology in 2018. She was awarded the Maulana Azad National Fellowship for Minority students during 2011–2016 by the University Grants Commission, New Delhi, India. She has worked at Sri Guru Granth Sahib World University, Fatehgarh Sahib, Punjab, India, as an Assistant Professor in the Department of Botany and Environmental Sciences for one semester. Then, she joined Sanmati Government College of Science Education and Research, Jagraon, Ludhiana, Punjab, India. She has been working as an Assistant Professor in Botany, Government College, Malerkotla, Punjab, India, and is actively engaged in publication work as an author, editor, and reviewer. So far, she has authored 12 paper publications in peer-reviewed national and international journals and one book titled, *Fungi for Human Health: Current Knowledge and Future Perspectives,* and book chapters with Springer Nature and Elsevier.

Khalid Rehman Hakeem, PhD
Professor, King Abdulaziz University, Jeddah, Saudi Arabia

Khalid Rehman Hakeem, PhD, is a Professor at King Abdulaziz University, Jeddah, Saudi Arabia. He is currently the Head of the Plant Biotechnology Research Unit) at Princess Dr. Najlaa Bint Saud Al Saud Center for Distinguished Research in Biotech, King Abdulaziz University, Jeddah. After completing his PhD (Botany; specialization in Plant Ecophysiology and Molecular Biology) from Jamia Hamdard, New Delhi, India, in 2011, he initially worked as Assistant Professor at the University of Kashmir, Srinagar. Later, he joined the University of Putra Malaysia, Selangor, Malaysia, and worked there as Post Doctorate Fellow in 2012 and Fellow

Researcher (Associate Prof.) from 2013–2016, respectively. He joined King Abdulaziz University in August 2016 as Associate Professor and was promoted to Professor in 2019.

Dr. Hakeem has more than 14 years of teaching and research experience in plant eco- physiology, biotechnology, molecular biology, medicinal plant research, plant-microbe-soil interactions, as well as in environmental studies. He is the recipient of several fellowships at both national and international levels. He was recently elected as a Fellow of the Royal Society of Biology, London. He has served as a visiting Scientist at Fatih University, Istanbul, Turkey, as well as at Jinan University, Guangzhou, China. Currently, he is involved with a number of international research projects with different government organizations. To date, Prof. Khalid has authored and edited more than 75 books with international publishers, including Springer Nature, Academic Press (Elsevier), CRC Press, etc. He also has more than 155 research publications in peer-reviewed international journals and 65 book chapters in edited volumes with international publishers to his credit. At present, he is serving as an editorial board member and reviewer for several high-impact international scientific journals from Elsevier, Springer Nature, Taylor and Francis, Cambridge Press, and John Wiley. He is included on the advisory board of Cambridge Scholars Publishing, UK. He is also a fellow of the Plantae group from the American Society of Plant Biologists, a member of the World Academy of Science, a member of the International Society for Development and Sustainability, Japan, and a member of the Asian Federation of Biotechnology, Korea.

Contents

Contributors

Shafat Ahmad Ahanger
Plant Pathology, FoA, Wadura, Sopore, SKUAST, Kashmir, India

Syed Kamran Ahmed
Department of Plant Protection, Aligarh Muslim University, Aligarh, Uttar Pradesh, India

Zahoor Ahmad
Department of Botany, University of Central Punjab, Bahawalpur Campus, Punjab Group of Colleges, Pakistan

Suhail Altaf
Department of Plant Pathology, SKUAST, Kashmir, India

Muhammad Ashar Ayub
Institute of Soil and Environmental Sciences, University of Agriculture, Faisalabad, Punjab Pakistan

Uzma Azeem
Government College Malerkotla, Malerkotla, Punjab, India

Arif Hussain Bhat
Department of Agriculture, BFC, Bathinda, Punjab, India

Mudasir Bhat
Department of Agriculture, SKUAST, Kashmir, India

Faiçal Brini
Laboratory of Biotechnology and Plant Improvement, Center of Biotechnology of Sfax, Sfax, Tunisia

M. S. Dar
Plant Pathology, MRCFC-Khudwani, SKUAST, Kashmir, India

Gurpaul Singh Dhingra
Department of Botany, Punjabi University, Patiala, Punjab, India

Hanan Ahmed Hashem
Department of Botany, Faculty of Science, Ain Shams University, Cairo, Egypt

Muatasim Jan
Centre of Research for Ethnobotany, Government Model Science College, Jiwaji University, Gwalior, Madhya Pradesh, India

Munisa Manzoor Khan
Department of Botany, University of Kashmir, Srinagar, Jammu and Kashmir, India

Nafeesa Farooq Khan
Department of Botany, University of Kashmir, Srinagar, Jammu and Kashmir, India

Rakesh Kumar Khare
Centre of Research for Ethnobotany, Government Model Science College, Jiwaji University, Gwalior, Madhya Pradesh, India

Sheikh Mehraj
Division of Fruit Science, SKUAST, Kashmir, India

Tawseef Ahmad Mir
Centre of Research for Ethnobotany, Government Model Science College, Jiwaji University, Gwalior, Madhya Pradesh, India

Neelofer Nabi
Department of Botany, University of Kashmir, Srinagar, Jammu and Kashmir, India

Muhammad Nadeem
Institute of Soil and Environmental Sciences, University of Agriculture, Faisalabad, Punjab, Pakistan

Subzar Ahmad Nanda
Department of Botany, University of Kashmir, Srinagar, Jammu and Kashmir, India

Maira Naveed
Institute of Soil and Environmental Sciences, University of Agriculture, Faisalabad, Punjab, Pakistan

Adil Nazir
Plant Pathology, MRCFC-Khudwani, SKUAST, Kashmir, India

Ashhad Parvez
Department of Botany, Aligarh Muslim University, Aligarh, Uttar Pradesh, India

Aaniqa Qayoom
Deptartment of Botany, University of Kashmir, Srinagar, Jammu and Kashmir, India

Irfan Rashid
Deptartment of Botany, University of Kashmir, Srinagar, Jammu and Kashmir, India

Ishfaq Ul Rehman
Deptartment of Botany, University of Kashmir, Srinagar, Jammu and Kashmir, India

Sumaya Rehman
Department of Botany, University of KashmirSrinagar, Jammu and Kashmir, India

Summia Rehman
Deptartment of Botany, University of Kashmir, Srinagar, Jammu and Kashmir, India

Muhammad Zia ur Rehman
Institute of Soil and Environmental Sciences, University of Agriculture, Faisalabad, Punjab, Pakistan

P. Q. Rizvi
Department of Plant Protection, Aligarh Muslim University, Aligarh, Uttar Pradesh, India

Peer Saffeullah
Department of Botany, Jamia Hamdard, New Delhi, India

Walid Saibi
Laboratory of Biotechnology and Plant Improvement, Center of Biotechnology of Sfax, Tunisia

Zahoor Ahmad Shah
KVK-Ganderbal, SKUAST, Kashmir, India

Anwar Shahzad
Department of Botany, Aligarh Muslim University, Aligarh, Uttar Pradesh, India

Ishfaq Ahmad Sheergojri
Deptartment of Botany, University of Kashmir, Srinagar, Jammu and Kashmir, India

Richa Shri
Department of Phramaceutical Sciences and Drug Research, Punjabi University, Patiala, Punjab, India

Seema Singh
Department of Botany, University of Kashmir, Srinagar, Jammu and Kashmir, India

Syed Sami Ullah
ICAR-CITH, Srinagar, Jammu and Kashmir, India

Shahid Umar
Department of Botany, Jamia Hamdard, New Delhi, India

Wajid Umar
Doctoral School of Environmental Science, Hungarian University of Agricultural and Life Sciences, Gödöllő, Hungary

Aadil Farooq War
Department of Botany, University of Kashmir, Srinagar, Jammu and Kashmir, India

Naqshe Zuhra
Institute of Soil and Environmental Sciences, University of Agriculture, Faisalabad, Punjab, Pakistan

Abbreviations

12, 13-EOT	12, 13(S)-epoxy-octadecatrieonic acid
12-OH-JA	12-hydroxyjasmonic acid
13-HOPT	13(S)-hydroperoxy-octadecatrienoic acid
13-LOX	13-lipoxygenase
24-EBL	24-epibrassinolide
2OG	2-oxoglutarate
ABA	abscisic acid
ACX	acyl-CoA oxidase
AH	amido hydrolase
AMI/AMI1	altered mersitem inflorescence/altered inflorescence meristem 1
AOC	allene oxide cyclase
AOS	allene oxide synthase
APX	ascorbate peroxidase
AsA	ascorbic acid
AsA-GSH	ascorbate-glutathione
AtOPR3	arabidopsis thaliana oxophytodienoate- reductase 3
Aux	auxin
Avr-gene	anti-virulence gene
bHLH	basic helix-loop-helix
BMV	Brome mosaic virus
BRs	brassinosteroids
CAT	catalase
CBF	C-repeat binding factor
CDPKs	calcium-dependent protein kinases
ChIP	chromatin immunoprecipitation
Chl	chlorophyll
CJ	cis-jasmone
COI1/ coi1	coronatine insensitive 1
CYP	cytochrome
CYP74A	cytochrome P450 enzyme
CYP94B3	cytochrome P450 94B3
CYP94C1	cytochrome P450 94C1

Dad1	delayed in anther dehiscence1
DGL	dongle
dn-OPDA	dinor-oxophytodienoic acid
DREB2A	dehydration-responsive element-binding protein 2A
EIN4	ethylene insensitive 4
ERF	ethylene response factor
ERS1	ethylene response sensor 1
ET	ethylene
ETI	effector-triggered immunity
FACs	fatty acid–amino acid conjugates
FAD	fatty acid desaturase
FR	far-red
GA/GAs	gibberellin/gibberellins
GF	griseofulvin
GLSs	glucosinolates
GLV	green leaf volatiles
GPX	guaiacol peroxidase
GSH	reduced glutathione
GUS	glucuronidase
GUS	promoter::glucuronidase
HAE	herbivore-associated elicitor
HDA6	histone deacetylase 6
HMs	heavy metals
HOPT	13(S)-hydroperoxy-octadecatrienoic acid
HSF	heat shock factor
HSPs	heat shock proteins
HT	high temperature
IAA	indole-3-acetic acid
IAH	ILR1-like amidohydrolase
ICE	inducer of cbf expression
IP5	pentakisphosphate
ISR	induced systemic resistance
JA	jasmonic acid
JA-ACC	jasmonoyl ACC
JA-Ile	jasmonoyl-isoleucine
JAO2	jasmonic acid oxidase 2
JAR1	jasmonate resistant 1
Jar1	jasmonic acid resistant 1

JAs	jasmonates
JAT1	jasmonates-associated transporter 1
JAZ	jasmonate ZIM-domain
jazD	jaz decuple
JIPs	jasmonate-induced proteins
JMT	jasmonic acid carboxyl methyltransferase
JOX	jasmonate-induced oxygenases
JRE4	jasmonate-responsive ethylene 4
KAT	3-ketoacyl-CoA thiolase
LOX/LOXs	lipoxygenase/lipoxygenases
LRRs	leucine-rich repeats
MAMPs	microbe-associated molecular patterns
MAPK	mitogen-activated protein kinase
MDA	malondialdehyde
MED25	mediator25
MeJA/MeJAs	methyl jasmonate/methyl jasmonates
MFP	multifunctional protein
MpJAZ	*M. polymorpha* JAZ gene
MPK4	mitogen-activated protein kinase 4
MT	microtubules
MTI	MAMP-triggered immunity
MYB	myeloblastosis
Myc	regulator genes and proto-oncogenes
NINJA	novel interactor of JAZ
NOS	nitric oxide synthase
NPR1	nonexpressor of PR genes 1
NR	nitrate reductase
NT domain	N-terminal domain
OCPs	organochlorine pesticides
OCs	organochlorines
O-Glc-JA	O-glucoside jasmonic acid
OPC	3-oxo-2-(2′(Z)-pentenyl)-cyclopentane-1-octanoic
OPCL	carboxyl-CoA ligase
OPCL1	OPC-8:0 CoA ligase1
OPDA/12-OPDA	12-oxo-phytodienoic acid
OPR	OPDA reductase
OPR3	OPDA reductase 3
PAHs	polyaromatic hydrocarbons

PCBs	polychlorinated biphenyls
PCD	programmed cell death
PCDDs	polychlorinated dibenzo-p-dioxins
PCDFs	polychlorinated dibenzofurans
PI	proteinase inhibitors
PIFs	phytochrome-interacting factors
PLA	phospholipase A
PLT	stem cell transcription factors plethora
PLTs	plethoras
POD	peroxidase
POPs	persistent organic pollutants
PRRs	pattern recognition receptors
PTSGMS	photothermosensitive genic male sterile
PUFA	polyunsaturated fatty acids
RNS	reactive nitrogen species
ROS	reactive oxygen species
Rubisco	ribulose 1,5-bisphosphate carboxylase/oxygenase
RuBP	ribulose-1,5-bisphosphate
SA	salicylic acid
SAR	systemic acquired resistance
SCF	Skp-Cullin-F-box
SGA/SGAs	steroidal glycoalkaloid/steroidal glycoalkaloids
SL	strigolactone
SOD	superoxide dismutase
SPLs	squamosa promoter-binding protein-like
SPP2	sucrose-phosphate phosphatase 2
Spr1	suppressor of prosystemin-response1
spr2	suppressor of pro-systemin-mediated responses2
SPS1-4	sucrose phosphate synthase 1–4
TA	tuberonic acid
TAG	tuberonic acid glucoside
TBZ	tiabendazole
TFs	transcription factors
TIAs	terpenoid indole alkaloids
TIR1	transport inhibitor response 1
Tom Lox-D	tomato lipoxygenase D
TPL	topless
TPRs	TPL-related proteins

TPS	terpene synthesis
TS1	tassel seed-1
UDP	uridine diphosphate
UPS	ubiquitin–proteasome system
UV	ultraviolet
Vir	virulence
VOC/VOCs	volatile organic compounds
VSP	vegetative storage proteins
ZIM	Zn-finger protein expressed in inflorescence meristem
α-DOXs	α-dioxygenases
α-LeA	A-linolenic acid

Preface

To date, a rapid change in environmental conditions is likely to override the adaptive efficiency of plants. These changes in the environment are either natural or anthropogenic, leading to soil and air pollution as well as other climatic changes. Plants are sessile and have limited mechanisms for avoiding various unpredictable and unfavorable environmental stresses, such as extremes of temperature, drought, flooding, deficiency and excess of nutrients, heavy metal toxicity, exposure to UV-rays, environmental pollution, and pathogen attack. They need flexible means for the acclimatization to the rapidly changing environment. To improve a plant's defense system, it is essential to be familiar with the mechanisms contributing toward stress tolerance.

Several plant metabolites including phytohormones, such as abscisic acid, salicylic acid, ethylene, auxins, brassinosteroids including jasmonates, play a role in plant defense against environmental stress. Jasmonates (jasmonic acid and its derivatives) represent a class of plant hormones with an array of physiological, biochemical, and molecular processes in a concentration-dependent manner (nanomolar to micromolar concentrations). These are ubiquitously distributed in the kingdom plantae and function as growth regulators or stress factors. Jasmonates have strong potential to ameliorate various abiotic and biotic stresses, such as exposure to mechanical stress, osmotic stress, herbivory, and exposure to UV-irradiation, ozone, and heavy metal stress.

Jasmonates help plants in mitigating the stress by various mechanisms such as by modulating the antioxidant defense system and up- and down-regulation of specific gene expression. The use of jasmonates in nano- to micromolar concentrations can be one of the solutions to mitigate the toxic effects of various biotic and abiotic stresses and to improve crop yield in an eco-friendly and sustainable manner. However, this needs more exploitation of jasmonates to understand their role and mechanism in plants to mitigate various kinds of stresses.

This book, *Jasmonates and Plant Defense: An Eco-Friendly and Sustainable Approach for Stress Tolerance,* puts the spotlight on the presence and roles of jasmonates in plant defense under biotic and abiotic stresses, genetic basis of jasmonates action, effects of jasmonates on the antioxidant defense system, and cross talk between jasmonates and other plant hormones.

—**Editors**

Acknowledgments

The editors convey thanks to all the authors for their precious ideas, time, and dedication to bringing this volume of book to its final form.

CHAPTER 1

Discovery and Biosynthesis of Jasmonates in Plants

TAWSEEF AHMAD MIR, MUATASIM JAN, and RAKESH KUMAR KHARE

Centre of Research for Ethnobotany, Government Model Science College, Jiwaji University, Gwalior, Madhya Pradesh, India

ABSTRACT

Jasmonates are oxylipins, which are basically the oxidation products of some unsaturated fatty acids. These substances are a complex group of compounds that play a role in a variety of biochemical changes and stressed physiological responses. Jasmonates are one of the well-studied types of plant oxylipins. Fatty acid derivatives known as jasmonates (JAs) include, among others, jasmonic acid (JA), methyl jasmonate (MeJA), and jasmonate isoleucine conjugate (JA-Ile). JA biosynthesis and signaling pathways in dicotyledonous plants, such as *Arabidopsis*, tobacco, and tomato, have received substantial research. There aren't many JA biosynthesis enzymes known to exist in monocotyledonous species. Trienoic fatty acids with a carbon atom 13 are peroxided by LOX, resulting in 13-hydroperoxide, which is subsequently changed into an allene oxide fatty acid and cyclized to create 12-oxo-phytodienoic acid (OPDA). More attention is being paid toward the biochemical functions of JA. Jasmonates have increasingly gained recognition as a protective and fertility hormone, modulating a variety of developmental and stress-related processes. In *Arabidopsis* and tomato, JAs have a direct impact on the development of stamen and trichomes, vegetative growth, cell

Jasmonates and Plant Defense: An Eco-Friendly and Sustainable Approach for Stress Tolerance.
Uzma Azeem, & Khalid Rehman Hakeem (Eds.)

cycle regulation, senescence, anthocyanin biosynthesis regulation, and responses to biotic and abiotic stressors. The JA signal collaborates with other hormones in plants. Numerous studies have also highlighted the connection between plant hormones and defense mechanisms.

1.1 INTRODUCTION

The oxidation products derived from the fatty acids are termed as oxylipins. These oxylipins are the compounds responsible for the production of a number of substances, which are in turn responsible for various developmental and stress response processes in the plant body.[1] By the process of initial oxidation by lipoxygenases (LOXs) or by α-dioxygenases (α-DOXs), oxylipins can be formed. However, they can also be formed by the nonenzymatic autoxidation process of unsaturated fatty acids.[2] These compounds provide a protective shield either as signaling molecules during developmental processes, pathogen attacks, and wounding or directly by antimicrobial substances that are harmful or lethal to attacking pathogens. Besides the progress in unveiling the major functions, most of the oxylipins are still unknown to the scientific community. Jasmonates (JAs), which include jasmonic acid (JA) and its other derivatives, including cis-jasmone, methyl jasmonate (MeJA), jasmonoyl ACC (JA-ACC), and a variety of other derivatives, are currently the plant oxylipins that have received the greatest attention in research. The most significant class of plant oxylipins are the green leaf volatiles (GLV). GLV releases local and systemic volatile organic compounds (VOC) in response to feeding on insects.[3] A number of volatile organic chemicals, including sesquiterpenes, monoterpenes, and carotenoid-like compounds, are biosynthesized by terpenic, lipidic, and shikimic pathways.[4] Most of the VOCs are the products of similar LOX pathways but not of the LOX pathway.[5] Phytoprostanes are a further class of plant oxylipins that have received the most attention. This oxylipin is produced through nonenzymatic action and is crucial for the OPDA-mediated plant stress response.[6] The biosynthesis and signaling pathways of JA have been extensively studied in plants like *Arabidopsis*, *Nicotiana tobaccum*, and *Solanum lycopersicum*. But in monocots, a small number of JA biosynthetic enzymes have been defined.[7,8] Trienoic fatty acids are peroxidized at carbon atom 13 to produce 13-hydroperoxide, which is then converted to make the JAs. This 13-hydroperoxide is converted to allene oxide fatty acid and then cyclized to another compound called 12-oxo-phytodienoic acid (OPDA). JA is formed from OPDA by the reduction in the double bond and side-by-side successive rounds of β-oxidation. It accepts C-18-PUFA and also C16-PUFA (hexadecatrienoic acid). In C16-PUFA, the intermediate formed is called dinor-OPDA, which may also be converted to JA after certain processes. Furthermore, JA can also be enzymatically modified into a number of derivatives and conjugates having well-defined bioactivities, such as JA, MeJA, cis-jasmone, and JA-Ile. Upon the formation of the SCFCOI1-JA-Ile-JAZ (jasmonate ZIM-domain) ternary complex, JA starts the signaling cascade,[9] where the repressors of JAZ are ubiquitinated and are then destroyed to release various transcription factors, such as MYC2, which causes the downstream transcription activation of defense or development responses.[10,11] The one and only jasmonate receptor that has been identified to date is the COI protein.[12,13] But the only ligand of the ubiquitin ligase complex SCFCOI1 E3 found was JA-Ile.[11] After being discovered in the oils of jasmine plant flowers in the 1960s as secondary metabolites, the biological function of JA has been the point of attraction for researchers. Besides being a good agent in terms of development and stress response in the plant body, JAs have also been described to be important as a fertility and defense hormone. In some plants, such as *Arabidopsis* and *Solanum lycopersicum*, JAs have been found to be involved in the development of trichomes and stamens, vegetative development, regulation of the cell cycle, biosynthesis of anthocyanin, and various biotic and abiotic stress responses.[14–16] There is not so much information regarding the JAs function in monocots, but it has been defined that they play a meager role in processes like sex determination, reproductive initiation, and many more.[7,8,17] The JA signal in plants acts in cooperation with other important hormones. Many of the studies are attracted toward this process and function of JAs. JA in *Arabidopsis* was shown to act cooperatively with ethylene,[18] and based on the stress condition, both cooperative and antagonistic actions were shown with salicylic acid (SA)[19] and abscisic acid.[20] Likewise, gibberellins also interact with JA and control the fertility of flowers in *Arabidopsis*. JA is also responsible for the synthesis of abscisic acid and ethylene in the leaf senescence of maize in a positive way.[8] Hence, JA is found to exert its role in various developmental processes in the plant body by interacting with some other beneficial hormones already present in the plant tissues. However, it still needs proper exploration.

1.2 DISCOVERY OF JASMONATES

The lipid-derived cyclopentanones, including JA and its other derivatives, are JAs. JAs are one of the more recent plant hormones, as opposed to the five classical ones (auxin, gibberellin, ethylene, cytokinin, and abscisic acid).[21] For the first time, a methyl ester of JA (MeJA) was isolated as a fragrance from the ethereal oil of *Jasminum grandiflorum* flowers.[22] Likewise, from the culture filtrates of the *Botryodiplodia theobromae* fungus, a free acid was isolated.[23] Later on, JA was also isolated from higher plant species such as *Cucurbita pepo*[24] and *Vicia faba*.[25] In 1980, some of the compounds structurally related to JA were synthesized, and then the biological activities were checked on the developmental process of rice seedlings.[26]

1.3 BIOSYNTHESIS OF JA

1.3.1 THE SCHEME OF JA BIOSYNTHESIS

Through the octadecanoid pathway, JA is derived from α-linolenic acid in the ratio of 18:3. Following the first proposal of the chemistry of the biosynthetic pathway of JA, studies conducted later in the field confirmed the chemistry and also added essential information about the enzymology, regulation, and cellular location of this pathway.[15,16,27,28] This pathway is started by LOX-catalyzed oxygenation of α-linolenic acid in the chloroplast (Figure 1.1). In order to produce the unstable chemical known as allene oxide, 12,13(S)-epoxy-octadecatrienoic acid (12,13-EOT), allene oxide synthase (AOS) is thought to use 13(S)-hydroperoxy-octadecatrienoic acid (13-HOPT) as a substrate. The 12,13-EOT is cyclized to the 9S, 13S 12-OPDA isomer, which is then transferred to the peroxisome. When it reaches the peroxisome, its cyclopentanone ring is reduced to 3-oxo-2-(2′(Z)-pentenyl)-cyclopentane-1-octanoic acid (OPC-8:0) with the help of OPDA reductase 3 (OPR3). From the carboxyl side chain for the completion of JA biosynthesis, the six carbons are removed by three cycles of β-oxidation.[29] The hexadecanoid pathway, which runs parallel to it, utilizes similar enzymes in the plastids, uses the same enzymes in the chloroplast, and forms dinor-oxophytodienoic acid (dn-OPDA) by acting upon hexadecatrienoic acid (16:3). If dn-OPDA is an OPR3 substrate, the peroxisome's oxidation enzymes may further convert this C16 molecule to JA.

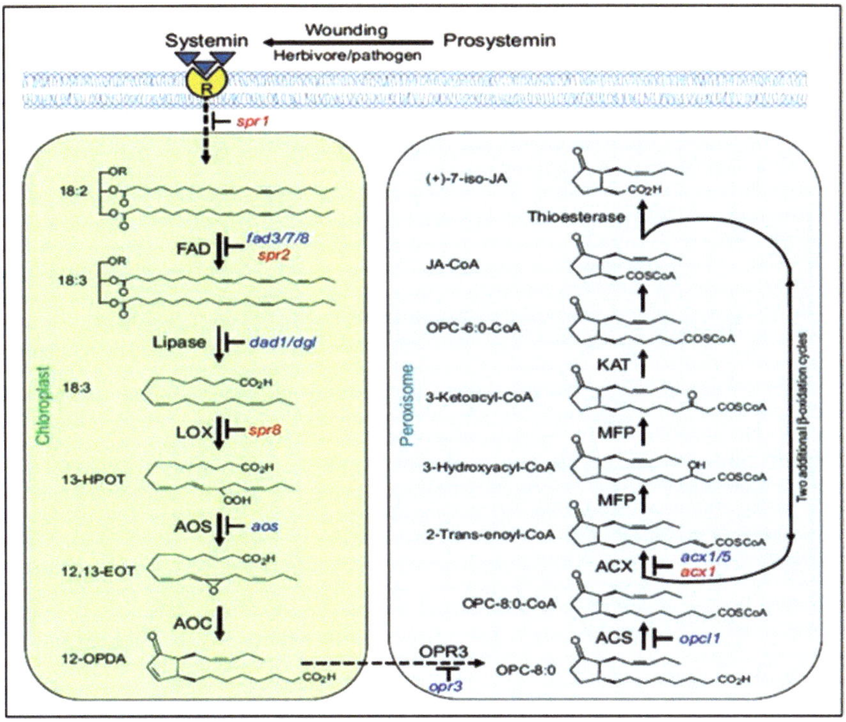

FIGURE 1.1 Jasmonic acid (JA) generation via the octadecanoid route in tomato and *Arabidopsis* [Modified from Zhai et al.[30]].

The fatty acid desaturase (FAD)-catalyzed desaturation process converts linoleic acid into the JA precursor, linolenic acid (18:3 and 18:2). Dienoic fatty acids (16:2 and 18:2) are transformed into trienoic fatty acids by three fatty acid desaturases (FAD3, FAD7, and FAD8) in the *Arabidopsis* genome (16:3 and 18:3). Since the fad3/7/8 triple mutant does not create any detectable 16:3 or 18:3, JA biosynthesis is not properly carried out by it.[31] Substantially, this triple mutant has a male sterile phenotype that can be easily recovered using exogenous 18:3 and JA.[31,32] The investigation of the triple mutant fad3/7/8 provided the first undeniable evidence that JA is required for *Arabidopsis* male reproductive development as a result. Additional knowledge on the function of FADs in JA biosynthesis has been gained from the analysis of the *Solanum lycopersicum* suppressor of prosystemin-response2 (spr2) mutant.[33] The spr2 mutation adversely affects wound-induced JA biosynthesis as well as the expression of protective genes. According to map-based cloning studies, the 3 chloroplast 3 fatty acid desaturase expressed by Spr2 is a homolog of the *Arabidopsis* FAD7 gene product. Surprisingly, spr2 plants developed normally and reproduced even though their levels of trienoic fatty acids were much lower. To learn more about how JAs function in different plant species, it is important to further describe the triple mutant of *Arabidopsis* fad3/7/8 and the tomato spr2 mutant.

1.3.2 RELEASE OF LINOLENIC ACID FROM GALACTOLIPIDS

It is believed that the process of releasing the 18:3 precursor from membrane lipids is essential for the beginning of JA production. This stage advanced significantly with the characterization of the mutant *Arabidopsis* delayed in anther dehiscence1 (dad1), which possesses a male sterile phenotype. The DAD1 gene codes for a chloroplast-specific phospholipase A1 (PLA1) enzyme that releases 18:3 from the sn1 location of chloroplast membrane glycerolipids.[34] There is a deficiency of dad1 mutant in anther dehiscence and pollen production, but JA accumulation after wounding is largely natural, meaning that DAD1-dependent JA biosynthesis is primarily involved in reproductive development.[34] DONGLE (DGL), a homolog of DAD1, has been shown to be present in abundance in the leaves in comparison to siliques, flowers, or roots, in contrast to DAD1.[35] Wounding induced DGL expression transiently, and a DGL-overexpressing mutant had increased basal JA aggregation and an increase in the expression of JA responsive genes, implying that DGL-dependent JA biosynthesis is primarily involved in the early wound reaction.[35] These findings support the idea that JA biosynthesis is differentially regulated in wound growth and reaction.

1.3.3 OXYGENATION OF A-LINOLENIC ACID BY 13-LOX

After the release of free 18:3 from galactolipids, it is converted to 13-HPOT, where the reaction is catalyzed by 13-LOX enzymes. All four 13-LOXs (LOX2, LOX3, LOX4, and LOX6) encoded in the *Arabidopsis* genome are capable of oxygenating 18:3 in vitro.[36] Wound-induced JA output is primarily regulated by LOX2.[37–39] The genes LOX3 and LOX4 play a key role in the growth of fertile flowers.[40] LOX6 is associated with wound-induced and drought-induced JA production, according to various genetic and physiological studies.[41,42] TomLoxD, a 13-LOX that facilitates the hydroperoxidation of linolenic acid in JA biosynthesis, is encoded by the Spr8 gene, according to the characterization of the spr8 mutation in tomato, which is itself a repressor of prosystemin-mediated reactions. Surprisingly, the Spr8 plant's basic growth and flower production are generally natural, implying that TomLoxD is engaged in the damage response directly.[43] The maize TASSELSEED1 (TS1) protein, which is involved in determining sex, and the TomLoxD proteins do have a great deal of similarities.[17,43] These findings back up the idea that various LOX isoforms have different physiological purposes. There is no "fitness penalty" associated with higher TomLoxD expression because it leads to enhanced production of genes that respond to wounds and increases tolerance to pests and biotrophic bacteria, but it has very minimal effects on overall plant development. This is important because ongoing defenses hinder plant development.[44] The TomLoxD gene may be utilized to defend crops because it does not have a significant "fitness expense" when it is overexpressed.[43]

1.3.4 DEHYDRATION OF 13-HPOT BY AOS

The conversion of 13-HPOT to the unstable 12,13-EOT is catalyzed by the plastid-localized 13-AOS, an unusual cytochrome P450 enzyme (CYP74A) that generally employs oxygenated fatty acid hydroperoxide substrates as an oxygen source.[45] There is only one AOS gene in the *Arabidopsis* genome, while there are two in the tomato genome.[46,47] Wounding induces the AOS genes in both *Arabidopsis* and tomato, implying that they would be engaged in the wound-induced JA biosynthetic pathway.[46,48] The analysis of the *Arabidopsis* aos mutant, which obstructs the function of the AOS gene, provided direct proof that AOS is essential for the biosynthesis of JA.[48] The mutated AOS is deficient in both basal and wound-induced JA deposition, as well as wound-induced defense gene expression. Furthermore, the aos mutants do have an extreme male sterile phenotype that does not regenerate during growth but can be saved by external MeJA spray.[48] Such findings show that AOS-dependent biosynthesis of JA is essential for *Arabidopsis* wound response and reproductive growth.

1.3.5 SYNTHESIS OF OPDA BY AOC

AOS produces allylic epoxides, which are water-insoluble. When allylic epoxides are hydrolyzed, α- and γ-ketols are formed. In order to create OPDA, which is often a racemic combination of the cis (+) and cis (−) enantiomers, allene oxide cyclase (AOC) participates in the cyclization of allylic epoxides.[49] The formation of allene oxide undergoes stereospecific cyclization to produce the cis (+) enantiomer OPDA exclusively catalyzed by AOC, which indicates that AOC has a greater contribution in the establishment of the JA precursor enantiomeric structure.[29] In *Arabidopsis*, AOC is expressed by four genes, while in tomato, it is expressed by a single gene.[50,51] The four *Arabidopsis* AOC genes were demonstrated utilizing the matching promoter::glucuronidase (GUS) fusion lines to demonstrate their robust and redundant expression in aerial tissues like flowers and leaves.[52] These findings, together with the assumption that neither of the single or double mutants of AOC1, ACO3, or AOC4 showed a JA-related phenotype, suggested that the four AOC genes function repetitively in the biosynthesis of JA. Surprisingly, this research discovered that AOC3 and AOC4 were both abundantly expressed in the root's meristematic region, suggesting that JA synthesis could occur effectively in plant root tissue.[52]

1.3.6 COMPLETION OF JA SYNTHESIS IN PEROXISOMES

OPDA is the end product of the plastid-localized portion of the JA biosynthetic pathway. The reduction of the cyclopentenone ring of OPDA and dnOPDA is catalyzed by peroxisomal OPDA reductase (OPR) to form OPC-8:0 and OPC-6:0, respectively.[29] Only *Arabidopsis* OPR3 among the six OPRs recognizes the (9S, 13S)-enantiomer of OPDA as a substrate, implying that OPR3 is associated with the production of JA. Evidently, the *Arabidopsis* opr3 mutant, which interrupts the expression of OPR3, exhibited compromised defensive responses and male sterility as a result of defective JA development, suggesting that OPR3 is needed for the biosynthesis of JA.[53,54] OPCL1, a carboxyl-CoA ligase, is responsible for activating OPC-8:0.[55] In addition to wounding, the *Arabidopsis* opcl1 null mutant accumulates more OPC-8:0 and less JA, showing that OPCL1 (OPC-8:0 CoA ligase 1) is still active in the biosynthesis of JA. Nevertheless, OPCL1 depletion only results in a 50% decrease in wound-induced JA aggregation, implying that other CoA-ligase isozymes are included in the stimulation of OPC-8:0 during the biosynthesis of JA.[55,56] The octanoic acid side chains of OPC-8:0 are shortened three times by oxidation, resulting in (+)-7-iso-JA. These latter processes in the synthesis of JA are catalyzed by the major enzymes of the oxidation cycle: acyl-CoA oxidase (ACX), multifunctional protein (MFP), and 3-ketoacyl-CoA thiolase (KAT). The mutant evaluation in *Arabidopsis* and tomato provided scientific evidence for the function of these enzymes in the biosynthesis of JA.[57,58] According to biochemical evidence, the first step in the β-oxidation of OPC-8:0-CoA is believed to be catalyzed by *Arabidopsis* ACX1A.[57] This finding is favored by the fact that the acx1 tomato mutant is defective in wound-induced biosynthesis of JA and has a reduced defensive reaction to wounds and insect infestation.[57] Likewise, degradation of altered inflorescence meristem 1 (AIM1), one of *Arabidopsis*'s two peroxisomal MFP proteins, impairs wound-induced JA deposition and protection gene transcription.[59]

1.3.7 THE BYPASS IN JA FORMATION: THE COI1-INDEPENDENT (CORONATINE INSENSITIVE 1) AND OPR3-INDEPENDENT ROUTE

As previously mentioned, the first discovered mechanism for the biosynthesis of JA entails the oxygenation of fatty acids, cyclization and reduction of the cyclopentenone ring, and shrinkage of the carboxylic acid side chain. The OPR3-catalyzed step that leads to cyclopentanone compounds is critical. The first mutant allele discovered, opr3-1, was indeed a variable mutant capable of producing OPR3 mRNA and JA in some situations.[60] Now, opr3-3, a new mutant allele with no OPR3 involvement, has been identified.[61] Full deficiency JA was therefore predicted. On the other hand, thorough genetic, biochemical, and analytical analyses showed a pathway of JA formation via dinor-OPDA and 4,5-ddh-JA.[61] After being injured, OPR3-3 plants absorb 4,5-ddh-JA and JA-Ile but not 4,5-ddh-JA-Ile, indicating that OPDA's carboxylic acid side chain is being oxidized to dinor-OPDA (Figure 1.2). In wild-type and OPR3-3 plants, 4,5-ddh-JA was found to have JA-like activity in normal JA responses, including the suppression of root growth, but not in coi1-30 and jar1-1 mutant plants, demonstrating the necessity of an effective COI1 and JAR1 protein. Initially, it was thought that OPR2 and OPR1 were involved in the degradation of several compounds with unsaturated carbonyls (conjugated enones).[29] Now, the recent biological behavior of OPR2 shows that various evolutionary mechanisms may exist based on the presence of different OPRs. As such, OPR3 orthologues are missing in lower plants like bryophytes, which is because of the JA absence in these plant species; however, these plant species do contain an OPR2-like genome.[62] This explains why the OPR3-dependent path is more common in vascular plants.[62]

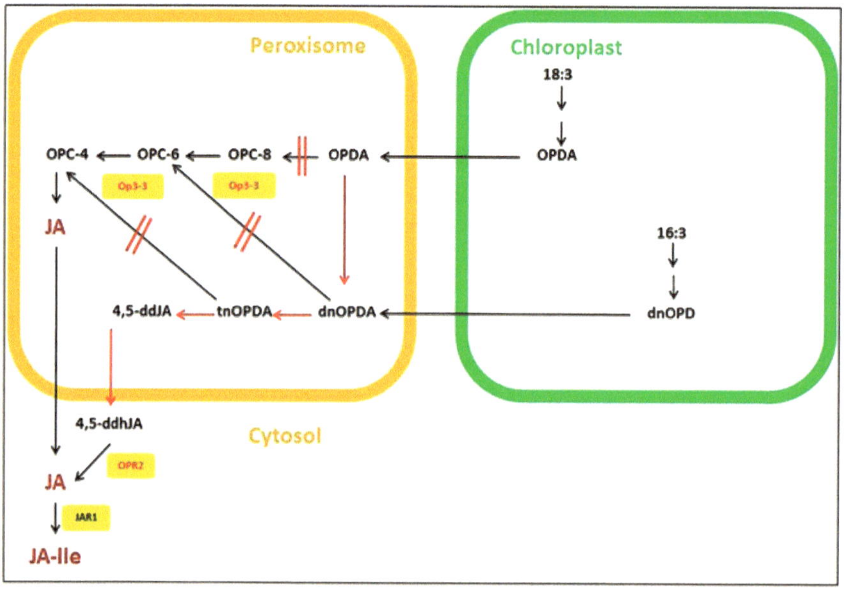

FIGURE 1.2 The byproduct of JA biosynthesis. OPDA and dnOPDA are formed in chloroplasts and transformed to JA and 4,5-ddh-JA in peroxisomes, respectively.

1.4 REGULATION OF JA BIOSYNTHESIS

1.4.1 *JASMONATE BURST*

It is very well understood that JA ratios differ considerably between organs, stages of development, and environmental factors.[14,15] Wounding and insect infestation both cause an increase in the biosynthesis of JA.[63,64] When there is any mechanical wound in *Arabidopsis*, it leads to a very fast (<5 min) rise in JA-Ile aggregation in systemic unharmed leaves, according to research from various laboratories.[65,66] These results provide evidence in favor of the hypothesis that a rapidly moving damage stimulus induces the production of JA-Ile in unharmed leaves, which then triggers the transcription of wound-responsive genes via the JA signaling pathway.

1.4.2 AMPLIFICATION OF JA PRODUCTION BY SYSTEMIN

It has been demonstrated that the 18-amino-acid peptide signals, systemin and JA, are both essential functional stimuli for tomato wound-induced systemic host defense.[67,68] It has been suggested that systemin and JA control systemic defense reactions via the same signaling pathway.[67–69] Grafting experimentation with mutants of tomato deficient in the biosynthesis of JA or signaling revealed that JA, not systemin, is the long-distance mobile signal for systemic resistance mechanisms.[70] Such a theory questioned an emerging paradigm while also raising an intriguing question about how systemin communicates with that of the JA cascade to cause structural protective reactions. The analysis of the tomato Spr1 mutation, which is immune to systemin, reveals that systemin primarily functions as a local intermediary for the aggregation of JA.[71] Spr1 differs from other tomato wound response mutants in that it mostly has deficiencies in the systemic defense response while exhibiting a largely normal local defense response. Moreover, in addition to systemin therapy, Spr1 is deficient in JA aggregation.[71] Spr1 is deficient in the processing, but not really the perception, of a long-distance damage signal, according to reciprocal grafting studies between Spr1 and wild-type plants, implying that Spr1 is involved in the incorporation of systemin perceived notions with the activation of the JA cascade.[71] Such findings indicate that systemin works at the injured parts to increase JA accumulation to a critical threshold, which then acts as a structural signal to trigger protective reactions in distal unharmed sections.[72]

1.4.3 POSITIVE FEEDBACK REGULATION OF BIOSYNTHETIC GENES OF JA

The swift induction of most JA biosynthetic genes by JA therapy, mechanical wounding, or other factors is an essential feature of the JA biosynthetic pathway, meaning that JA synthesis is actively controlled by JA signaling. MYC2, a basic-helix-loop-helix (bHLH) transcription factor that binds to the JA-responsive G-box (CACGTG) or G-box-like motif in *Arabidopsis* regulates a full variety of JA responses and functions as a master TF of JA signaling.[73] MYC2 has been shown to bind directly to the promoter of LOX2 and control its expression. SlMYC2, the *Arabidopsis* homolog of MYC2, also regulates the wound-induced transcription of the TomLoxDin tomato.[43] Several excellent reviews have identified additional regulation of biosynthesis of the JA, including substrate abundance and tissue specificity.[16]

1.5 THE PHYSIOLOGICAL ROLES OF JA IN PLANT DEVELOPMENT AND DEFENSE

1.5.1 JAs SERVE AN IMPORTANT ROLE AGAINST ABIOTIC STRESSES

One of the well-studied abiotic stresses in plants is salinity. Salinity stress effects in soybean saplings can be efficiently alleviated by the foliar application of MeJA.[74] Exogenous jasmonate can help a salt-sensitive grapevine cell line expand, and JA-signaling components like JAZ proteins control the salt stress response.[75] Plant tolerance to ozone (O_3) stress relies heavily on JA. The fad3/7/8 JA biosynthesis mutant and the jar1 JA-signaling mutant are more responsive to O_3. MeJA pretreatment also reduces O_3-induced H_2O_2 concentration and SA levels, as well as totally eliminating O_3-induced cell death.[76]

1.5.2 JAs ARE REQUIRED SIGNALS FOR POLLEN DEVELOPMENT IN DICOTS AND FOR SEX DETERMINATION IN MONOCOTS

In addition to its protective function, jasmonate has been demonstrated to have a significant role in plant reproduction. *Arabidopsis* JA biosynthesis pathway variations such as fad3/7/8, aos, opr3, and JA signaling mutant coi1 are male-sterile, demonstrating that JA is a crucial signal for the development of the male organ in the flowers of bisexual plants.[16] The three differences that characterize this JA-dependent male-sterile phenotype are: (a) at flowering time, mutant anthers lost their ability to dehisce to shed pollen; (b) in the anthers, the pollen grains are often inviable. The filaments of the stamens are noticeably thinner in mutants, meaning that the anthers do not stretch to the stigma level. Exogenous JA therapy will rescue the fertility defect phenotype of JA biosynthetic mutants; however, JA procedures cannot rescue the signaling mutant coi1.[16] The formation of stamen has also been linked to certain JA signaling components. Myeloblastosis-related transcription factors MYB21 and MYB24 were isolated from OPR3 stamens and used as JA-dependent transcription factors for flower growth.[77] Myb21 mutants showed shorter anther filaments, delayed anther dehiscence, and markedly reduced male fertility. While a myb24 mutant was morphologically a wild type, a myb21myb24 double mutant showed that the myb21 mutation aggravated all three facets of the myb24 morphology. Exogenous jasmonate had no effect on the ability of the myb21 and myb21myb24 mutant plants to survive. These results demonstrate that MYB21 and MYB24 are JA signaling elements that mediate the JA response during stamen development.[77] Some JA signaling mutants, on the other hand, are male-fertile rather than sterile. Myc2 mutants, for example, do not have a male-sterile morphology[78]; the jar1 mutant is a male-fertile mutant.[79] It's probable that an alternate part or ligand for JA signaling occurs in these mutants. The tomato (*Solanum lycopersicum*) JA vision mutant jar1 is male-fertile but female-sterile, unlike *Arabidopsis*, indicating JA's function in plant reproductive differentiation is thought to be species-dependent.[80] Rice is a monocotyledonous plant with bisexual leaves, similar to *Arabidopsis*. Male sterility was observed in a JA-deficient rice hebiba mutant, indicating that JA is needed for male organ development in bisexual flower species.[81] Maize is another monocot that belongs to monoecious plants, which have separate male and female inflorescences on the same plant (called tassel and ear). Monosexual florets develop from bisexual floret primordia of the top or axillary meristems in the tassel or ears through a sex determination pathway mediated by a range of sex-determining genes.[82] According to a recent publication on ts1 (tasselseed1), a mutant in which male inflorescence (tassel) turns a female-fertile structure that can be pollinated to yield seeds,[17] JAs are an essential phytohormone that initiates the sex determination program of tassels. In our most recent study, the JA-deficient mutant opr7/8 displayed a 100% feminized tassel, demonstrating that the JA signal is required for tassel development in maize.[8] The 13-lipoxygenase (LOX8) encoded by TS1 is destroyed, causing JA dysfunction in the tassel meristem. OPR isoforms opr7 and opr8 are double mutants, and as a result, the plant suffers systemic JA depletion as a result of the mutation. Numerous studies have connected gibberellin (GA) to the development of the ear. GA is another crucial phytohormone for maize sex determination, as demonstrated by the dwarfism and masculinized ears (i.e., the formation of male florets), revealed by the GA biosynthesis mutants an1, d1, d2, d3, and d5, as well as by the GA vision mutants D8 and D9.[83]

1.5.3 SOME PLANT SPECIES FEMALE ORGANOGENESIS IS INFLUENCED BY JAs

The ortholog of coi1 and JA signaling mutants in tomato, jar1, showed seed-bearing sterility. A few viable seeds were retrieved from the fruit, but the great majority (>99%) of the fertilized eggs of the mutant fruit were not viable during fruit development. The size and mass of fully ripened jai1-1 fruit were much lower than those of mature wild-type fruit. Less than 0.1% of the viable seed output of plant species grown under comparable conditions was found to be produced by the jai1-1 plants.[80] The JA-deficient mutant opr7/8 in maize showed outgrowth of multiple female reproductive buds as well as excessive elongation of ear shanks, suggesting that JA is an important signal for female organ growth.[8]

1.5.4 JAs REGULATE VEGETATIVE GROWTH

In response to biotic and abiotic stresses, JA defensive signaling is activated, available resources are depleted, and plant growth is severely limited. It is very well understood that JAs function as growth promoters in plant shoots and roots.[84] By inhibiting cell mitosis, wound-induced aggregation of endogenous JAs suppresses plant development of shoot and root.[85] JA vision mutant coi1 eased JA inhibition to shoot and root.[86] JA-signaling mutants like jin1/myc2, jin4/jar1, and jai3 have significantly decreased JA-induced root and leaf growth inhibition.[78] To control growth processes and defensive responses, the JA signal combines other plant hormones, such as CK (cytokinins), GA (gibberellic acid or gibberellin), IAA (indole acetic acid), ABA (abscisic acid), and ET (ethylene).[20]

1.5.5 JAs PROMOTE FRUIT/SEED RIPENING

The tomato mutant jar1 produces far smaller fruits than the wild variety, and the young mutant fruit seeds have a significant increase in seed abortion (>99%), meaning that the JA signal is important during the early stages of fruit growth and seed development in tomatoes.[80] Apples (*Malus sylvestris*) and sweet cherries (*Prunus avium*) fruits and seeds collected endogenous JA during the early stages of ripening, indicating that JA is crucial for fruit and seed growth.[87]

1.6 CONCLUSION

Jasmonate biosynthesis, regulation, and signaling pathways have all improved dramatically in recent years. Despite the advances made, certain issues remain unsolved and require additional examination. This includes the topic of what substrates are used and how they are fed into the biosynthetic pathway for JAs. Are esterified and/or free fatty acids used in the first phases of jasmonate biosynthesis? The lipases involved in jasmonate biosynthesis, except DAD1, have yet to be discovered, and the processes and signals that trigger lipase activation and the release of jasmonate precursors under various situations (e.g., in reaction to injury) are still unknown. The aforementioned question will most likely be answered by determining the lipase(s) and their substrate specificity.

KEYWORDS

- **jasmonic acid**
- **biosynthesis**
- **plants**
- **oxylipins**
- **systemin**

REFERENCES

1. Andersson, M. X.; Hamberg, M.; Kourtchenko, O.; Brunnstrom, A.; McPhail, K. L.; Gerwick, W. H.; Gobel, C.; Feussner, I.; Ellerstrom, M. Oxylipin Profiling of the Hypersensitive Response in *Arabidopsis thaliana*. Formation of a Novel Oxo-phytodienoic Acid Containing Galacto Lipid, Arabidopside E. *J. Biol. Chem.* **2006,** *281,* 31528–31537.

2. Göbel, C.; Feussner, I. Methods for the Analysis of Oxylipins in Plants. *Phytochem.* **2009,** *70,* 1485–1503.
3. Farag, M. A.; Paré, P. W. C-6 Green Leaf Volatiles Trigger Local and Systemic VOC Emissions in Tomato. *Phytochemistry* **2002,** *61,* 545–554.
4. Fons, F.; Froissard, D.; Bessière, J. M.; Buatois, B.; Rapior, S. Biodiversity of Volatile Organic Compounds From Five French Ferns. *Nat. Prod. Commun.* **2010,** *5,* 1655–1658.
5. Müller, C. Host Finding and Oviposition Behavior in a Chrysomelid Specialist—The Importance of Host Plant Surface Waxes. *J. Chem. Ecol.* **2001,** *27,* 985.
6. Eckardt, N. A. Oxylipin Signaling in Plant Stress Responses. *Plant Cell* **2008,** *20,* 495–497.
7. Tani, T.; Sobajima, H.; Okada, K.; Chujo, T.; Arimura, S.; Tsutsumi, N.; Nishimura, M.; Seto, H.; Nojiri, H.; Yamane, H. Identification of the OsOPR7 Gene Encoding 12-oxophytodienoatereductase Involved in the Biosynthesis of Jasmonic Acid in Rice. *Planta* **2008,** *227,* 517–526.
8. Yan, Y.; Christensen, S.; Isakeit, T.; Engelberth, J.; Meeley, R.; Hayward, A.; Emery, R. J.; Kolomiets, M. V. Disruption of OPR7 and OPR8 Reveals the Versatile Functions of Jasmonic Acid in Maize Development and Defense. *Plant Cell* **2012,** *24,* 1420–1436.
9. Sheard, L. B.; Tan, X.; Mao, H.; Withers, J.; Ben-Nissan, G.; Hinds, T. R.; Kobayashi, Y.; Hsu, F. F.; Sharon, M.; Browse, J.; He, S. Y.; Rizo, J.; Howe, G. A.; Zheng, N. Jasmonate Perception by Inositol-Phosphate-Potentiated COI1-JAZ Co-receptor. *Nature* **2010,** *468,* 400–405.
10. Chini, A.; Fonseca, S.; Fernandez, G.; Adie, B.; Chico, J. M.; Lorenzo, O.; Garcia-Casado, G.; Lopez-Vidriero, I.; Lozano, F. M.; Ponce, M. R.; Micol, J. L.; Solano, R. The JAZ Family of Repressors is the Missing Link in Jasmonate Signaling. *Nature* **2007,** *448,* 666–671.
11. Thines, B.; Katsir, L.; Melotto, M.; Niu, Y.; Mandaokar, A.; Liu, G.; Nomura.; He, S. Y.; Howe, G. A.; Browse, J. JAZ Repressor Proteins are Targets of the SCF(COI1) Complex During Jasmonate Signaling. *Nature* **2007,** *448,* 661–665.
12. Katsir, L.; Schilmiller, A. L.; Staswick, P. E.; He, S. Y.; Howe, G. A. COI1 is a Critical Component of a Receptor for Jasmonate and the Bacterial Virulence Factor Coronatine. *Proc. Natl Acad. Sci. U. S. A.* **2008,** *105,* 7100–7105.
13. Yan, J.; Zhang, C.; Gu, M.; Bai, Z.; Zhang, W.; Qi, T.; Cheng.; Peng.; Luo, H.; Nan, F.; Wang.; Xie, D. The Arabidopsis CORONATINE INSENSITIVE1 Protein is a Jasmonate Receptor. *Plant Cell* **2009,** *21,* 2220–2236.
14. Creelman, R. A.; Mullet, J. E. Biosynthesis and Action of Jasmonates in Plants. *Plant Mol. Biol.Annu. Rev. Plant Physiol. Plant Mol. Biol.* **1997,** *48,* 355–381.
15. Wasternack, C. Jasmonates: An Update on Biosynthesis, Signal Transduction and Action in Plant Stress Response, Growth and Development. *Ann. Bot.* **2007,** *100,* 681–697.
16. Browse, J. The Power of Mutants for Investigating Jasmonate Biosynthesis and Signaling. *Phytochem* **2009,** *70,* 1539–1546.
17. Acosta, I. F.; Laparra, H.; Romero, S. P.; Schmelz, E.; Hamberg, M.; Mottinger, J. P.; Moreno, M. A.; Dellaporta, S. L. Tasselseed1 is a Lipoxygenase Affecting Jasmonic Acid Signaling in Sex Determination of Maize. *Science* **2009,** *323,* 262–265.

18. Xu, D.; Kohli, R. M.; Massey, V. The Role of Threonine 37 in Flavin Reactivity of the Old Yellow Enzyme. *Proc. Natl Acad. Sci. U. S. A.* **1999,** *96,* 3556–3561.
19. Beckers, G. J.; Spoel, S. H. Fine-tuning Plant Defense Signaling: Salicylate Versus Jasmonate. *Plant Biol. (Stuttg.)* **2006,** *8,* 1–10.
20. Anderson, J.; Badruzsaufari, E.; Schenk, P. M.; Manners, J. M.; Desmond, O. J.; Ehlert, C.; Maclean, D. J.; Ebert, P. R.; Kazan, K. Antagonistic Interaction Between Abscisic Acid and Jasmonate-Ethylene Signaling Pathways Modulates Defense Gene Expression and Disease Resistance in *Arabidopsis*. *Plant Cell* **2004,** *16,* 3460–3479.
21. Kende, H.; Zeevaart, J. The Five "Classical" Plant Hormones. *Plant Cell* **1997,** *9,* 1197–1210.
22. Demole, E.; Lederer, E.; Mercier, D. Isolementetdétermination de la Structure du Jasmonate de Méthyle, Constituant Odorant Caractéristique de Lessence de Jasmin. *Helv. Chim. Acta* **1962,** *45,* 675–685.
23. Aldridge, D. C.; Galt, S.; Giles, D.; Turner, W. B. Metabolites of *Lasiodiplodia theobromae*. *J. Chem. Soc.* **1971,** 1623–1627.
24. Fukui, H.; Koshimizu, K; Usuda, S.; Yamazaki, Y. Isolation of Plant-growth Regulators from Seeds of *Cucurbita-pepo* L. *Agric. Biol. Chem. Tokyo* **1977,** *41,* 175–180.
25. Dathe, W.; Ronsch, H.; Preiss, A.; Schade, W.; Semdbner, G.; Schreiber, K. Endogenous Plant Hormones of the Broad Bean *Vicia faba* L. (−)-jasmonic Acid, a Plant Growth Inhibitor in Pericarp. *Planta* **1981,** *153,* 530–535.
26. Yamane, H.; Sugawara, J.; Suzuki, Y.; Shimamura, E.; Takahashi, N. Syntheses of Jasmonic Acid Related-Compounds and their Structure-Activity-Relationships on the Growth of Rice Seedings. *Agric. Biol. Chem. Tokyo* **1980,** *44,* 2857–2864.
27. Vick, B. A.; Zimmerman, D. C. The Biosynthesis of Jasmonic Acid: A Physiological Role for Plant Lipoxygenase. *Biochem. Biophys. Res. Commun.* **1983,** *111,* 470–477.
28. Wasternack, C.; Hause, B. Jasmonates: Biosynthesis, Perception, Signal Transduction and Action in Plant Stress Response, Growth and Development. An Update to the 2007 Review in Annals of Botany. *Ann. Bot.* **2013,** *111,* 1021–1058.
29. Schaller, A.; Stintzi, A. Enzymes in Jasmonate Biosynthesis—Structure, Function, Regulation. *Phytochem* **2009,** *70,* 1532–1538.
30. Zhai, Q. Z.; Zhang, X.; Wu, F. M.; Feng, H.; Deng, L.; Xu, L.; Zhang, M.; Wang, Q.; Li, C. Transcriptional Mechanism of Jasmonate Receptor COI1-Mediated Delay of Flowering Time in Arabidopsis. *Plant Cell* **2015,** *27,* 2814–2828.
31. McConn, M.; Browse, J. The Critical Requirement for Linolenic Acid is Pollen Development, Not Photosynthesis, in an *Arabidopsis* mutant. *Plant Cell* **1996,** *8,* 403–416.
32. McConn, M.; Creelman, R. A.; Bell, E.; Mullet, J. E.; Browse, J. Jasmonate is Essential for Insect Defense in *Arabidopsis*. *Proc. Natl Acad. Sci. U. S. A.* **1997,** *94,* 5473–5477.
33. Li, C.; Liu, G.; Xu, C.; Lee, G.I; Bauer, P.; Ling, H.; Ganal, M. W.; Howe, G. A. The Tomato Suppressor of Prosystemin-Mediated Responses2 Gene Encodes a Fatty Acid Desaturase Required for the Biosynthesis of Jasmonic Acid and the Production of a Systemic Wound Signal for Defense Gene Expression. *Plant Cell* **2003,** *15,* 1646–1661.
34. Ishiguro, S.; Kawai-Oda, A.; Ueda, J.; Nishida, I.; Okada, K. The DEFECTIVE IN ANTHER DEHISCENCE Gene Encodes a Novel Phospholipase A1 Catalyzing the Initial Step of Jasmonic Acid Biosynthesis, which Synchronizes Pollen Maturation,

Anther Dehiscence, and Flower Opening in *Arabidopsis*. *Plant Cell* **2001,** *13,* 2191–2209.
35. Hyun, Y.; Choi, S.; Hwang, H. J.; Yu, J.; Nam, S.; Ko, J.; Park, J.; Seo, Y. S.; Kim, E. Y.; Ryu, E. B.; Kim, W. T.; Lee, Y.; Kang, H.; Lee, I. Cooperation and Functional Diversification of Two Closely Related Galactolipase Genes for Jasmonate Biosynthesis. *Dev. Cell* **2008,** *14,* 183–192.
36. Bannenberg, G.; Martinez, M.; Hamberg, M.; Castresana, C. Diversity of the Enzymatic Activity in the Lipoxygenase Gene Family of *Arabidopsis thaliana. Lipids* **2009,** *44,* 85–95.
37. Bell, E.; Creelman, R. A.; Mullet, J. E. A Chloroplast Lipoxygenase is Required for Wound-induced Jasmonic Acid Accumulation in Arabidopsis. *Proc. Natl Acad. Sci. U. S. A.* **1995,** *92,* 8675–8679.
38. Glauser, G.; Dubugnon, L.; Mousavi, S. A.; Rudaz, S.; Wolfender, J.; Farmer, E. E. Velocity Estimates for Signal Propagation Leading to Systemic Jasmonic Acid Accumulation in Wounded Arabidopsis. *J. Biol. Chem.* **2009,** *284,* 34506–34513.
39. Schommer, C.; Palatnik, J. F.; Aggarwal, P.; Chételat, A.; Cubas, P.; Farmer, E. E.; Nath, U.; Weigel, D. Control of Jasmonate Biosynthesis and Senescence by miR319 Targets. *PLoS Biol.* **2008,** *6,* e230.
40. Caldelari, D.; Wang, G.; Farmer, E. E.; Donget, X. Arabidopsis lox3 lox4 Double Mutants are Male Sterile and Defective in Global Proliferative Arrest. *Plant Mol. Biol.* **2011,** *75,* 25–33.
41. Chauvin, A.; Caldelari, D.; Wolfender, J. L.; Chauvin, A.; Caldelari, D.; Wolfender, J.; Farmer, E. E. Four 13-lipoxygenases Contribute to Rapid Jasmonate Synthesis in Wounded *Arabidopsis thaliana* Leaves: A Role for Lipoxygenase 6 in Responses to Long-Distance Wound Signals. *New Phytol.* **2013,** *197,* 566–575.
42. Grebner, W.; Stingl, N. E.; Oenel, A.; Mueller, M. J.; Berger, S. Lipoxygenase6-Dependent Oxylipin Synthesis in Roots is Required for Abiotic and Biotic Stress Resistance of *Arabidopsis*. *Plant Physiol.* **2013,** *161,* 2159–2170.
43. Yan, L. H.; Zhai, Q. Z.; Wei, J. N.; Li, S.; Wang, B.; Huang, T.; Du, M.; Sun, J.; Kang, L.; Li, C.; Li, C. Role of Tomato Lipoxygenase D in Wound-induced Jasmonate Biosynthesis and Plant Immunity to Insect Herbivores. *PLoS Genet.* **2013b,** *9,* e1003964.
44. Bostock, R. M. Signal Crosstalk and Induced Resistance: Straddling the Line between Cost and Benefit. *Annu. Rev. Phytopathol.* **2005,** *43,* 545–580.
45. Howe, G. A.; Schilmiller, A. L. Oxylipin Metabolism in Response to Stress. *Curr. Opin. Plant Biol.* **2002,** *5,* 230–236.
46. Howe, G. A.; Lee, G. I.; Itoh, A.; Li, L.; DeRocher, A. E. Cytochrome P450-dependent Metabolism of Oxylipins in Tomato. Cloning and Expression of Allene Oxide Synthase and Fatty Acid Hydroperoxide Lyase. *Plant Physiol.* **2000,** *123,* 711–724.
47. Laudert, D.; Pfannschmidt, U.; Lottspeich, F.; Holländer-Czytko, H.; Weiler, E. W. Cloning, Molecular and Functional Characterization of *Arabidopsis thaliana* Allene Oxide Synthase (CYP74), the First Enzyme of the Octadecanoid Pathway to Jasmonates. *Plant Mol. Biol.* **1996,** *31,* 323–335.
48. Park, J. H.; Halitschke, R.; Kim, H. B; Baldwin, I. T.; Feldmann, K. A.; Feyereisen, R. A Knock-out Mutation in Allene Oxide Synthase Results in Male Sterility and Defective Wound Signal Transduction in Arabidopsis due to a Block in Jasmonic Acid Biosynthesis. *Plant J.* **2002,** *31,* 1–12.

49. Brash, A. R.; Baertschi, S. W.; Ingram, C. D.; Harris, T. M. Isolation and Characterization of Natural Allene Oxides: Unstable Intermediates in the Metabolism of Lipid Hydroperoxides. *Proc. Natl Acad. Sci. U. S. A.* **1988,** *85,* 3382–3386.
50. Ziegler, J.; Stenzel, I.; Hause, B.; Maucher, H.; Hamberg, M.; Grimm, R.; Ganal, M.; Wasternack. C. Molecular Cloning of Allene Oxide Cyclase. The Enzyme Establishing the Stereochemistry of Octadecanoids and Jasmonates. *J. Biol. Chem.* **2000,** *275,* 19132–19138.
51. Stenzel, I.; Hause, B.; Miersch, O.; Kurz, T.; Maucher, H.; Weichert, H.; Ziegler, J.; Feussner, I.; Wasternack, C. Jasmonate Biosynthesis and the Alleneoxidcyclase Family of *Arabidopsis thaliana. Plant Mol. Biol.* **2003b,** *51,* 895–911.
52. Stenzel, I.; Otto, M.; Delker, C.; Kirmse, N.; Schmidt, D.; Miersch, O.; Hause, B.; Wasternack, C. ALLENE OXIDE CYCLASE (AOC) Gene Family Members of Arabidopsis Thaliana: Tissue- and Organ-specific Promoter Activities and In Vivo Heteromerization. *J. Exp. Bot.* **2012,** *63,* 6125–6138.
53. Sanders, P. M.; Lee, P. Y.; Biesgen, C.; Boone, J. D.; Beals, T. P.; Weiler, E. W.; Goldberg, E. B. The Arabidopsis DELAYED DEHISCENCE1 Gene Encodes an Enzyme in the Jasmonic Acid Synthesis Pathway. *Plant Cell* **2000,** *12,* 1041–1061.
54. Stintzi, A.; Browse, J. The Arabidopsis Male-sterile Mutant, opr3, Lacks the 12-oxophytodienoicacid Reductase Required for Jasmonate Synthesis. *Proc. Natl Acad. Sci. U. S. A.* **2000,** *97,* 10625–10630.
55. Koo, A. J.; Chung, H. S.; Kobayashi, Y.; Howe, G. A. Identification of a Peroxisomal Acyl-activating Enzyme Involved in the Biosynthesis of Jasmonic Acid in Arabidopsis. *J. Biol. Chem.* **2006,** *281,* 33511–33520.
56. Kienow, L.; Schneider, K.; Bartsch, M.; Stuible, H.; Weng, H.; Miersch, O.; Wasternack, C.; Kombrink, E. Jasmonates Meet Fatty Acids: Functional Analysis of a New Acyl-coenzyme Asynthetase Family from Arabidopsis Thaliana. *J. Ex. Bot.* **2008,** *59,* 403–419.
57. Li, C.; Schilmiller, A. L.; Liu, G.; Lee, G. I.; Jayanty, S.; Sageman, C.; Vrebalov, J.; Giovannoni, J. J.; Yagi, K.; Kobayashi, Y.; Howe, G. A. Role of Beta-oxidation in Jasmonate Biosynthesis and Systemic Wound Signaling in Tomato. *Plant Cell* **2005,** *17,* 971–986.
58. Schilmiller, A. L.; Koo, A. J.; Howe, G. A. Functional Diversification of Acyl-coenzyme A Oxidases in Jasmonic Acid Biosynthesis and Action. *Plant Physiol.* **2007,** *143,* 812–824.
59. Delker, C.; Zolman, B. K.; Miersch, O.; Wasternacke, C. Jasmonate Biosynthesis in Arabidopsis Thaliana Requires Peroxisomal Beta-oxidation Enzymes—Additional Proof by Properties of pex6 and aim1. *Phytochem* **2007,** *68,* 1642–1650.
60. Chehab, E. W.; Kim, S.; Savchenko, T.; Kliebenstein, D.; Dehesh, K.; Braam, J. Intronic t-DNA Insertion Renders Arabidopsis opr3 a Conditional Jasmonic Acid-Producing Mutant. *Plant Physiol.* **2011,** *156,* 770–778.
61. Chini, A.; Monte, I.; Zamarreño, A. M.; Hamberg, M.; Lassueur, S.; Reymond, P.; Weiss, S.; Stintzi, A.; Schaller, A.; Porzel, A.; García-Mina, J. M.; Solano, R. An opr3-Independent Pathway Uses 4,5-didehydrojasmonate for Jasmonate Synthesis. *Nat. Chem. Biol.* **2018,** *14,* 171–178.
62. Han, G. Z. Evolution of Jasmonate Biosynthesis and Signaling Mechanisms. *J. Exp. Bot.* **2017,** *68,* 1323–1331.

63. Blechert, S.; Brodschelm, W.; Holder, S.; Kammerer, L.; Kutchan, T. M.; Mueller, M. J.; Xia, Z. Q.; Zenket, M. H. The Octadecanoic Pathway: Signal Molecules for the Regulation of secondary pathways. *Proc. Natl Acad. Sci. U. S. A.* **1995,** *92,* 4099–4105.
64. McCloud, E. S..; Baldwin, I. T. Herbivory and Caterpillar Regurgitants Amplify the Wound-induced Increases in Jasmonic Acid but Not Nicotine in *Nicotiana Sylvestris. Planta* **1997,** *203,* 430–435.
65. Glauser, G.; Grata, E.; Dubugnon, L.; Rudaz, S.; Farmer, E. E.; Wolfender, J. Spatial and Temporal Dynamics of Jasmonate Synthesis and Accumulation in Arabidopsis in Response to Wounding. *J. Biol. Chem.* **2008,** *283,* 16400–16407.
66. Koo, A. J.; Gao, X.; Jones, AD.; Howe, G. A. A Rapid Wound Signal Activates the Systemic Synthesis of Bioactive Jasmonates in *Arabidopsis. Plant J.* **2009,** *59,* 974–986.
67. Farmer, E. E.; Johnson, R. R.; Ryan, C. A. Regulation of Expression of Proteinase Inhibitor Genes by Methyl Jasmonate and Jasmonic Acid. *Plant Physiol.* **1992,** *98,* 995–1002.
68. Ryan, C. A. The Systemin Signaling Pathway: Differential Activation of Plant Defensive Genes. *Biochim. Biophys. Acta* **2000,** *1477,* 112–121.
69. Ryan, C. A.; Pearce, G. Systemin: A Polypeptide Signal for Plant Defensive Genes. *Annu. Rev. Cell Dev. Biol.* **1998,** *14,* 1–17.
70. Li, L.; Li, C.; Lee, G. I.; Howe, G. A. Distinct Roles for Jasmonate Synthesis and Action in the Systemic Wound Response of Tomato. *Proc. Natl Acad. Sci. U. S. A.* **2002,** *99,* 6416–6421.
71. Lee, G. I.; Howe, G. A. The Tomato Mutant spr1 is Defective in Systemin Perception and the Production of a Systemic Wound Signal for Defense Gene Expression. *Plant J.* **2003,** *33,* 567–576.
72. Stenzel, I.; Hause, B.; Maucher, H.; Pitzschke, A.; Miersch, O.; Ziegler, J.; Ryan, C. A.; Wasternack, C. Allene Oxide Cyclase Dependence of the Wound Response and Vascular Bundle-specific Generation of Jasmonates in Tomato—Amplification in Wound Signalling. *Plant J.* **2003a**, 33, 577–589.
73. Kazan, K.; Manners, J. M.; MYC2: The Master in Action. *Mol. Plant* **2013,** *6,* 686–703.
74. Yoon, J. Y.; Hamayun, M.; Lee, S. K.; Lee, I. J.; Methyl Jasmonate Alleviated Salinity Stress in Soybean. *J. Crop Sci. Biotechnol.* **2009,** *12,* 63.
75. Ismail, A.; Riemann, M.; Nick, P. The Jasmonate Pathway Mediates Salt Tolerance in Grapevines. *J. Exp. Bot.* **2012,** *63,* 2127–2139.
76. Rao, M. V.; Lee, H.; Creelman, R. A.; Mullet, J. E.; Davi, K. R. Jasmonic Acid Signaling Modulates Ozone-induced Hypersensitive Cell Death. *Plant Cell* **2000,** *1,* 1633–1646.
77. Mandaokar, A.; Thines, B.; Shin, B.; Lange, B. M.; Choi, G.; Koo, Y. J.; Yoo, Y. J.; Choi, Y. D.; Choi, G.; Browse, J. Transcriptional Regulators of Stamen Development in Arabidopsis Identified by Transcriptional Profiling. *Plant J.* **2006,** *46,* 984–1008.
78. Lorenzo, O.; Chico, J. M.; Sanchez-Serrano, J. J.; Solano, R. JASMONATE-INSENSITIVE1 Encodes a MYC Transcription Factor Essential to Discriminate Between Different Jasmonate-regulated Defense Responses in Arabidopsis. *Plant Cell* **2004,** *16,* 1938–1950.
79. Staswick, P. E.; Tiryaki, I. The Oxylipin Signal Jasmonic Acid is Activated by an Enzyme that Conjugates it to Isoleucine in Arabidopsis. *Plant Cell* **2004,** *16,* 2117–2127.

80. Li, L.; Zhao, Y.; McCaig, B. C.; Wingerd, B.; Wang, J.; Whalon, M. E.; Pichersky, E.; Howe, G. A. The Tomato Homolog of CORONATINE-INSENSITIVE1 is Required for the Maternal Control of Seed Maturation, Jasmonate-signaled Defense Responses, and Glandular Trichome Development. *Plant Cell* **2004,** *16,* 126–143.
81. Riemann, M.; Muller, A.; Korte, A.; Furuya, M.; Weiler, E. W.; Nick, P. Impaired Induction of the Jasmonate Pathway in the Rice Mutant Hebiba. *Plant Physiol.* **2003,** *133,* 1820–1830.
82. Bortiri, E.; Hake, S. Flowering and Determinacy in Maize. *J. Exp. Bot.* **2007,** *58,* 909–916.
83. Chuck, G. Molecular Mechanisms of Sex Determination in Monoecious and Dioecious Plants. *Adv. Bot. Res.* **2010,** *54,* 53–83.
84. Staswick, P. E.; Su, W.; Howell, S. H. Methyl Jasmonate Inhibition of Root Growth and Induction of a Leaf Protein are Decreased in an *Arabidopsis thaliana* Mutant. *Proc. Natl Acad. Sci. U. S. A.* **1992,** *89,* 6837–6840.
85. Zhang, Y.; Turner, J. G. Wound-induced Endogenous Jasmonates Stunt Plant Growth by Inhibiting Mitosis. *PLoS One* **2008,** *3,* e3699.
86. Xie, D. X.; Feys, B. F.; James, S.; Nieto-Rostro, M.; Turner, J. G. COI1: An Arabidopsis Gene Required for Jasmonate-regulated Defense and Fertility. *Science* **1998,** *280,* 1091–1094.
87. Kondo, S.; Tomiyama, A.; Seto, H. Changes of Endogenous Jasmonic Acid and Methyl Jasmonate in Apples and Sweet Cherries during Fruit Development. *J. Am. Soc. Hortic. Sci.* **2000,** *125,* 282–287.

CHAPTER 2

Presence and Roles of Jasmonates in Plant Growth and Metabolic Processes

MUNISA MANZOOR KHAN, AADIL FAROOQ WAR, SUMAYA REHMAN, SUBZAR AHMAD NANDA, and ISHFAQ UL REHMAN

Department of Botany, University of Kashmir, Srinagar, Jammu & Kashmir, India

ABSTRACT

Plants must continually observe external cues in order to change their growth, development, and defensive capacities to maximize their reproductive success, especially when environmental conditions are unusual. Under these conditions, plant hormones, including jasmonates, play an active role in mediating plant growth and developmental processes. These jasmonates, which are byproducts of oxygenated fatty acids, are involved in a number of metabolic processes for regulating development and growth in plants while preparing for defense responses against various abiotic and biotic stresses. The jasmonates comprise jasmonic acid (JA), an endogenous hormone, and its derivatives, methyl ester (MeJA), which is a volatile compound mediating interplant communication, and isoleucine conjugate (JA-Ile). These phytohormones are well-documented in higher plants, and their occurrences have been intensely investigated in the modal plant *Arabidopsis thaliana.* The jasmonates play diverse roles in plants, ranging from defense to cell death and leaf senescence, in addition to their involvement in symbiosis and signaling gene expression. One of their prominent and most studied roles is detected in wound response of plants against herbivores, and their signaling mechanisms frequently interact

Jasmonates and Plant Defense: An Eco-Friendly and Sustainable Approach for Stress Tolerance.
Uzma Azeem, & Khalid Rehman Hakeem (Eds.)

with those of other signaling pathways to enhance the plant's defenses against pathogens and herbivores. Glucosinolates (GLSs), terpenoids, and alkaloids, which are involved in coordinating the defense response and plant development, are secondary metabolites produced as a result of jasmonate signaling. The present chapter will offer some crucial insights regarding the distribution of jasmonates in plants and their function in plant growth and metabolic processes under stress conditions, in light of the fact that abiotic and biotic pressures are a constant threat to plants.

2.1 INTRODUCTION

To maximize their reproductive success, plants need to constantly notice external stimuli in order to modify their growth, development, and defense capacities, particularly when they face abiotic and biotic stresses. They respond to these stimuli by activating specific metabolic pathways involving a slew of molecular actors who interact in complicated networks.[1] Amid several molecular signals involved in defense responses, the phytohormone, jasmonic acid, or jasmonate (JA), is considered to play a significant role.[2] JAs, which include the hormone jasmonic acid (JA) and its derivatives, are involved in regulating plant growth and developmental processes, such as nitrogen and phosphorus uptake, stimulating stomatal opening, and impeding Rubisco biosynthesis,[3] besides their key role in wound response against pathogen attack.[4] Plants contain a variety of JAs, most notably jasmonic acid, which is widely distributed in nature.[5] The name and molecular structure of JAs were revealed with the characterization and isolation of methyl jasmonate from the jasmine oil of *Jasminum grandiflorum*.[6] The JAs intervene in their responses during biotic and abiotic stress conditions by stimulating a sequence of gene expression[7] and acting in association with other hormone signaling pathways, for instance, auxin, salicylic acid (SA), ethylene (ET), brassinosteroids (BRs), abscisic acid (ABA), and gibberellin (GA) in a network manner[8] (Figure 2.1). Some JAs are also released as volatile organic compounds (VOCs), allowing plants to communicate and alert one another to potential threats.[9]

One of the prominent roles of JAs is that they act as master signals in activating a variety of triggered defense responses during herbivory attack[4] and attack from necrotrophic pathogens,[11] regulating the plant metabolism under these biotic stress conditions. In many cases, systemin signals the wound response to external signals and triggers the intracellular increase of JA, which then acts to regulate the stability between defense and growth responses.[10]

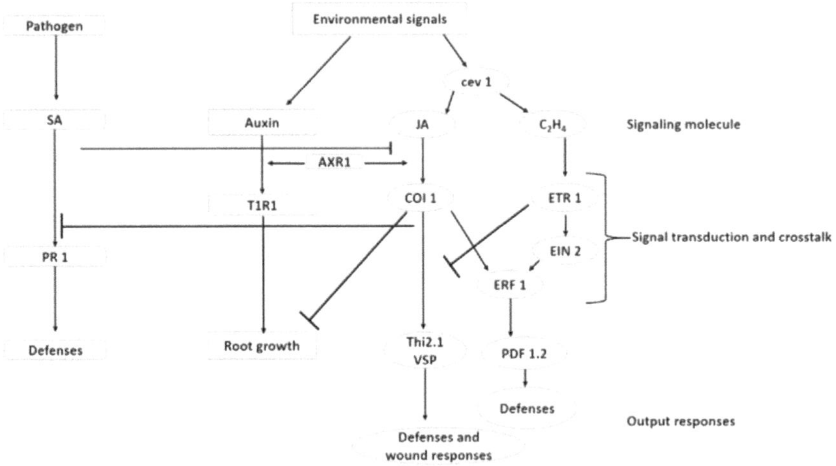

FIGURE 2.1 The interaction of jasmonate, ethylene, auxin, and salicylic acid.[10] The interaction of jasmonate, ethylene, auxin, and salicylic acid-dependent pathways is depicted in this figure. JA via COI1 and ethylene via ETR1 and EIN2 work in concert to induce the expression of PDF1.2 in a way that is dependent on ERF1. CEV1 is capable of constitutive JA and ethylene signaling. A wound signal may stimulate the production of JA, which in turn stimulates the expression of JA-responsive genes such as Thi2.1 and VSP. AXR-dependent signaling mechanisms may be used by jasmonate and auxin. Antagonism between SA and JA signaling is represented by COI1-dependent PR1 repression. Positive and negative interactions are indicated by arrows and bars, respectively.

Source: Adapted from Ref. [10]

At a molecular level, the JAZs form an important component of the classical JA signaling cascade, where they act as negative regulators.[12–13] JAZ proteins can bind with both the R2R3-MYBs and the JA-inducible bHLH (basic helix-loop-helix) TFs (transcription factors) of the MBW complex in the absence of bioactive JAs, preventing the MBW complex from forming and decreasing its transcriptional activity. The JAZ proteins are targeted for proteolysis in response to JA perception, allowing the MBW complex to assemble and release to initiate JA-induced anthocyanin biosynthesis.[14]

2.2 PRESENCE OF JASMONATES

All kinds of JAs find their source of origin in plant species, with the principal JA, that is, jasmonic acid, occurring universally in the plant kingdom.[15–16] Being a natural plant growth regulator, JA is extensively dispersed in plants. For the first time, it has been isolated from the culture of *Lasiodiplodia theobromae*, a major JA member.[17] The jasmonic acid derivative, MeJA, is mainly isolated from jasmine oil, which is harvested from *Jasminum grandiflorum.*[18] In addition to JA and MeJA, more JAs with multiple biological roles have been described, especially cis-jasmone, jasmonoyl ACC (JA-ACC), and jasmonoyl isoleucine (JA-Ile).[18–21] In *Solanum tuberosum* leaves, two types of JA isolates are found, namely tuberonic acid (TA) and its glucoside (TAG), which are supposed to induce tuber formation.[22–23] Many plant species synthesize these compounds by hydroxylation and subsequent glycosylation.[22] A JA-like compound, cucurbic acid, in the *Cucurbita pepo* seeds[24] has been discovered to be aggressively involved in tuberization.[25] The production of JA in monocotyledons and dicotyledons, particularly in *Arabidopsis*, has been extensively studied among all JAs. At least two pathways are involved in the production of JA in *Arabidopsis*, comprising the 43 α-linolenic acid (18:3) initial octadecane pathway and the hexadecatrienoic acid (16:3) initial 44 hexadecane pathway (Figure 2.2).[26–28]

To begin JA biosynthesis, linolenic acid is released from the chloroplast membrane. The 12-oxo-PDA is then produced from this linolenic acid. This 12-oxo-PDA is carried into the peroxisome, where a number of enzymes convert it to JA before it is exported to the cytoplasm. Depending on the carboxylic acid group, the pentenyl side chain, or the pentanone ring are chemically modified. JA can be metabolized into a variety of different compounds, including methyl jasmonates, jasmonic acid, 12-HSO4-JA, jasmonyl isoleucine, 12-hydroxyjasmonic acid sulfate, 12-oxo-phytodienoic acid (12-OPDA), and 12-oxo-PDA.[29] In the octadecane pathway, 45 unsaturated fatty acids (18:3) are converted to 12-OPDA in the chloroplast, correspondingly. After that, the peroxisome repeatedly oxidizes 12-OPDA to create JA. JA is finally transformed into 48 different JA structures in the cytoplasm, including JA-Ile, MeJA, and 49 12-hydroxyjasmonic acids (12-OH-JA) (Figure 2.2). One of these JAs, 50 JA-Ile, is the form of JA that is biologically active in plants.[30] Plants also produce JAs in response to signaling from wounds brought on by herbivore or pathogen attacks, which later trigger the expression of genes involved in defense. It has been shown that within about 2 h after mechanical wounding or herbivory, proteinase inhibitors (PIs) encoding genes are expressed systemically in the well-studied tomato (*Lycopersicon esculentum*) system upon JAs action.[31–33]

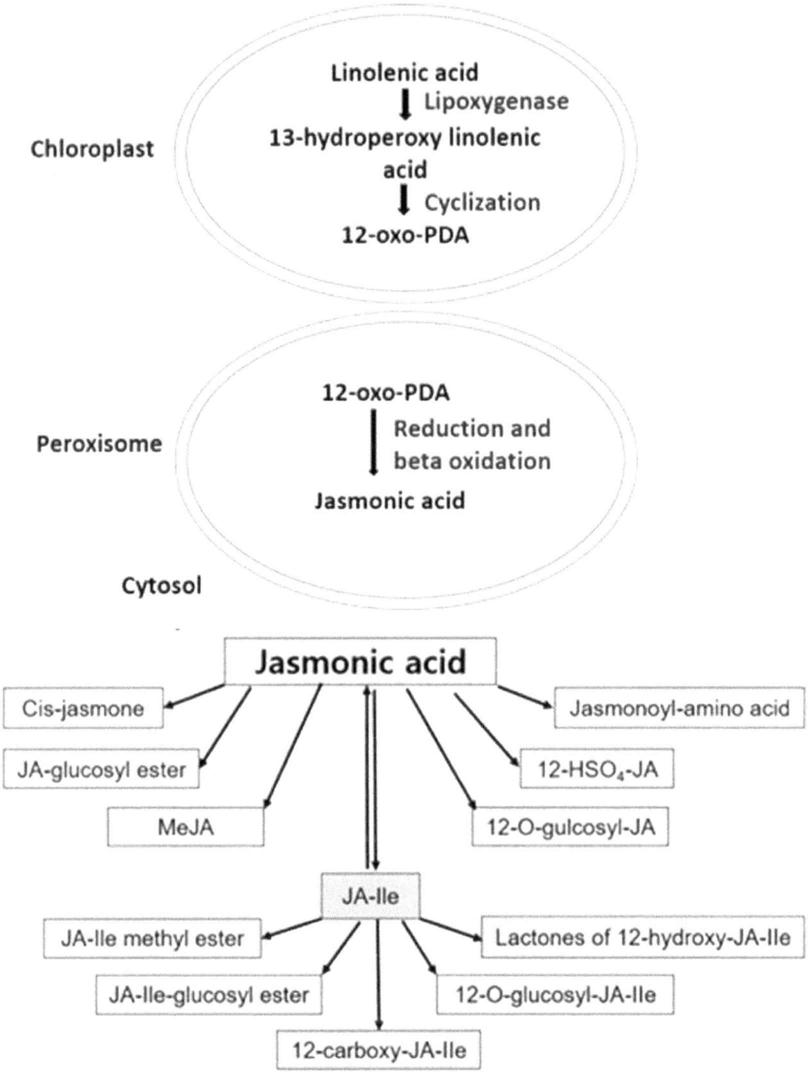

FIGURE 2.2 In response to abiotic stress, jasmonic acid is biosynthesized and metabolized.
Source: Adapted from Ali et al., 2020

2.3 ROLES OF JASMONATES IN PLANT GROWTH AND METABOLIC PROCESSES

2.3.1 *ROLE OF JASMONATES IN PLANT GROWTH AND DEVELOPMENT*

JAs are well-known for activating specific metabolic pathways[34–35] and are essential for controlling the biosynthesis of specialized metabolites like glucosinolates, alkaloids, and terpenoids.[36] The regulation of these metabolites and the transcription factors involved is reported in various plant species (Table 2.1). Each plant species has typically evolved distinct and specific JA-controlled metabolic pathways, demonstrating how evolution efficiently exploited a highly conserved hormonal signaling cascade to drive species-specific cellular and metabolic networks.[34,35,37] Recently, it has been shown that plants from the Solanaceae family produce steroidal glycoalkaloids (SGAs) as protective metabolites through the involvement of JA and GA (gibberellin) crosstalk, which act on MYC1 and MYC2 transcription factors, in which JA acts to elevate while GA acts to reduce the synthesis of SGAs.[38] A shotgun proteomics approach, followed by biochemical profiling in the diatom *Stauroneis* sp. revealed enhancements in biomass yields and bioactive metabolite synthesis upon exposure to methyl jasmonates,[39] which upregulate almost 197 proteins, including those of carbohydrate metabolism, photosynthetic processes, chloroplast biogenesis, and lipid biosynthesis. In the case of *Hypericum perforatum*, a medicinally important plant, the exposure to MeJA and FR (far-red) lighting increased secondary compound accumulation (hypericin, flavonols, flavan-3-ols, and phenolic acid) without interfering with plant growth.[40] During recent times, JAs have been seen actively involved in several plant growth and metabolic processes in various plant species under different circumstances.[39–43] In *Arabidopsis*, JAs mediate many responses under salt stress conditions through the regulation of the MYC2 transcription factor and jasmonate ZIM-domain (JAZ) repressors.[43] In the case of *Fragaria* × *ananassa*, JA, in conjunction with ABA, gives salt stress protection through an antioxidant mechanism.[44] During postharvest ripening of tomatoes, exogenously applied methyl jasmonate regulates sucrose metabolism via upregulation of sucrose-phosphate synthase1-4 (SPS1-4) and sucrose-phosphate phosphatase2 (SPP2), which are associated with sucrose biosynthesis, and downregulation of those related to sucrose degradation.[42]

TABLE 2.1 Jasmonate-Induced Synthesis of Several Metabolites and the Associated Transcription Factors (TFs) in Different Plant Species.[40]

Plant species	Metabolite biosynthesis	JA-induced TFs	Regulated genes or pathways	References
Arabidopsis thaliana	Flavonoids	MYC2	MYB75/PAP1, EGL3	45
Artemisia annua	Artemisinin	MYC2	CYP71AV1, DBR2	46
Catharanthus roseus	Terpenoid indole alkaloids (TIAs)	MYC2	TDC	47
Gossypium arboreum	Gossypol	WRKY1	CAD1-A	48
Nicotiana attenuata	Volatile terpenes	WRKY3, WRKY6	upstream of JA biosynthesis	49
Nicotiana tabacum	Nicotine	MYC2a/MYC2b	PMT1a	50
Nicotiana benthamiana	Nicotine	NbbHLH1, NbbHLH2	PMT	51
Panax quinquefolius	Ginsenoside	WRKY1	Ginsenoside biosynthesis	52
Taxus cuspidata	Taxol	JAMYC1, JAMYC2, JAMYC4	Paclitaxel biosynthetic pathway	53
Taxus chinensis	Taxol	WRKY1	DBAT	54
Salvia miltiorrhiza	Tanshinones and phenolic, salvianolic acids	MYC2a, MYC2b	HCT6, CYP98A14, CYP98A14	55

2.3.2 JASMONATES IN BIOTIC STRESS AMELIORATION

Jasmonic acid is well known for mediating plant defense responses to biotic stresses such as herbivore attacks or pathogen infections.[4] JA signaling begins with the perception of molecular patterns (PAMPs/DAMPs/HAMPs) from biotic agents by plant pattern-recognition receptors (PRRs), followed by JA biosynthesis and action[56] mediated by various intracellular signals such as mitogen-activated protein kinase (MAPK) cascades, calcium-dependent protein kinases (CDPKs), ROS, and calcium ions.[57] In the JA signaling pathway, the main repressors are jasmonate ZIM-domain (zinc finger inflorescence meristem) JAZs, which halt JA signaling via binding to transcription factors comprising MYCs (MYC2, MYC3, and MYC4).[36] When environmental stimuli cause a spike in jasmonic acid-isoleucine (herbivore attack or pathogen infection), the JAZs are targeted for ubiquitination and deprivation in the 26S proteasome by coronatine insensitive1 (COI1), an associate of the ubiquitin E3 ligase family.[58] Degradation of JAZs causes transcription factors to be derepressed, which in turn causes the expression of defense-related genes to be activated.[36] JA signaling maintains the balance between defense and growth as well as helps reprogram primary metabolism during the stressed conditions.[59] For instance, the production of jaz decuple (jazD) mutant deficient in JAZ1-7, -9, -10b, and -13 shows high resistance to necrotrophic fungal pathogenic infections and caterpillar feeding but has a harshly underdeveloped growth phenotype because an increase in the defense system is supplemented by a decrease in the growth of the phenotype.[60] The results of omics analyses showed that jazD significantly induces certain metabolic pathways, including the biosynthesis of indole glucosinolates, and because excessive metabolite accumulation can be harmful to plant cells, a JAZ-repressive transcriptional hierarchy is thus essential for successful plant growth.[36] In the regulation of plant growth and defense, JAs, particularly jasmonic acid, have to coordinate with other hormonal signals to maintain this balance[8] (Figure 2.1). The importance of these relationships among JA and other plant hormones in regulating plant stress reactions has received a lot of attention.[13,27,30] In plants, transcription factors, such as SQUAMOSA promoter-binding protein-like (SPLs), interact with a specific group of JAZ proteins, impairing COI1-dependent JAZ degradation. The miR156-SPL-JAZ module acts by targeting these SPLs during the transition from juvenile to adult stages of growth development. With the progression of plant growth, miR156 decays and SPL levels rise, regulating a variety of features of growth, development, and adaptation in plants.[61] With the miR156-SPL-JAZ module, the JA response gradually decays. Because the plant has few concentrations of defensive compounds during its early stages, an active JA response is critical for defense. In *Arabidopsis*, defense substances like GLSs are constitutively accumulated during plant growth, making adult plants further resistant to insect herbivores. The age-dependent decay of JA signaling is one tactic that plants can employ to balance defense and growth.

2.3.3 ROLE OF JASMONATES IN TEMPERATURE STRESS

JAs play an active role under cold and high temperature stress conditions in different types of cultivable plants.[62–65] The ascorbate-glutathione (AsA-GSH) cycle and the SlICE-SlCBF-SlCOR (ICC) signaling pathway are regulated by the transcription factor SlMYC2, which MeJA activates during the postharvest period to prevent chilling injury in tomato fruits stored in cold storage.[62] Surely, MeJA treatment increases the accumulation of internal ascorbate (AsA) and glutathione (GSH) bodies in tomato fruits by increasing the actions of enzymes involved in the AsA-GSH cycle. In the meantime, MeJA-treated tomato fruit had expressively higher levels of expression of C-repeat-binding factor 1 (SlCBF1), cold-regulated gene (SlCOR413), and inducer of SlCBF expression (SlICE1 and SlICEa), all of which are part of the SlICE-SlCBF-SlCOR (ICC) signaling pathway.[62] JA treatment to postharvest peach fruits prior to storage at 4°C develops their chilling tolerance by regulating the ethylene and sugar metabolism.[65] Melatonin-induced cold tolerance of grafted watermelon plants is mediated by methyl jasmonates, which, when applied, resulted in H_2O_2 accumulation and cold tolerance, but inhibiting JA biosynthesis eliminated rootstock- or melatonin-induced H_2O_2 accumulation and cold tolerance.[63] In rice, JAs play a unique role in reducing the spikelet-opening impairment caused by high temperature (HT) stress during anthesis, probably through improvement in the activities of several antioxidant enzymes such as guaiacol peroxidase, superoxide dismutase (SOD), catalase (CAT), and glutathione peroxidase, and further increasing the protective compounds such as HSPs and ascorbic acid (AsA), thereby quashing the ROS generation and decreasing programmed cell death under stress.[66] By increasing the antioxidant capacity, JAs also lessen the harm that HT stress has on the stigma vitality of rice lines that are photothermosensitive genic male sterile (PTSGMS).[64] Using methyl JAs during anthesis on rice panicles reduced the accumulation of reactive oxygen species and increased stigma vitality, increasing fertilization and seed-setting rates in HT-stressed hybrids of PTSGMS rice lines.[64]

2.3.4 ROLE OF JASMONATES TO COMBAT DAMAGE CAUSED BY UV RAYS

JAs are not only involved in regulating plant growth and development but also activate the defensive system against various forms of biotic and abiotic stresses, including UV radiations in the range of 280–320 nm (UV-B),[67] which are known to affect plant health worldwide.[68] The impacts of UV-B differ between plant species and cultivars of the same species, depending on their sensitivity as well as the intensity and duration of UV-B exposure.[69,70] The absence of UV-B-induced antiherbivore defense in the JA-deficient variety of tobacco proposed that JA signaling brought resistance mainly against insect herbivory, and its incidence resulted in a high accumulation of phenylpropanoid derivatives in wild-type plants, which improved the plant defense mechanism and increased the expression of wound-response and JA-inducible genes such as trypsin proteinase inhibitor (TPI).[70] It has been shown that UV-B amplified the defense against *Spodoptera litura* by increasing secondary metabolites via the JA and/or JA-Ile pathways, as well as the herbivory-induced defense metabolite glucosinolate content in *Arabidopsis thaliana*.[71] It is revealed that under UV-B stress, an inverse relationship between JA and SA is well-known in soybean, pea, and mung bean cultivars. Increased JA content improved plant resistance, whereas increased SA levels increased oxidative stress in plants when exposed to high UV-B levels.[72] In mung bean and pea cultivars, higher endogenous JA accumulation is associated with a better tolerance mechanism to UV-B stress. For instance, in these cultivars, it was observed that better nodulation, as well as nitrogenase, nitrate reductase (NR), and nitrite reductase (NiR) activity, occurs with higher JA accumulation when compared to cultivars with low endogenous JA levels under UV-B stress.[73,74] Exogenously applied JA at concentrations of 1.0 and 2.5 mM mitigated the negative effects of UV-B radiation in wheat seedlings.[67] This induction of UV-B tolerance in wheat by JA was synchronized with the accumulation of substances linked to osmotic adjustments in plants and anthocyanin deposition, as well as increased superoxide dismutase (SOD) activities. Moreover, exogenously applied JA improved UV-B tolerance in plants by increasing proline content, and it has been proposed that this change in proline metabolism is primarily responsible for the expression of genes involved in redox control.[67]

KEYWORDS

- **jasmonic acid**
- **biotic stresses**
- ***Arabidopsis thaliana***
- **crosstalk**
- **α-linolenic acid**

REFERENCES

1. Lacchini, E.; Goossens, A. Combinatorial Control of Plant Specialized Metabolism: Mechanisms, Functions, and Consequences. *Annu. Rev. Cell Dev. Biol.* **2020,** *36,* 291–313.
2. Betsuyaku, S.; Katou, S.; Takebayashi, Y.; Sakakibara, H.; Nomura, N.; Fukuda, H. Salicylic Acid and Jasmonic Acid Pathways are Activated in Spatially Different Domains Around the Infection Site During Effector-Triggered Immunity in *Arabidopsis Thaliana. Plant Cell Physiol.* **2018,** *59* (1), 8–16.
3. Wasternack, C.; Hause, B. Jasmonates: Biosynthesis, Perception, Signal Transduction and Action in Plant Stress Response, Growth and Development. An Update to the 2007 Review in Annals of Botany. *Ann. Bot.* **2013,** *111* (6), 1021–1058.
4. Howe, G. A. Jasmonates as Signals in the Wound Response. *J. Plant Growth Regul.* **2004,** *23* (3), 223–237.
5. Ahmad, P.; Rasool, S.; Gul, A.; Sheikh, S. A.; Akram, N. A.; Ashraf, M.; Kazi, A.; Gucel, S. Jasmonates: Multifunctional Roles in Stress Tolerance. *Front. Plant Sci.* **2016,** *7,* 813.
6. Demole, E.; Lederer, E.; Mercier, D. Isolement et Détermination de la Structure du Jasmonate de Méthyle, Constituant Odorant Caractéristique de L'essence de Jasmin. *Helv. Chim. Acta* **1962,** *45* (2), 675–685.

7. Gupta, A.; Hisano, H.; Hojo, Y.; Matsuura, T.; Ikeda, Y.; Mori, I. C.; Senthil-Kumar, M. Global Profiling of Phytohormone Dynamics During Combined Drought and Pathogen Stress in *Arabidopsis Thaliana* Reveals ABA and JA as Major Regulators. *Sci. Rep.* **2017,** *7* (1), 1–13.
8. Yang, J.; Duan, G.; Li, C.; Liu, L.; Han, G.; Zhang, Y.; Wang, C. The Crosstalks Between Jasmonic Acid and Other Plant Hormone Signaling Highlight the Involvement of Jasmonic Acid as a Core Component in Plant Response to Biotic and Abiotic Stresses. *Front. Plant Sci.* **2019,** *10,* 1349.
9. Baldwin, I. T.; Halitschke, R.; Paschold, A.; Von Dahl, C. C.; Preston, C. A. Volatile Signaling in Plant-Plant Interactions: "Talking Trees" in The Genomics Era. *J. Sci.* **2006,** *311* (5762), 812–815.
10. Devoto, A.; Turner, J. G. Regulation of Jasmonate-Mediated Plant Responses in *Arabidopsis*. *Ann. Bot.* **2003,** *92* (3), 329–337.
11. Browse, J. Jasmonate: An Oxylipin Signal with Many Roles in Plants. *Vitam. Horm.* **2005,** *72,* 431–456.
12. Goossens, J.; Mertens, J.; Goossens, A. Role and Functioning of BHLH Transcription Factors in Jasmonate Signalling. *J. Exp. Bot.* **2017,** *68* (6), 1333–1347.
13. Wasternack, C.; Song, S. Jasmonates: Biosynthesis, Metabolism, and Signaling by Proteins Activating and Repressing Transcription. *J. Exp. Bot.* **2017,** *68* (6), 1303–1321.
14. Qi, T.; Song, S.; Ren, Q.; Wu, D.; Huang, H.; Chen, Y.; Fan, M.; Peng, W.; Ren, C.; Xie, D. The Jasmonate-ZIM-Domain Proteins Interact with the WD-Repeat/bHLH/MYB Complexes to Regulate Jasmonate-Mediated Anthocyanin Accumulation and Trichome Initiation in *Arabidopsis Thaliana*. *Plant Cell* **2011,** *23* (5), 1795–1814.
15. Schaller, M. The Behavioural Immune System and the Psychology of Human Sociality. *Philos. Trans. R. Soc. Lond., B, Biol. Sci.* **2011,** *366* (1583), 3418–3426.
16. Pirbalouti, A. G.; Mirbagheri, H.; Hamedi, B.; Rahimi, E. Antibacterial Activity of the Essential Oils of Myrtle Leaves Against Erysipelothrix Rhusiopathiae. *Asian Pac. J. Trop. Biomed.* **2014,** *4,* S505–S509.
17. Tsukada, K.; Takahashi, K.; Nabeta, K. Biosynthesis of Jasmonic Acid in a Plant Pathogenic Fungus, *Lasiodiplodia Theobromae*. *Phytochem* **2010,** *71* (17–18), 2019–2023.
18. Avanci, N.; Luche, D.; Goldman, G.; Goldman, M. Jasmonates are Phytohormones with Multiple Functions, Including Plant Defense and Reproduction. *Genet. Mol. Res.* **2010,** *9* (1), 484–505.
19. Rohwer, C. L.; Erwin, J. E. Horticultural Applications of Jasmonates. *J. Hortic. Sci. Biotechnol.* **2008,** *83* (3), 283–304.
20. Wasternack, C.; Kombrink, E. Jasmonates: Structural Requirements for Lipid-Derived Signals Active in Plant Stress Responses and Development. *ACS Chem. Biol.* **2010,** *5* (1), 63–77.
21. Koo, A. J.; Howe, G. A. Catabolism and Deactivation of the Lipid-derived Hormone Jasmonoyl-Isoleucine. *Front. Plant Sci.* **2012,** *3,* 19.
22. Wakuta, S.; Hamada, S.; Ito, H.; Matsuura, H.; Nabeta, K.; Matsui, H. Identification of A β-Glucosidase Hydrolyzing Tuberonic Acid Glucoside in Rice (*Oryza Sativa* L.). *Phytochemistry* **2010,** *71* (11–12), 1280–1288.

23. Seto, Y.; Hamada, S.; Ito, H.; Masuta, C.; Matsui, H.; Nabeta, K.; Matsuura, H. Tobacco Salicylic Acid Glucosyltransferase is Active Toward Tuberonic Acid (12-hydroxyjasmonic Acid) and is Induced by Mechanical Wounding Stress. *Biosci. Biotechnol. Biochem.* **2011,** *75* (12), 2316–2320.
24. Nam, K.; Yoshihara, T. Bioorganic Chemistry of The Induction of Potato Tuber Formation: A Review. *Fruit Veg. Cereal Sci. Biotech.* **2008,** *2,* 69–81.
25. Honda, I.; Seto, H.; Turuspekov, Y.; Watanabe, Y.; Yoshida, S. Inhibitory Effects of Jasmonic Acid and Its Analogues on Barley (Hordeum Vulgare L.) Anther Extrusion. *Plant Growth Regul.* **2006,** *48* (3), 201–206.
26. Engelberth, J.; Koch, T.; SchüLer, G. D.; Bachmann, N.; Rechtenbach, J.; Boland, W. Ion Channel-Forming Alamethicin is a Potent Elicitor of Volatile Biosynthesis and Tendril Coiling. Cross Talk Between Jasmonate and Salicylate Signaling in Lima Bean. *Plant Physiol.* **2001,** *125* (1), 369–377.
27. Kazan, K. Diverse Roles of Jasmonates and Ethylene in Abiotic Stress Tolerance. *Trends Plant Sci.* **2015,** *20* (4), 219–229.
28. Wasternack, C. Action of Jasmonates in Plant Stress Responses and Development—Applied Aspects. *Biotechnol. Adv.* **2014,** *32* (1), 31–39.
29. Ali, M.; Baek, K.-H. Jasmonic Acid Signaling Pathway in Response to Abiotic Stresses in Plants. *Int. J. Mol. Sci.* **2020,** *21* (2), 621.
30. Wasternack, C.; Strnad, M. Jasmonate Signaling in Plant Stress Responses and Development–Active and Inactive Compounds. *New Biotechnol.* **2016,** *33* (5), 604–613.
31. Farmer, E. E.; Ryan, C. A. Octadecanoid Precursors of Jasmonic Acid Activate the Synthesis of Wound-Inducible Proteinase Inhibitors. *Plant Cell* **1992,** *4* (2), 129–134.
32. Ryan, C. A. The Systemin Signaling Pathway: Differential Activation of Plant Defensive Genes. *Biochim. Biophys. Acta, Protein Struct. Mol. Enzymol.* **2000,** *1477* (1–2), 112–121.
33. Howe, G. A.; Lee, G. I.; Itoh, A.; Li, L.; Derocher, A. E. Cytochrome P450-Dependent Metabolism of Oxylipins in Tomato. Cloning and Expression of Allene Oxide Synthase and Fatty Acid Hydroperoxide Lyase. *Plant Physiol.* **2000,** *123* (2), 711–724.
34. Shoji, T. The Recruitment Model of Metabolic Evolution: Jasmonate-Responsive Transcription Factors and a Conceptual Model for the Evolution of Metabolic Pathways. *Front. Plant Sci.* **2019,** *10,* 560.
35. Wasternack, C.; Strnad, M. Jasmonates are Signals in the Biosynthesis of Secondary Metabolites-Pathways, Transcription Factors and Applied Aspects-A Brief Review. *New Biotechnol.* **2019,** *48,* 1–11.
36. Chen, X.; Wang, D.-D.; Fang, X.; Chen, X.-Y.; Mao, Y.-B. Plant Specialized Metabolism Regulated by Jasmonate Signaling. *Plant Cell Physiol.* **2019,** *60* (12), 2638–2647.
37. De Geyter, N.; Gholami, A.; Goormachtig, S.; Goossens, A. Transcriptional Machineries in Jasmonate-Elicited Plant Secondary Metabolism. *Trends Plant Sci.* **2012,** *17* (6), 349–359.
38. Panda, S.; Jozwiak, A.; Sonawane, P. D.; Szymanski, J.; Kazachkova, Y.; Vainer, A.; Vasuki, H.; Almekias-Siegl, E.; Dikaya, V.; Bocobza, S. Steroidal Alkaloids Defense Metabolism and Plant Growth are Modulated by the Joint Action of Gibberellin and Jasmonate Signaling. *New Phytol.* **2021.**

39. Mc Gee, D.; Archer, L.; Parkes, R.; Fleming, G. T.; Santos, H. M.; Touzet, N. The Role of Methyl Jasmonate in Enhancing Biomass Yields and Bioactive Metabolites in *Stauroneis* sp. (Bacillariophyceae) Revealed by Proteome and Biochemical Profiling. *J. Proteom.* **2021,** *249,* 104381.
40. Paponov, M.; Antonyan, M.; Slimestad, R.; Paponov, I. A. Decoupling of Plant Growth and Accumulation of Biologically Active Compounds in Leaves, Roots, and Root Exudates of *Hypericum Perforatum* L. by The Combination of Jasmonate and Far-red Lighting. *Biomolecules* **2021,** *11* (9), 1283.
41. Zhu, T.; Herrfurth, C.; Xin, M.; Savchenko, T.; Feussner, I.; Goossens, A.; De Smet, I. Warm Temperature Triggers JOX and ST2A-Mediated Jasmonate Catabolism To Promote Plant Growth. *Nat. Commun.* **2021,** *12* (1), 1–8.
42. Tao, X.; Wu, Q.; Li, J.; Cai, L.; Mao, L.; Luo, Z.; Li, L.; Ying, T. Exogenous Methyl Jasmonate Regulates Sucrose Metabolism in Tomato During Postharvest Ripening. *Postharvest Biol. Technol.* **2021,** *181,* 111639.
43. Delgado, C.; Mora-Poblete, F.; Ahmar, S.; Chen, J.-T.; Figueroa, C. R. Jasmonates and Plant Salt Stress: Molecular Players, Physiological Effects, and Improving Tolerance by Using Genome-Associated Tools. *Int. J. Mol. Sci.* **2021,** *22* (6), 3082.
44. Jamalian, S.; Truemper, C.; Pawelzik, E. Jasmonic and Abscisic Acid Contribute to Metabolism Re-adjustment in Strawberry Leaves Under NaCl Stress. *Int. J. Fruit Sci.* **2020,** *20* (Suppl 2), S123–S144.
45. Dombrecht, B.; Xue, G. P.; Sprague, S. J.; Kirkegaard, J. A.; Ross, J. J.; Reid, J. B.; Fitt, G. P.; Sewelam, N.; Schenk, P. M.; Manners, J. M. MYC2 Differentially Modulates Diverse Jasmonate-Dependent Functions in *Arabidopsis*. *Plant Cell* **2007,** *19* (7), 2225–2245.
46. Shen, Q.; Lu, X.; Yan, T.; Fu, X.; Lv, Z.; Zhang, F.; Pan, Q.; Wang, G.; Sun, X.; Tang, K. The Jasmonate-Responsive Aa MYC 2 Transcription Factor Positively Regulates Artemisinin Biosynthesis in *Artemisia Annua*. *New Phytol.* **2016,** *210* (4), 1269–1281.
47. Paul, P.; Singh, S. K.; Patra, B.; Sui, X.; Pattanaik, S.; Yuan, L. A Differentially Regulated AP 2/ERF Transcription Factor Gene Cluster Acts Downstream of A MAP Kinase Cascade to Modulate Terpenoid Indole Alkaloid Biosynthesis in *Catharanthus Roseus*. *New Phytol.* **2017,** *213* (3), 1107–1123.
48. Xu, Y.-H.; Wang, J.-W.; Wang, S.; Wang, J.-Y.; Chen, X.-Y. Characterization of GaWRKY1, a Cotton Transcription Factor that Regulates the Sesquiterpene Synthase Gene (+)-δ-Cadinene Synthase-A. *Plant Physiol.* **2004,** *135* (1), 507–515.
49. Skibbe, M.; Qu, N.; Galis, I.; Baldwin, I. T. Induced Plant Defenses in the Natural Environment: *Nicotiana Attenuata* WRKY3 and WRKY6 Coordinate Responses to Herbivory. *Plant Cell* **2008,** *20* (7), 1984–2000.
50. Zhang, H.-B.; Bokowiec, M. T.; Rushton, P. J.; Han, S.-C.; Timko, M. P. Tobacco Transcription Factors NTMYC2a and NTMYC2b Form Nuclear Complexes With The NtJAZ1 Repressor and Regulate Multiple Jasmonate-Inducible Steps in Nicotine Biosynthesis. *Mol Plant.* **2012,** *5* (1), 73–84.
51. Todd, A. T.; Liu, E.; Polvi, S. L.; Pammett, R. T.; Page, J. E. A Functional Genomics Screen Identifies Diverse Transcription Factors that Regulate Alkaloid Biosynthesis in *Nicotiana Benthamiana*. *Plant J.* **2010,** *62* (4), 589–600.
52. Sun, Y.; Niu, Y.; Xu, J.; Li, Y.; Luo, H.; Zhu, Y.; Liu, M.; Wu, Q.; Song, J.; Sun, C. Discovery of WRKY Transcription Factors Through Transcriptome Analysis and

Characterization of A Novel Methyl Jasmonate-Inducible PqWRKY1 Gene From Panax Quinquefolius. *Plant Cell Tissue Organ Cult.* **2013,** *114* (2), 269–277.
53. Lenka, S. K.; Nims, N. E.; Vongpaseuth, K.; Boshar, R. A.; Roberts, S. C.; Walker, E. L. Jasmonate-Responsive Expression of Paclitaxel Biosynthesis Genes in *Taxus Cuspidata* Cultured Cells is Negatively Regulated by The bHLH Transcription Factors TcJAMYC1, TcJAMYC2, and TcJAMYC4. *Front. Plant Sci.* **2015,** *6,* 115.
54. Li, S.; Zhang, P.; Zhang, M.; Fu, C.; Yu, L. Functional Analysis of a WRKY Transcription Factor Involved in Transcriptional Activation of the DBAT Gene in Taxus Chinensis. *Plant Biol.* **2013,** *15* (1), 19–26.
55. Zhou, Y.; Sun, W.; Chen, J.; Tan, H.; Xiao, Y.; Li, Q.; Ji, Q.; Gao, S.; Chen, L.; Chen, S. Sm MYC2a and Sm MYC2b Played Similar but Irreplaceable Roles in Regulating the Biosynthesis of Tanshinones and Phenolic Acids in *Salvia Miltiorrhiza. Sci. Rep.* **2016,** *6* (1), 1–12.
56. Wu, J.; Baldwin, I. T. New Insights into Plant Responses to the Attack From Insect Herbivores. *Annu. Rev. Genet.* **2010,** *44,* 1–24.
57. Zhang, L.; Zhang, F.; Melotto, M.; Yao, J.; He, S. Y. Jasmonate Signaling and Manipulation by Pathogens and Insects. *J. Exp. Bot.* **2017,** *68* (6), 1371–1385.
58. Sheard, L. B.; Tan, X.; Mao, H.; Withers, J.; Ben-Nissan, G.; Hinds, T. R.; Kobayashi, Y.; Hsu, F.-F.; Sharon, M.; Browse, J. Jasmonate Perception by Inositol-Phosphate-Potentiated COI1–JAZ Co-receptor. *Nature* **2010,** *468* (7322), 400–405.
59. Colinas, M.; Goossens, A. Combinatorial Transcriptional Control of Plant Specialized Metabolism. *Trends Plant Sci.* **2018,** *23* (4), 324–336.
60. Guo, Q.; Yoshida, Y.; Major, I. T.; Wang, K.; Sugimoto, K.; Kapali, G.; Havko, N. E.; Benning, C.; Howe, G. A. JAZ Repressors of Metabolic Defense Promote Growth and Reproductive Fitness in *Arabidopsis. Proc. Natl. Acad. Sci. U. S. A.* **2018,** *115* (45), E10768–E10777.
61. Yu, Z.-X.; Wang, L.-J.; Zhao, B.; Shan, C.-M.; Zhang, Y.-H.; Chen, D.-F.; Chen, X.-Y. Progressive Regulation of Sesquiterpene Biosynthesis in *Arabidopsis* and *Patchouli* (Pogostemon Cablin) by The MiR156-Targeted SPL Transcription Factors. *Mol Plant.* **2015,** *8* (1), 98–110.
62. Li, Z.; Min, D.; Fu, X.; Zhao, X.; Wang, J.; Zhang, X.; Li, F.; Li, X. The Roles of SLMYC2 in Regulating Ascorbate-Glutathione Cycle Mediated by Methyl Jasmonate in Postharvest Tomato Fruits Under Cold Stress. *Sci. Hortic.* **2021,** *288,* 110406.
63. Li, H.; Guo, Y.; Lan, Z.; Xu, K.; Chang, J.; Ahammed, G. J.; Ma, J.; Wei, C.; Zhang, X. Methyl Jasmonate Mediates Melatonin-induced Cold Tolerance of Grafted Watermelon Plants. *Hortic. Res.* **2021,** *8* (1), 1–12.
64. Chen, J.; Miao, W.; Fei, K.; Shen, H.; Zhou, Y.; Shen, Y.; Li, C.; He, J.; Zhu, K.; Wang, Z. Jasmonates Alleviate the Harm of High-temperature Stress During Anthesis To Stigma Vitality of Photothermosensitive Genetic Male Sterile Rice Lines. *Front. Plant Sci.* **2021,** *12,* 412.
65. Zhao, Y.; Song, C.; Brummell, D. A.; Qi, S.; Lin, Q.; Duan, Y. Jasmonic Acid Treatment Alleviates Chilling Injury in Peach Fruit by Promoting Sugar and Ethylene Metabolism. *Food Chem.* **2021,** *338,* 128005.
66. Yang, J.; Miao, W.; Chen, J. Roles of Jasmonates and Brassinosteroids in Rice Responses to High Temperature Stress—A Review. *Crop J.* **2021**.

67. Liu, X.; Chi, H.; Yue, M.; Zhang, X.; Li, W.; Jia, E. The Regulation of Exogenous Jasmonic Acid on UV-B Stress Tolerance in Wheat. *J. Plant Growth Regul.* **2012,** *31* (3), 436–447.
68. Yao, Y.; You, J.; Ou, Y.; Ma, J.; Wu, X.; Xu, G. Ultraviolet-B Protection of Ascorbate and Tocopherol in Plants Related with Their Function on the Stability on Carotenoid and Phenylpropanoid Compounds. *Plant Physiol. Biochem.* **2015,** *90,* 23–31.
69. Kakani, V.; Reddy, K.; Zhao, D.; Sailaja, K. Field Crop Responses to Ultraviolet-B Radiation: A Review. *Agric. For. Meteorol.* **2003,** *120* (1–4), 191–218.
70. Demkura, P. V.; Abdala, G.; Baldwin, I. T.; Ballaré, C. L. Jasmonate-Dependent and-Independent Pathways Mediate Specific Effects of Solar Ultraviolet B Radiation on Leaf Phenolics and Antiherbivore Defense. *Plant Physiol.* **2010,** *152* (2), 1084–1095.
71. Qi, J.; Zhang, M.; Lu, C.; Hettenhausen, C.; Tan, Q.; Cao, G.; Zhu, X.; Wu, G.; Wu, J. Ultraviolet-B Enhances the Resistance of Multiple Plant Species to Lepidopteran Insect Herbivory Through the Jasmonic Acid Pathway. *Sci. Rep.* **2018,** *8* (1), 1–9.
72. Choudhary, K. K.; Singh, S.; Agrawal, M.; Agrawal, S. Role of Jasmonic and Salicylic Acid Signaling in Plants under UV-B Stress. In *Jasmonates and Salicylates Signaling in Plants*; Springer, 2021; pp 45–63.
73. Choudhary, K. K.; Agrawal, S. Cultivar Specificity of Tropical Mung Bean (*Vigna Radiata* L.) to Elevated Ultraviolet-B: Changes in Antioxidative Defense System, Nitrogen Metabolism and Accumulation of Jasmonic and Salicylic Acids. *Environ. Exp. Bot.* **2014,** *99,* 122–132.
74. Choudhary, K. K.; Agrawal, S. Ultraviolet-B Induced Changes in Morphological, Physiological and Biochemical Parameters of Two Cultivars of Pea (*Pisum Sativum* L.). *Ecotoxicol. Environ. Saf.* **2014,** *100,* 178–187.

CHAPTER 3

Genetic and Molecular Basis of Jasmonate Metabolism and Interactions with Other Phytohormones

NEELOFER NABI[1], SEEMA SINGH[1], PEER SAFFEULLAH[2], and SHAHID UMAR[2]

[1]*Department of Botany, University of Kashmir, Srinagar, Jammu and Kashmir, India*

[2]*Department of Botany, Jamia Hamdard, New Delhi, India*

ABSTRACT

Jasmonates are known to play multifunctional roles in crop growth and development. The genetic and molecular basis of jasmonate metabolism has been summarized in this chapter, which have shown substantial role in controlling plant metabolism under stress and non-stress conditions. During stress conditions, the regulation of jasmonates synthesis is impaired, which is associated with a cascade of metabolic events like signal transduction. Although mitogen-activated protein kinases (MAPKs) are vital components of jasmonic acid (JA) signaling and biosynthesis pathways, phytohormones like abscisic acid, auxin, cytokinin, ethylene, gibberellin, salicylic acid, and brassinosteroids are also crucial mediators of crop growth and development during JA synthesis and signal transduction. This chapter also discusses crosstalk of JA with other phytohormones in regulating balance between crop growth and defense responses, which involves proteins like jasmonate ZIM domain protein (JAZ) and

Jasmonates and Plant Defense: An Eco-Friendly and Sustainable Approach for Stress Tolerance.
Uzma Azeem, & Khalid Rehman Hakeem (Eds.)

MYC2, which are key players in JA signaling. The crosstalk with these phytohormones affects JA synthesis synergistically or antagonistically. The investigation of different signaling molecules could be helpful to inspect the intricacies of signal transduction pathway of JA at molecular level. Further research is required to properly address the queries like actions of metabolites and receptors involved in JA signaling. The role of transcription factors involving MAPKs, SIPK, WIPK, and WRKY need to be fully explored to comprehend mechanism of action of JAs.

3.1 INTRODUCTION

Jasmonates are a class of phytohormones needed for crop growth and development. Jasmonic acid (JA) is the primary jasmonate which is metabolized into a wide variety of other derivatives.[1] Enzymes such as allene oxide synthase, allene oxide cyclase, and lipoxygenase regulate JA biosynthesis via octadecanoid pathway. The biological activities of all other metabolites of JA are not clear except JA-isoleucine (JA-Ile) conjugate, whose role is well established. The derivatives of JA have different biological roles in comparison to classic JA functions. For example, induction of tuber formation in potatoes is attributed to hydroxyl derivative of JA (12OH-JA) and its glucosylated derivative (O-glucoside JA), also called tuberonic acid.[2] Nyctinastic leaf movements of various plants such as rain trees (*Samaneasaman*) and legumes (*Medicagotruncatula*) are attributed to 12-O-Glc-JA, known as a leaf-closing factor. 12OH-JA is produced by fungal monooxygenase in response to rice blast by the infection of fungus *Magnaportheoryza*, thus acting as a virulent factor.[3] In tomatoes, 12OH-JA inhibits various typical JA responses which include inhibition growth of roots and seed germination or expression of JA-inducible genes.[4] Aubert et al.[5] reported indirect evidence suggesting the signaling roles of 12OH-JA-Ile and 12COOH-JA-Ile in imparting host resistance against fungal pathogen *Botrytis cineria*. Jasmonate methyl transferase and jasmonate-amidosynthetase catalyze biosynthesis of methyl jasmonate and JA-Ile, respectively. These derivatives are formed by conjugation, hydroxylation, carboxylation, decarboxylation, sulfation, etc. Various enzymes are required for these reactions to occur. The endogenic JA levels differ with the prevailing environmental conditions, as the genes involved in JA modulation are kept under regulation during stress conditions; hence, JA is main plant stress hormone.[6] Two inducible, crosslinked catabolic pathways are required to turn over JA-Ile in *Arabidopsis*: The oxidative pathway is catalyzed by cytochrome P450 enzymes belonging to subfamily 94 (CYP94), and the deconjugation is catalyzed by amidohydrolases. The genetic inactivation of cytochrome P450 enzymes and amidohydrolases has immense influence on JAs (jasmonates) homeostasis, including JA-Ile accretion, which although correlates with increased tolerance to biotic stress in absence of over induction of negative signaling regulators. However, the jasmonic acid (JA) oxidase 2 (jao2) mutant leads to the impairment of JA oxidation which induces constitutive defense reactions without enriching JA-Ile levels in naive leaves and amplifies resistance to subsequent biotic stress.[7]

3.2 ENZYMES INVOLVED IN JA SYNTHESIS AND THEIR GENETIC CONTROL

3.2.1 CYP94S

In *Arabidopsis,* cytochrome P450 (CYP94 family proteins) comprises six members that catalyze ω-oxidation pathway. These are exactly homologous to the CYP86 clade of P450s that were earlier specified as FA ω-hydroxylases.[8] Three CYP94 genes (CYP94B1, CYP94B3, and CYP94C1) were recognized by gene co-expression with JA-inducible marker genes, and, besides their expression, was eradicated in the coi1 mutant, which is in line with the reduced 12OH-JA-Ile levels and impaired turnover of JA-Ile in coi1.[9] Mutational change such as functional loss in CYP94B3 resulted in strong decrease in 12-OH-JA-Ile accretion and simultaneous enhancement in JA-Ile.[10–12] In in vivo studies, overexpression of CYP94B3 led to the opposite effects, confirming the precursor and product association between JA-Ile and 12OH-JA-Ile.[10] Similar enzymatic functions are carried by CYP94B1, homolog of CYP94B3, which contributes in JA-Ile 12-hydroxylation activity in wounded leaves and variation in expression in different tissues.[13] However, these two CY94Bs approximately account for 95% in hydroxylation reaction of JA-Ile to 12-OH-JA-Ile in wounded leaves. Further catalyzation of 12OH-JA-Ile to 12COOH-JA-Ile is mostly acted by CYP94C1. But, the defective hindrance of 12COOH-JA-Ile production in the cyp94c1 single loss-of-function and total abolition of 12COOH-JA-Ile in the cyp94b3cyp94c1 double mutant, and the in vitro enzyme assay observations put forward the overlapping roles of CYP94B3 and CYP94C1 in the two consecutive oxidation steps. The enhanced catalysis of JA-Ile in the CYP94B1 and CYP94B3 overexpressing plants led to typical JA-deficient phenotypes implying that the ω-oxidation pathway leads to JA signal inactivation.[10] CYP94C1 overexpression, on the other hand, did not result in strong depletion of JA-Ile yet caused attenuated defense response against *Botrytis cinerea* infection, leading authors to propose the importance of hormones to the oxidation ratio rather than absolute changes in JA-Ile levels for the resistance against this pathogen.[5] However, enhanced oxidation of JA-Ile is correlated inversely to molecule's capacity to enhance the COI1-JAZ receptor complex formation as revealed by in vitro protein pull-down experiments,[9] suggesting the formation of 12OH-JA-Ile from JA-Ile, and to 12COOH-JA-Ile, probably attenuate the COI1-dependent signaling pathway of JA. Although possessing three to fourfold enhanced amount of JA-Ile as compared to wild type, the double loss-of-function mutant cyp94b1cyp94b3 possess a wide array of phenotypes that have resemblance that of the JA-Ile-deficiency phenotypes.[14] For example, these phenotypes possess enhanced resistant to wound-induced restraintment of growth, accreted less anthocyanin, and little trichomes, and also such phenotypes were more liable to insect herbivory. In addition, global metabolite investigation found that wounded cyp94b1cyp94b3 mutants drastically reduced number of known metabolites that are required for positive regulation of JA-Ile. These observations are contradictory to plants over-accumulating JA-Ile; however, these phenotypes can be well explained in plants that are not capable to sense JA-Ile. However, JA signaling pathway did not reflect any defectiveness in mutants, because these mutants responded same as wild-type in response to exogenous JA. Thus, reveals complexity of the plants long-term response to prevailing conditions that can chronically uprise endogenous JA levels. Besides, hydroxylation of JA-Ile CYP94B1 and CYP94B3 also control other functions. In such scenario, the another hypothetical signaling molecule produced by CYP94Bs have huge effect as JA-Ile in modulating wound response, and, at the same time, should have capacity of overriding JA-Ile signaling in JA-Ile overaccumulating mutant background.

3.2.2 ILL6 AND IAR3

The hydrolytic pathway in *Arabidopsis* is metabolized by IAR3 and ILL6 of the ILR1-like amidohydrolase (IAH) family. Many members of this seven-gene family are reported, possessing capacity of executing cleavage reactions for several auxin conjugates.[15] In *N. attenuata*, an IAR3 homologue was first recognized to act as a JA-Ile hydrolase[16] soon followed by lots of reports about the corresponding role of *Arabidopsis* genes.[17–18] The wound responses and JA inducible feature of gene expression helped in their recognition.[17] Transcripts of these genes in the IAH family increase in response to wounding and JA treatment.[19] On the contrary, ILR1 was later left out from being responsible in JA metabolism.[18] Recombinant IAR3 and ILL6 enzymes showed broad substrate specificity against several amino acid-substituted conjugates of JA besides JA-Ile, including -Leu, -Val, -Gln, -Thr, and -Phe conjugates. Also, both enzymes were found to hydrolyze 12OH-JA-Ile[18]; besides, further amendment of the –OH group, such as oxidation (12COOH-JA-Ile), glucosylation (12-O-Glc-JA-Ile), or sulfation (12-HSO_4-JA-Ile) hindered the hydrolytic activity of these enzymes.[18] The enhanced sensitivity of these hydrolases is by the change at the C12 position and is in agreement with the proposal that these derivatives might be the by-product of JA metabolism; also, the pentenyl side chain amendment is an efficient way to inactivate JA-Ile. The main variation between the two enzymes is their catalytic property for auxin amino acid conjugates, as only IAR3, is able to hydrolyze auxin conjugates, but not ILL6. It has been found, that plants either mutated in or impaired in expressing ILL6 and IAR3, the endogenous JA level found in such plants are in agreement with the observations of the in vitro enzyme assays. An enhanced JA-Ile profile was noted in ill6 loss-of-function mutants and was found to be very prominent in ill6 iar3 double mutant suggesting their role in JA-Ile turnover in plants. Additionally, overexpressing of ILL6 or IAR3 reduced JA-Ile, thus further suggesting their importance in vivo to catalyze JA-Ile. 12OH-JA-Ile amount in the loss-of-function and overexpressing plants followed the scheme of JA-Ile apart from that the results were even more stunning.[17–19] Hydrolysis of 12OH-JA-Ile is the only known metabolic pathway in plants to produce 12-hydroxy-JA (12OH-JA). However, persevering levels of 12OH-JA in the iar3 ill6 double mutant indicated that there may be alternate routes: for example, by direct 12-hydroxylation of JA. A monooxygenase performing the direct 12-hydroxylation of JA has been found recently in rice blast fungus.[3] The enzymes, named JASMONATE-INDUCED OXYGENASES (JOXs), belong to the 2-oxoglutarate (2OG) Fe(II)-dependent oxygenase family and were able to catalyze 12-hydroxylation of JA in vitro. The quadruple loss of function mutants of the four JOX genes displayed phenotypes consistent with reduced turnover of JA. Similar to mutants of CYP94Bs, plants overexpressing ILL6 or IAR3 possess the same JA-Ile-deficient effect, providing evidence that the hydrolysis of JA-Ile is not a futile only to be reactivated to JA-Ile by JAR1. It is likely that there are distinctive pools of JA: one aiding as a substrate for JA-Ile biosynthesis, and the other, being metabolized, and the two pools are spatially separated by different subcellular compartments in which metabolic enzymes are confined.[18] Apart from JA-related phenotypes, IAH-overexpressing lines exhibited auxin-related phenotypes, that is, IAR3-over-expressing plants were hypersensitive to auxin conjugates. The reason for this could be that IAR3 released free auxin, indole-3-acetic-acid (IAA), which is the bioactive form of auxin. Consequently, a remarkable mechanism of hormonal crosstalk was anticipated between the JA and auxin signaling pathways during stress conditions.

3.2.3 JMT

Methyl jasmonate (MeJA), the chemical form of JA, was identified and purified from flowers of jasmine (*Jasminum grandiflorum*).[20] JA methylation is acted upon by S-adenosyl-L-methionine at the C1 carboxyl group; however, in *Arabidopsis*, it is catalyzed by JA carboxyl methyltransferase (JMT).[21] MeJA possessing the property of volatility and effectively crossing cell membranes, thus MeJA acts as a long-distance inter- and intracellular mobile JA signal. Although, to induce the JA response by MeJA, there should be demethylation of MeJA by an esterase[22] to JA followed by conjugation to JA-Ile which then can interrelate with the COI1-JAZ receptor system.[23] Further proof for such a scenario was obtained by the ectopic expression of JMT in *N. attenuata* which gave rise in amassing of MeJA at the expense of JA-Ile and, more importantly, the decrease of the defense gene expression.[24] MeJA is formed by expression of *JMT*.[21] Earlier, a floral nectary-specific gene *NTR1* was extracted from

Brassica campestris, which structurally resembles *SAMT* which encodes *S*-adenosyl-l-methionine:JA carboxyl methyl transferase.

3.2.4 *ST2A*

12-hydroxyjasmonic acid sulfate (12HSO$_4$-JA) is sulfated derivative of JA that occurs predominantly in *Arabidopsis* and many plant species. In vitro studies have shown that the hydroxyjasmonatesulfo-transferase (ST2A) catalyzes the formation of 12HSO$_4$-JA found in *Arabidopsis.*[4,25] ST2A uses phosphoadenosine 5/phosphosulfate (PAPS) as the activated sulfate donor and displays stringent substrate specificity for hydroxylated JAs, 12OH-JA and 11OH-JA. 12OH-JA is likely to be precursor of 12HSO$_4$-JA, and exhibits similar kinetics as its precursor molecule. The transcripts of ST2A gene can be provoked by wounding in a COI1-dependent manner they can also be provoked by JA-Ile or 12OH-JA. Unlike JA-Ile, 12HSO$_4$-JA is not bioactive as it could not impede root growth or seed germination and was unable to prompt expression of JA-responsive marker gene.[4,19,25] These observations also reveal that 12HSO$_4$-JA is not reverted to JA or JA-Ile. Similar is the case with 12OH-JA, which also does not induce the classical JA-Ile response. Gidda et al.[25] treated *Arabidopsis* with 12OH-JA and found that there is no enhancement of JA after 36 h of incubation.

3.3 METABOLISM OF JA

JA is produced from cis-(+)-12-oxo-phytodienoic acid (OPDA) using fatty acids, either free or esterified to membrane glycerolipids in the plastid. In peroxisomes, OPDA is converted to JA, (+)-7-iso-JA by three rounds of ß oxidation cycle. JA is metabolized into inactive, partially active, and active compounds (Figure 3.1) by various metabolic pathways.[26–28] Methylation to JA-methyl ester, JA glucosyl ester formation, decarboxylation to cis-jasmone, hydroxylation to 12-hydroxyJA, sulfation of 12-hydroxy-JA, O-glucosylation of 12-hydroxy-JA to 12-O-glucosyl-JA, conjugation of JA with amino acids, specially isoleucine to provide JA-Ile, methylation to JA-Ile methyl ester, 12-hydroxylation of JA-Ile, carboxylation of 12-OH-JA-Ile, O-glucosylation of JA-Ile to 12-O-glucosyl-JA-Ile, and formation of JA-Ile glucosyl ester are specified. These reactions require enzymes like jasmonic acid oxidase2 (JAO2), jasmonate-induced oxidase 1,2,3,4 (JOX1,2,3,4),

JAR1, jasmonoyl-isoleucine synthetase, JA-Ile-12-hydroxylase CYP94B3, 12-OH-JA-Ile carboxylase CYP94C1, amidohydrolases IAR3 and ILL6, JMT, JA methyl transferase, ST2a, and 12-OH-JA sulfotransferase. In cytosol, an active form of the - (+)-7-iso-JA-L-Ile [also known as (3R,7S)-JA-L-Ile], is synthesized by the conjugation to L-isoleucine catalyzed by JAR1. JA-Ile can either be acted upon by CYP94 enzymes or IAHs and is thus oxidized to 12OH-JA-Ile and 12COOH-JA-Ile and hydrolyzed to JA, respectively. 12OH-JA-Ile is hydrolyzed by IAH enzymes to 12OH-JA, which then can be further acted by either ST2A orglucosyltransferase for formation of sulfated or glucosylated derivatives of JA. Both the CYP94s and IAHs are present in ER. MeJA, volatile methyl derivative of JA, is catalyzed by a JA carboxyl methyltransferase (JMT). The reactions and enzymes/genes involved in JA metabolism are discussed in details below.

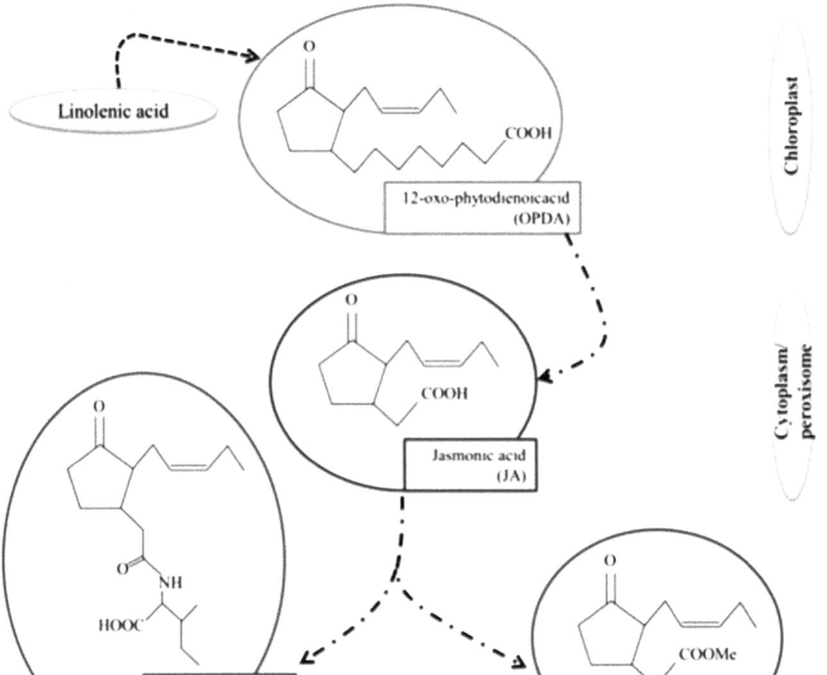

FIGURE 3.1 The pathway involved in synthesis of jasmonic acid (JA) and its derivatives in different cell organelles.

3.3.1 CONJUGATION

In cytosol, bioactive JA is produced by JASMONATE RESISTANT 1 (JAR1) by the formation of an amide-linked conjugate with amino acids, Jasmonoyl-isoleucine synthetase, which is a member of the GH3 gene family (AtGH3.11) in *Arabidopsis*, also found in soyabean for auxin induction,[28] leads to JA-Ile production, an ultimate step in JA perception.[29] The GH3 protein family comprises acyl acid-amidosynthetases which activate diverse substrates such as indole-3-acetic acid (IAA), benzoic acids, and JA through the Mg ATP-dependent adenylation in their carboxyl groups.[30] The function of JA-Ile became clear by recognizing it as the most biologically active JA compound,[31] by SCF COI1–JAZ co-receptor complex formation mediated by JA-Ile[23,33] and JA-Ile as the ligand.[34] This layout has been utilized to model a ligand-based antagonist of JA-Ile by modifying coronatine[35] or for creation of inhibitors of JAR1, for example, jarin1 through chemical screening.[36] jar1 Plants are resistant to JA-induced inhibition of root growth, decrease in the plants own expression of JA responsive-marker genes, and are more liable to the soil fungus *Pythiumirregulare* and insect herbivory.[37,38] Also, pharmacological studies have revealed the value of this catalytic step for JA signaling.[39] Treatment of plants with jarin-1 resulted in hampered degradation in JA-induced JAZ proteins, JA-responsive gene expression, and anthocyanin accretion, signifying that inhibition of JAR1 blocks JA signaling. Crystallization of JAR1 showed detailed explanation for the formation of epimer (+)-7-iso-JA-Ile.[30] However, (+)-7-iso-JA-Ile is predominantly produced in response to wounding. The amount of JA-Ile plays an imperative regulatory role for all JA-dependent progressions and is continued by JAR1 activity and ensuing metabolic reactions such as hydroxylation, hydrolysis, and carboxylation of conjugates. Recently, (+)-7-iso-JA-Ala, (+)-7-iso-JA-VAL, (+)-7-iso-JA-Leu, and (+)-7-iso-JA-Met have also been recognized as endogenous derivatives of JA derivatives via respective coronafacic acid conjugates.[40] These conjugates depicted distinctive binding in the COI1–JAZ interaction assay. The internally originated JA is a (3R,7S) isomer (or (+)-7-iso-JA) but this is converted into thermodynamically more stable isomer (3R,7R) spontaneously, at an equilibrium.[31,39] JA-Ile(3R,7S) is more active in stimulating complex establishment with the COI1-JAZ receptors.[31] There is further need for investigation on JAR1's molecular mechanism to impact the stereochemistry and epimerization of JA-Ile.[31] JA-Ile levels persist in jar1, being approximately 10–20% of wild type in seedlings and wounded mature leaves.[29,36] DFL2, a member of the GH3 gene family member, possesses the closest sequence homology to JAR1, and shows a vital role in red light-specific hypocotyl elongation,[40] but its function in JA metabolism remains unclear. Conjugation of JA to Ile is a unique modification which leads to deactivation of the signal for the purpose of temporary storage or catabolic turnover according to the other known cases of plant hormones but JA-Ile activates the signal.[15,41] Crystallographic studies of JAR1 did not unveil obvious candidate motifs to surmise post translational modification.[31] Protein–protein interaction studies have revealed a protein (FIP1) interacting with JAR1 that may significantly regulate phytochrome A-mediated signaling.[42]

3.3.2 CATABOLISM OF JA-ILE

The amounts of growth hormones in plants are regulated by both biosynthesis and catabolism. The accumulation of JA-Ile in leaves by wounding is spontaneously declined with time confirming the presence of metabolic pathways for conversion of JA-Ile. There are two major pathways for JA-Ile catabolism. Subcellular localization investigations have revealed ER localization of the enzymes catalyzing the two pathways; hence, JA-Ile catabolism occurs in ER.[13,18,43] In ω-oxidation pathway, there is oxidation of the terminal carbon at the methyl end of the pentenyl side chain of JA-Ile, thus forming 12-hydroxy-JA-Ile (12OH-JA-Ile) and its further oxidation to 12COOH-JA-Ile.[10,12,44] Hydrolytic pathway is the second catabolic pathway where JA-Ile is hydrolyzed at the amide bond to form JA and Ile.[16–19] Both pathways are controlled by COI1-dependent JA signaling, and hence, the coi1 mutant is impaired in JA-Ile turnover.[9,10] The ω-oxidation pathway is operated by cytochrome P450 of the CYP94 family: CYP94B3 and B1 are JA-Ile o-hydroxylases producing 12OH-JA-Ile, and CYP94C1 catalyzes a more complete JA-Ile oxidation to 12COOH-JA-Ile. Various in vitro pull-down assays and feeding experiments have revealed that 12OH-JA-Ile hold on some co-receptor-assembling and gene-inducing activities relative to JA-Ile, and that 12COOH-JA-Ile is fully inactive in both the above assays.[5,10,13] Ectopic overexpression of CYP94 enzymes resulted in decrease in JA transcriptional responses due to JA-Ile oxidation, and diminished resistance to JA-defended insect or fungus invasions with reduced male fertility, due to enhanced JA-Ile turnover.[5,10,11,13] Loss-of-function *Arabidopsis* cyp94 mutants have resulted in gene redundancy. In case of single or double cyp94 mutant lines, it has been reported that upon wounding there is more consistent accumulation of JA-Ile, along with depleted oxidation products.[10–12] Deconjugation by amido hydrolase (AH) is the second major JA-Ile catabolic pathway, first investigated in *N. attenuata*[16] and then categorized as the IAR3 and ILL6 enzymes in *Arabidopsis.*[19] In JA pathway, the bioactive form is formed by JA conjugation to isoleucine. This leads to the probability that JA-Ile could be readily turnovered by cleavage reaction or JAR1-reverse reaction. Systematically, coregulated gene lists include genes encoding ILL1-related family of amido-hydrolases, like ILL6. In vitro examinations on regulated JA expression have revealed that recombinant IAR3 and newly characterized ILL6 cleave JA-Ile.[19] IAR3 acts as a bifunctional enzyme which mediates cross-talk between JA and auxin signaling.[18] Also, iar3 and ill6 mutants have impaired wound-induced JA profiles, consistent with a role in JA-Ile turnover. Although increased JA-Ile levels were confined to iar3 mutants, strong increase in 12OH-JA-Ile and 12COOH-JA-Ile levels were observed in both iar3- and ill6-deficient lines. A possible interpretation is that excessive JA-Ile accumulation is due to impaired cleavage capacity directed into the oxidation pathway, as proved by hyper-induction of CYP94 genes in amido-hydrolase mutants.[19] The IAR3 and ILL6 enzymes also hydrolyze CYP94-generated 12OH-JA-Ile, explaining indirect biosynthetic route for 12OH-JA formation.[18,19] As for CYP94, AH-overexpressing-lines shows deficiency symptoms of JA-Ile,[18] hence showing that the foremost function of both pathways is the attenuation of JA responses. In vitro studies revealed that NaJIH, an herbivore-inducible IAR3 homolog, was found to equally hydrolyze JA-Ile and IAA-Ala. Besides gene non-functioning within CYP94 and AH families, balancing mechanism happen when defective mutant(s) of one pathway over express genes or enzymes of another pathway.[19] Marquis et al.[44] carried out a study where JA-Ile catabolic pathways were observed. When higher-order mutant lines, fully deficient in either one (triple cyp94b1b3c1, named 3cyp or double amido hydrolase, named 2ah) or both (quintuple mutant, named 5ko) unexpected results were found prior being subjected to mechanical wounding or to infection by *B. cinerea.* CYP94 and AH pathways were observed to contribute similarly to shape JA-Ile dynamics upon wounding. In 5ko mutant, JA-Ile accumulation occurred to exceptionally high levels, specifying that hormone removal was hampered in this line. Contrastingly, in *B. cinereal*-infected leaves, 3cyp majorly repressed JA-Ile oxidation, with no alteration in JA-Ile levels in comparison to wild. The JA-Ile levels in infected 2ah and 5ko leaves revealed that turnover JA-Ile was predominated by AH pathway, upon fungal infection. Moreover, the level of engineered JA-Ile poorly correlated with stress-induced defense systems; the impaired JA-Ile catabolism lead to enhanced pest tolerance and defense responses in absence of over expression of JAZ and JAM genes which occurs in stress- and pathway-specific way. JAM, atypical subclass of bHLH factors, hampers the activity of transcription factors MYC.[45] In most cases, exaggerated JA-Ile accumulation clears the way for expression of repressors especially in 5ko. The 2ah and 3cyp lines failed to display enhanced expression of JAZ and JAM repressors, and showed more resilience to pest and fungal attacks.[44] Data advocates that in *Arabidopsis*, the signaling is controlled by a negative feedback loop. In order to take advantage of decreased levels of JA-Ile, a proper understanding of this feedback mechanism like determinants of transcriptional repression responses leads to be worked out. JA-Ile catabolism is related to other factors like abiotic stress management. Over expression of OsCYP94C2 leads to enhanced survival during salinity stress in rice[43]; however, an allele leads reduced expression of OsCYP94B4 (HAN1), and hence hampers JA-Ile catabolism, which inconsequently leads to its mitigation of temperature stress.[46] Also, an unusual expression of ZmCYP94B1 causes sexual differentiation defect in maize flowers that disrupts JA signaling in Tassleseed5 (Ts5) mutant.[47] A detailed characterization of CYP94 and AH gene families in crops is in budding stage[48] and will decipher the further intricacies of JA-Ile catabolism regulation of JA processes.

3.3.3 HYDROXYLATION AND CARBOXYLATION

Direct hydroxylation of JA to 12-OH-JA has not been identified hitherto, except for putative activity of a fungal monooxygenase.[3] Quite a few members of the CYP94 gene family have been recognized to hydroxylate JA-Ile (CYP94B3, CYP94C1) and to carboxylate 12-OH-JA-Ile (CYP94C1).[11] This work was of late stretched to other gene families like CYP94B1 and CYP94B2.[49] Hydroxylation and carboxylation contribute via ω-oxidation pathway to JA catabolism. Data on mutants and overexpression lines of CYP94 gene family members disclosed connection in JA-Ile signaling in *Botrytis cinerea* infection.[11] Nevertheless, a complex homeostasis of JA-Ile ensues in which hydroxylation and carboxylation eliminate the active JA-Ile, while the amido-hydrolases ILL6 and IAR3 cleave JA-Ile and 12-hydroxy-JA-Ile.[28]

3.3.4 DECARBOXYLATION

Decarboxylation of JA leads to formation of Cis-Jasmone (CJ), though an additional pathway encompassing cis-OPDA conversion to iso-OPDA has been suggested. Numerous essential roles of CJ on plant and insect interactions have been perceived[50,51] but an experiment on transcript profiling disclosed that CJ activates genes that are unlike those induced by classical JA application.[52] CJ could act as a catabolic sink of JA even though only an inadequate quantity of JA metabolite flux into CJ in *N.attenuata* leaves damaged by insects.[53] The volatile CJ has been identified as an integral component of the floral bouquet appealing insects for pollination. Multitrophic interfaces of plants with aphids and their parasitoids and wounding also leads to CJ formation.[51] CJ-treated plants express a different set pf genes unlike.[52,53] Moreover, decarboxylation of JA, isomerization of cis-(+)-OPDA into iso-OPDA, followed by ß-oxidation to 3,7-didehydro-JA and decarboxylation were suggested for synthesis of CJ.[54]

3.3.5 METHYLATION AND ESTERIFICATION

Although MeJA has long been recognized as a component of flower fragrance that aids in pollination and as a signaling compound. In most of the processes where JA or JA-Ile are active, MeJA is not active. Transgenic approaches revealed that MeJA is split by esterases and afterwards converted to JA-Ile by JAR1.[24] These findings are consistent with the JA-Ile-receptor complex's inability to bind MeJA.[34] MeJA transferases by sustaining JA profile is responsible for many plant stress responses and other various plant processes using clone from tomato, strawberry, rice, and *Arabidopsis*.[27] (+)-7-iso-JA was first isolated from *Lasiodiplodia theobromae*.[55] On the other hand, many esters of JA known as lasiojasmonates, which were purified from *Lasiodiplodia* sp., responsible for grapevine infection.[56]

However, these derivatives have been found biologically inactive by various phytotoxic and zootoxic assays.

3.3.6 O-GLYCOSYLATION

Glycosylation, preferably glucosylation, is a typical modification of phytohormones and secondary. Tuber-inducing factors of potato 12-OH-JA also known as tuberonic acid and its glucosylderivative 12-O-Glucosyl-JA[27] abundantly accumulate JA in wounded leaves of tomato or *Arabidopsis*.[4] In filaments of *Zea mays* and leaves of *Glycine max*, constitutive accumulation of 12-O-Glucosyl-JA can occur in three times higher magnitude in comparison to JA.[4] The leaf closing factor 12-O-glucosyl-JA present in *Albizzia* and *Samanea* species possesses similar structural features of the 12-OH-JA.[57] Very few findings are known on characteristic properties of glucosyltransferase of 12-OH-JA and a 12-OH-JA-glucoside hydrolase.[58,59] Similar to 12-HSO_4-JA, 12-O-Glc-JA accumulates in wounded leaves with a short delay after the initial burst of JA.[4] Both 12-HSO_4-JA and 12-O-Glc-JA accumulation is simultaneously reduced in mutants disrupted in 12OH-JA biosynthesis.[19] On the other hand, an increase in 12-O-Glc-JA in the adenosine 5′phosphosulfate kinase double mutant, apk1apk2, where 12-HSO_4-JA accumulation is impaired.[60] These findings exhibit that enzymes are liable for either 12-HSO_4-JA or 12-O-Glc-JA production competing for a common precursor 12OH-JA. In *Arabidopsis,* till now the enzymes that catalyze the JA 12-O-glucosylation production is unknown. In *Arabidopsis*, in spite, a wound-inducible glucosyl-transferase that possesses catalytic activity toward both 12OH-JA and SA has been purified from rice cell cultures and tobacco cultures.[59] Recently, a maize gene called silkless1 (sk1) encoding uridinediphosphate (UDP) glycosyltransferase was cloned, which is likely to function in the JA metabolic pathway.[60,61] Sex determination in maize depends upon the JA; it helps in development of bisexual flowers into male and female florets via a JA regulated pistil elimination method. Constitutive expression of sk1 transgene deprived both JA and OPDA in tassels leads to full feminization of the plant. The genetics of other sugar-attached JA derivatives like jasmonoyl-1-b-glucose, jasmonoyl-1-b-gentiobiose, hydroxyjasmonoyl-1-b-glucose from tobacco BY-2 cells,[62] and 12-O-glucosyl-JA-Ile from *Arabidopsis*[12] remain to be unrecognized.

3.4 JASMONATES AND CROSSTALK WITH OTHER PHYTOHORMONES

In natural environmental conditions, plants share a complex nexus with the biotic and abiotic components. Plants display wide number of interactions with various microbial pathogens and their infection strategies. In response to microbial infections/attack, plants have programmed a defense system/immune response to combat their damage. There are mainly two defined ways of innate immunity against pathogens.[63] The first type is prompted by the identification of highly conserved microbe-associated molecular patterns (MAMPs) *via* host cell transmembrane proteins that act as pattern recognition receptors (PRRs), which further trigger MAMP-triggered immunity (MTI).[63] This provokes a high-level defense against non-pathogenic microbes and, also, some pathogens. However, adapted pathogens have developed the resistance by introduction of virulence effector proteins, thus make plants susceptible to infection.[63] The type second defense responses distinguish pathogen infection by microbial effectors within the plant cell through nucleotide-binding site-leucine-rich repeat (NB-LRR) resistance (R) proteins.[63] This lead to programmed cell death called as hypersensitive response by activation of effector-triggered immunity (ETI).[63] Phytohormones are key elements in the modulation of the defense network that interprets the pathogen-induced early signaling array into cascade of ameliorative responses.[64] Phytohormones regulate the plant growth and various developmental processes and are required in minute amounts.[65] The amount, type, and timing of plant hormone produced in response to stress greatly depend upon the plant species and pathogen invader.[66] Major plant hormones include cytokinins (CKs), gibberellins (GAs), ethylene (ET), and abscisic acid (ABA), and also other phytohormones which include jasmonates (JAs), salicylic acid (SA), and brassinosteroids (BRs).[64] The JA, SA, and other phytohormones (Figure 3.2) are leading regulators of the plant's defense mechanism.[64,65,67] Basically, JA pathway is predominantly provoked by and potent in arbitrate resistance in response to herbivore and necrotrophic pathogen attack, However, the SA pathway is mainly provoked in response to biotrophic pathogens.[68] The raised resistance against necrotropic pathogens is always correlated with raised biotrophic attack and vice versa as JA and SA defense pathways generally oppose each other.[69] The success of defense response mostly depends on the combined contribution and amount

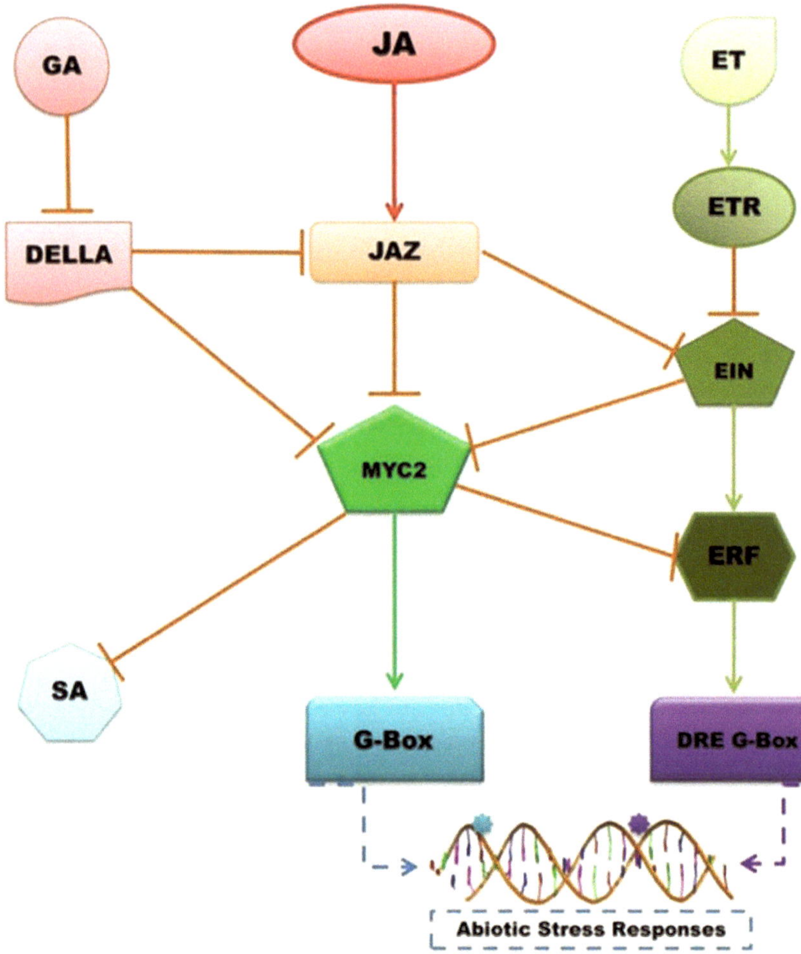

FIGURE 3.2 An overview of molecular mechanism of signal transduction induced by phytohormones like jasmonic acid (JA), in synergism with gibberellic acid (GA), ethylene (ET), and salicylic acid for conferring abiotic stress tolerance to crops. ETR, ethylene receptor; EIN, ethylene insensitive; JAZ, jasmonatezim-domain.

of these hormones during microbial invasion. Signal translocation and perception depends on receptor–ligand interaction.[70] In several hormonal signaling like SA, ABA, and ET, the signal is perceived by receptors, transmitted through cytoplasm to nucleus.[22,64,71] But, in case of JA, auxins, and GAs signaling hormone receptors are present in the nucleus.[64,72] In plant hormone signaling, the 26S ubiquitin-proteasome system (UPS) performs a key role and almost investigated for every hormone.[72] Nuclear proteins linked to transcriptional repression that bear an ethylene response factor-associated amphiphillic repression (EAR) domain are targeted by UPS.[72] Examples include auxin/indole-3-acetic acid (Aux/IAA), some jasmonate-ZIM domain (JAZ), and the BRASSINAZOLE RESISTANT1 (BZR1) repressors.[73] Hence, nuclear protein turnover is an important component in hormone signaling to produce exact defense responses for a particular microbial pathogen. In this chapter, we have summarized examples of JA with few other phytohormones and their interactions at the molecular level.

3.4.1 JA–GA INTERACTION

Phytohormone gibberellin is involved in regulating plant growth under different environmental cues.[74] DELLA proteins are the principle constituents of GA signaling and negative regulators of plant growth-promoting TFs inside nucleus, for example, phytochrome interacting factors (PIFs).[75] There are five DELLA proteins found in *Arabidopsis*: gibberellic acid insensitive (GAI), repressor of GA1–3 (RGA), RGA-like1 (RGL1), RGL2, and RGL3.[76,77] The activation of PIFs depends upon the degradation of DELLAs by the 26S proteasome. GA attaches to its receptor GA insensitive dwarF1 (GID1) which boosts the GID1–DELLA interface, which consecutively, invokes the interaction between DELLAs and the specific E3 ubiquitin ligase SLY1/GID2 complex.[76–78] DELLAs regulate plant defense pathway by regulating the equilibrium of JA/SA. The JA and GA signals operate synergistically or antagonistically in many defense and developmental functions of plant. In *Arabidopsis*, JA and GA have cumulative effects in the formation of trichomes and filaments (stamen)[79,80] and JAZs/DELLAs interaction with the WD-repeat/bHLH/MYC complex formation is responsible for cumulative effect of JA/DELLAs interaction.[80] DELLA proteins enhance plant resistance to necrotrophs by provoking JA signaling and enhance plants susceptibility to virulent biotrophs by obstruction in the SA pathway and vice versa.[81] So, loss-of-function mutants of DELLA attenuates JA modulated pathogen attack whereas defense genes are overexpressed by JA in persistent active DELLA variants.[81] Also, a GA-deficient variant ga1 induces overexpression of JA-responsive genes[82]; this ratifies that DELLA proteins have positive gene regulation with JA. Numerous recent reports suggest that DELLA interrelate with JAZ proteins and control JA signaling by the way competitive binding with transcription factors like MYC.[82] In accord with this model, JA-responsive genes which relies on MYC2 are found highly provoked in response to JA treatment in mutants accumulating DELLAs like that of GA biosynthetic mutant ga1-3 in comparison to mock subjected controls.[81] Indeed, upregulation of RGL3 induces MYC2-dependent JA-i gene expression, whereas rgl3 mutation inhibits it.[82] Similarly, RGL3 upregulates JA-mediated resistance to the necrotroph *B. cinerea* and are prone to the hemibiotroph *Pseudomonas syringae*.[82] JA prefers defense over growth by intervention with GAs signaling pathway by the COI1–JAZ–DELLA–PIF signaling module.[83] In response to herbivore attacks, enhanced levels of JA leads to proteasomal degradation of JAZs and induction of MYC2 plant defense, while DELLAs are alleviated of JAZs inhibition, and interact with phytochrome-interacting factors (PIFs) to reduce the growth. By contrast, GA provokes DELLA degradation, induces the PIFs to activate growth and on the other hand, the JAZ repressors are liberated to impede MYC2-mediated plant defense responses.[81,83] So, DELLAs and JAZs appear to interpret environmental signals that make plants to acclimate their physiological processes in accord with prevailing environment. In rice, JA and GA signals also interact antagonistically. For example, the reduced height of plant by JA is not observed in the *slender rice 1 (slr1)* variant, lacking the activity of *SLR1*, a rice DELLA. This represents that *SLR1* modulates the opposite crosstalk between JA and GA by which rice growth is greatly affected.[83] The expression pattern of *OsDREB1s*, which act downstream of the JA–responsive gene *OsPILI14*, supports the opposite effects of JA and GA, as *OsbHLH148* promotes *OsDREB1s* expression and *OsPIL14*suppresses*OsDREB1s* expression.[59,84] A latest study used knock-out mutants of OsJAZ9 postulated that OsJAZ9 is a chief JAZ protein that intercedes the opposite effects of JA and GA.[85]

3.4.2 JA–ET INTERACTION

Ethylene in combination with JA regulate the immune response against the necrotropic pathogens which indeed is multiplex processes depending upon the environment.[86] Ethylene signaling is modulated by 4 receptors which include ethylene response 1 (ETR1), ethylene response sensor 1 (ERS1), ethylene response 2 (ETR2), ethylene response sensor 2 (ERS2), and ethylene insensitive 4 (EIN4).[87] Necrotropic pathogens are acted upon by synergistic approach of ET and JA.[88] In *Arabidopsis*, the wound response caused by herbivory is dealt by MYC branch, while the increased resistance to necrotrophic pathogens is dealt by ERF branch.[88,89] Marker genes like vegetative storage protein2 (VSP2) are responsible for JA defense modulated by MYC-type Tfs. The ERF-type response is regulated by members of the AP2/ERF family of TFs, for example, ethylene response factor 1 (ERF1) and ORA59,[90] that modulate the expression of JA-responsive marker genes like PDF1.2. Overexpression of ERF1 increases resistance against *B. cinereal* and other necrotrophic pathogens, and enhances susceptibility to the hemibiotroph *Pseudomonas*.[89] ET and JA pathways modulate the ERF branch by their synergistic action, while they modulate MYC branch by antagonistic way.[90] JAZ proteins inhibit the transcriptional activity of ET TFs EIN3 and EIL1, by hindering their DNA-binding capacity by histone deacetylase6 (HDA6), a transcriptional co-repressor.[91] EIN3/EIL1 TFs remain active when JAZ proteins are degraded in presence of JA. The result is the synergistic induction of ERF1 and its downstream target genes like PDF1.2.[91] Besides, EIN3 and EIL1 are identified repressors of salicylic acid induction deficient 2 (SID2), a gene coding an isochorismate synthase necessary for SA biosynthesis.[79] So, JAZ repressors also regulate JA–SA antagonistic crosstalk by the repression on EIN3 and EIL1.

3.4.2.1 JA–ET SYNERGISTIC INTERACTION

JA and ET rely upon each other for synergistic induction of resistance against necrotropic pathogens via expression of pathogen-responsive genes, such as plant defensin 1.2 (PDF1.2). For the expression of PDF1.2, the factors which include apetala 2/ethylene response factor (AP2/ERF) domain transcription factors, ERF1 and octadecanoid-responsive arabidopsis AP2/ERF 59 (ORA59), bind GC box in the PDF1.2 promoter.[92,93] There are no molecular evidences that ERF1 and ORA59 are vital molecules for interaction of JA and ET signaling in modulating the expression of pathogen-responsive genes.[92] Zhu et al.[91] found that in the induction of pathogen-responsive gene expression and root hair development ein3 eil1 double mutants are non-responsive when subjected to ET and JA treatments. EIN3/EIL1 is another wing of JAZ-interacting transcription factors as EIN3/EIL1 physically associate with minimum of three JAZ members (JAZ1/3/9). JAZ protein inhibits EIN3 functions in plants by overexpression of stabilized JAZ1 protein that causes suppression of EIN3ox (EIN3 overexpression). Also, JA or ET treatment increases EIN3 DNA-binding capacity as revealed by chromatin immunoprecipitation (ChIP) assay. It was also noted that histone deacetylase 6 (HDA6) binds with JAZs and EIN3/EIL1 as a co-repressor, while JA treatment hinders the HDA6-EIN3 binding to release its inhibition. Next, it was observed that after JA treatment, histone acetylation increased on ERF1 promoter.[91] Hence, it can be concluded JA directly provokes EIN3/EIL1 by suppression pathway, while ET activates EIN3/EIL1 via the proved post-transcriptional regulation mechanism. Both of the above methods are necessary for complete activation of EIN3/EIL1. In ein2 mutants, EIN3/EIL1 protein concentration is very low, while there is hyperaccumulation of JAZ repressors in coi1, so neither ET nor JA treatment could induce pathogen-responsive gene expression in coi1 or ein2 due to the lack of functional EIN3/EIL1. Among mitogen-activated protein kinase (MAPK) cascade, JA provokes the MKK3–MPK6 cascade[94] to modulate upregulation of pathogen-responsive gene expression.[27]

3.4.2.2 JA–ET ANTAGONISTIC INTERACTION

JA/ET antagonistic interaction has been reported by two independent researchers and their observations are almost similar, that is, JA–ET antagonistic regulation relies on EIN3/EIL1. It is known that ET-induced apical hook emergence is produced because of induction of HOOKLESS1 (HLS1) expression, responsible for a protein similar to N-acetyltransferase. There is a direct combination of EIN3 with the HLS1 promoter and promotes its transcription.[95] As HLS1 has an important role in formation of apical hook, it is assumed that JA antagonizes ET's additive task in hook formation by regulation of HLS1. It is reported that the hls1 mutant is non-responsive to JA treatment, which confirms that HLS1 is must for JA antagonism.[96,97] Analysis by quantitative reverse transcription–PCR and HLS1 promoter-driven glucuronidase expression shows that JA downregulates HLS1 mRNA expression even in presence of ET. From earlier observations, HLS1 is a direct target gene of EIN3, the expression of HLS1 in ein3 eil1 is in reduced levels in comparison to the wild type, and In ein3 and eil1, JA was unable to reduce its expression. Thus, it is concluded that JA antagonizes ET-induced hook formation by downregulation of EIN3/EIL1-controlled HLS1 expression.[96,97] ET-induced hook formation is inhibited by JA promoted by MYC2 and its closest homologues MYC3 and MYC4 (MYC3 and MYC4 JAZ-interacting transcription factors).[98] HLS1 is expressed at a higher level in myc234 (the myc2 myc3 myc4 triple mutant) than in the wild type, and JA was unable to inhibit HLS1 expression in myc234.[96] Thus, MYC2 is a negative modulator of HLS1 expression and JA negatively modulates HLS1 expression by MYC2 family proteins. It has been MYC2 physically interacts with EIN3 and EIL1 as revealed by dual luciferase reporter assay.[96] Zhang et al.[97] also found that MYC2 interacts with EIN3/EIL1 and demonstrated that MYC2 abrogates EIN3 DNA binding capacity in an with the help of electrophoretic mobility shift assay (EMSA). In support of these in vitro studies, in vivo co-immunoprecipitation and ChIP assay also showed similar results.

3.4.3 JA–SA INTERACTION

Plant defenses against pathogens and herbivory are operated at the expense of allocation and ecological costs.[64] When plants are infected by biotroph pathogens, SA is produced which suppresses JA-dependent immune response of plant,[99] which will evidently prioritize the investment of resources in SA-dependent defense over JA-dependent responses.[100] Also, SA responses are repressed when plants are subjected to JA elicitation.[101] It has been observed that SA downregulate JA signaling downstream of the COI1-JAZ receptor complex by acting on GCC-box motifs in JA-responsive promoters through opposite effect on the accretion of the APETALA2/ethylene response factor (AP2/ERF)-type transcriptional activator ORA59. SA suppresses JA-responsive promoters in which GCC-box is overexpressed and thus this promoter motif is enough for SA-mediated downregulation of JA-induced gene expression.[102] SA induces antagonistic effect on JA at transcriptional level by lowering the accretion of GCC-box binding TF like ORA59.[99] Many proteins are well known in modulating SA-mediated reduction of JA expression, including MAPKs, redox regulators, NPR1, and nuclear TGA and WRKY TFs.[64] In *Arabidopsis*, MAPK 4 (MPK4) are reported to be positive modulator of JA pathway and negative modulator of JA pathway.[64] In *Arabidopsis*, mutants of mpk4 result in enhanced SA levels, consistent expression of PR genes of SA and increased resistance to *Pseudomonas syringae*. Also, it has been found that impaired expression of JA responsive genes and the resistance against *Alternaria brassicicola* in the mutants of mpk.[103]

In addition to above molecules, glutaredoxins (GRXs), like GRX480 and many of the ROXY class are other vital modulators which determine the opposite effects of SA and JA-modulated signaling by redox regulation, as they are able to catalyze disulfide bonds.[104] These GRXs act on TGA TFs responsible for the modulation of SA responsive PR genes and oppose JA-responsive genes such as plant defensin1.2 (PDF1.2) and ORA59, for example, interaction of GRX480 with TGA2 and TGA6.[104] The tga2356 quadruple mutant and tga256 triple mutant are found to be impaired in SA-mediated inhibition of the JA pathway[104,105] represents that TGAs efficiently modulate SA–JA crosstalk. SA-responsive PR genes are activated by TGA TFs by the intertranscription[99] action of NPR1 which is indeed regulated by SA itself.[106] For SA-responsive defense gene expression nuclear localization of NPR1 is vital, but not for SA-mediated suppression of the JA pathway,[107] suggesting that SA–JA interaction may be regulated by cytosolic NPR1. Despite of this, NPR1 modulates various SA-dependent nuclear TFs or cofactors necessary for down-regulated expression of JA which includes TGA and WRKY TFs.[65] Thus, WRKY TFs like WRKY50, WRKY51, WRKY70, and WRKY62 are also an important point of interaction of SA and JA signaling.[64] Finally, the JA nuclear TF MYC2 acts as a negative modulator of SA signaling in *Arabidopsis* as myc2 variants possess enhanced accretion of SA, increase expression of PR genes and enhance resistance to *Pseudomonas syringae* in contrast to wild type plants.[108] However, in spite of ample information of about the molecules that help in interplay of SA and JA,[99] the molecular mechanism of interaction between SA and JA remains unclear.

3.4.4 THE JA–CYTOKININ INTERACTION

There is an antagonistic relationship between cytokinin (CK) and JA, for example, JA inhibits cytokinin-induced soybean (*Glycine max*) callus proliferation, and revokes the effect of CK on chlorophyll formation.[109] The opposite effect of JA and CK is proved also by their differential regulation of gene expression responsible for chlorophyll formation. In *Arabidopsis*, the xylem differentiation in roots is modulated by JA and an opposite interplay of JA and CK is very vital for JA-dependent xylem formation.[110] Xylem is originated from procambial responsible for development of plant vascular system.[111] It has been observed in *Arabidopsis* that CK plays as negative modulator of xylem differentiation by maintaining stem cell identity as revealed by exogenous application of CK. In addition to this, the wooden leg mutants that are impaired in CK signaling are lacking procambial cells in their roots. Besides, additional xylem formation takes place in mutants that lack transcription of Type-B ARRs, for example, ARR12, ARR10, or recombinant plants with more expressive AHP6 genes (negative regulator of CK).[110] In another experiment, when wild-type plants and JA-deficient OPDA reductase 3 (*opr3*) variants were subjected to exogenous JA application; it resulted, in an extra xylem formation. However, JA signaling variants, such as *coi1* and jasmonate resistant 1 (*jar1*), did not result in extra xylem formation.[110] From these observations, it can be evident that JA causes inhibition of the procambium-specific CK response, and also the effect of JA on extra xylem formation is abrogated by CK, validates that JA antagonistically interacts with CK in xylem development in roots of *Arabidopsis*. The reason is that JA inhibits the expression of the cytokinin-responsive pin-formed 7 (PIN7) gene, responsible for xylem development. In addition, myc2 mutant did not lead to the emergence of additional xylem,[111] when subjected to exogenous JA treatment and the expression of AHP6, encoding a CK signaling inhibitor, was reduced in exogenous JA, confirming that the JA-responsive MYC2 transcription factor mediates this process. Thus, it is obvious that an opposite relationship between JA and CK is also responsible for the modulation of JA-dependent stress responses.[90] Nitschke et al.[91] observed that plants with low CK levels or impaired CK signaling led a JA-dependent cell death phenotype against circadian stress, in contrast to wild-type plants,[91] indicating that JA and CK have opposite effects against circadian stress.

3.4.5 THE JA–AUXIN INTERACTION

The relationship between JA and auxin in crop growth and development acts in various activities like secondary metabolite production, cell elongation, and tendril coiling[112]; however, this relationship has not been illustrated at the molecular level. The discoveries of genes responsible for JA and auxin metabolism and signaling pathways have found that JA and auxin interact to regulate plant development. The crosstalk occurring in JA and auxin in the modulation of root development is well established. JA hinders apical root growth; JA-treated wild-type plants have much shorter roots in compar ison to untreated wild-type plants, while mutant plants with defective JA signaling produce roots of same in length than wild-type plants even in JA-treated conditions.[113] In comparison to JA, auxin is important for root growth. So, auxin deficiency or impaired in signaling factors, like (trp2-12) and auxin resistant 3 (arx3-1), produce very short roots as compared to wild-type plants.[114] Chen et al.[115] demonstrated that root growth inhibition is regulated by an interaction of JA with auxin. In this case study, they observed that the cause of JA-mediated inhibition of root growth is by inhibition meristematic activity of root and external application JA inhibits the expression of the auxin-responsive transcription factors plethoras (PLTs), which are responsible for maintenance of stem cell and cell proliferation.[115] But, the expression profile of PLTs was not downregulated in impaired JA-signaling types, for example, coi1-1 and myc2, indicating that JA signaling which relies on COI1 regulates the JA-promoted root phenotype, and the MYC2 transcription factor downregulates the expression of PLTs. As, MYC2 interacts with the promoters of PLTs, it indicates that JA-induced root growth inhibition by JA-responsive MYC2 is due to downregulation of auxin-responsive PLTs, and hints that JA and auxin have opposite effect in the formation of root growth. The JA–auxin interactivity is involved in development of root floral organs, for instance, stamens and petals.[116] The key regulators for floral organ development are R2R3 and transcription factors like MYB21 and MYB24 which in turn are modulated by auxin-responsive transcription factors ARF6 and ARF8. Hence, it indicates that auxin mediate with JA to modulate the development of floral organs.[116] JA–auxin interaction also aids in regulation of leaf senescence. Auxins bring about the suppression of leaf senescence.[117] Several studies depict an inverse correlation between auxins and senescence wherein auxin concentration drops with leaf age.[118] Auxin-modulated JA signaling is crucial for crop development. Recently, a systematic understanding of how interaction of JA and auxin signaling pathways are calibrated by WRKY57 throughout JA-induced leaf senescence.[119] These results propose that JAZ4/8 and IAA29 exhibit antagonistic roles in JA-induced leaf senescence, thus is in line with the opposing roles of JA and auxin in JA-induced leaf senescence. Interestingly, the JA-induced leaf-senescence phenotypes in wild-type leaves are relieved by the use of exogenous auxin and are reliant on WRKY57. Together, these results validate that the JA-induced leaf-senescence is opposed by auxin via WRKY57.[119]

3.4.6 BR–JA INTERACTION

Brassinosteroids (BR) regulate many plant growth processes and regulate JA signaling and JA-dependent growth inhibition. DWARF4 produces an important enzyme meant for BR biosynthesis and a leaky mutation of DWARF4 rehabilitate JA sensitivity in the coi1 mutants and manifested JA hypersensitivity in the wild-type. Moreover, expression of DWARF4 was downregulated by JA in a COI1-dependent manner, and by external application of BR resulted in reduction on inhibition of root growth by JA.[119] These observations confirm that a BR–JA interaction is responsible for regulation of JA signaling. BR acts as a main hormone in defense downregulating the JA pathway.[120] Opposite results were obtained in a study where the JA-insensitive *jai1-1* and the BR biosynthesis-defective *dpy* mutants were used as experimental scenario. These observations from such experimental setup revealed that JA enhanced the establishment of anti-herbivory characters while as BR impeded it. For example, reduction in trichome number in *jai1-1*, highlighting the role of JA in trichome production.[121] On the other hand, *dpy* mutants produced high trichome density revealing that BR negatively controls its production. Similar, results were also observed in *Arabidopsis* which are BR-insensitive mutant *bri1*.[122] Similarly, opposite effect were noted between BR and JA for zgb (Zgb, a sesquiterpene associated with tomato resistance against many insects) contents, as few amount were found in *jasmonic acid insensitive1-1(jai1-1)* and maximum amount in dumpy *(dpy). The dpy×jai1-1* double mutant showed the jai1-1 phenotype, that is, BR negatively controls anti-herbivory traits in tomato (an opposite effect to that of JA). However, BR action resembles that of SA by opposing JA responses. The role of ethylene in host-pathogen interactions earlier proved that SA is not important for BR functioning.[123,124] Interaction between BRs and JA may modulate the resource investment as BRs are required in growth of plants[125] while JAs in defense. In tomato, positive modulation of the *jai1-1*-dependent jasmonate pathway was found, resulting in enhanced expression of defensive traits when mutant which were defective in biosynthesis of BR, *dpy* (impaired in growth), were used as in an experiment. Thus, these observations reflect that BR might be the plant hormone that modulates the resource allocation by modifying the responseness to JAs. There is no doubt, that JAZs modulation is now considered as a path present in plants to solve this dilemma.[126]

3.5 CONCLUSIONS

In conclusion, JA is under strict genetic control and interacts synergistically or antagonistically with a variety of phytohormones, such as gibberellic acid, cytokinin, auxin, and brassinosteroids, to modulate plant growth and development. These interactions may help optimize growth and development of plant during stress. The cross talk between different growth hormones rely on the plant species. Example like JA and gibberellic acid have opposite effect on height of the plant; however, in stem development they are having additive effect. Similarly, JA and auxin have opposite effect in apical root growth, but have additive effect in lateral root development. The exploration of other signaling molecules can be helpful to inspect the details of underlying molecular mechanisms of JA signal transduction. Much work is to be done in near future to find the proper answers of the queries like action of JA-related metabolites, and identification of universal JA receptors, etc.

KEYWORDS

- **jasmonic acid**
- **metabolism**
- **methyl jasmonate**
- **phytohormones crosstalk**

REFERENCES

1. Wang, J.; Song, L.; Gong, X.; Xu, J.; Li, M. Functions of Jasmonic Acid in Plant Regulation and Response to Abiotic Stress. *Int. J. Mol. Sci.* **2020,** *4,* 1446.
2. Yoshihara, T.; Omir, E. S. A.; Koshino, H.; Sakamura, S.; Kkuta, Y.; Koda, Y. Structure of a Tuber-inducing Stimulus from Potato Leaves (*Solanumtuberosum* L.). *Agric. Biol. Chem.* **1989,** *53* (10), 2835–2837.
3. Patkar, R. N.; Benke, P. I.; Qu, Z.; Chen, Y. Y. C.; Yang, F.; Swarup, S.; Naqvi, N. I. A Fungal Monooxygenase-derived Jasmonate Attenuates Host Innate Immunity. *Nat. Chem. Biol.* **2015,** *11* (9), 733–740.
4. Miersch, O.; Neumerkel, J.; Dippe, M.; Stenzel, I.; Wasternack, C. Hydroxylated Jasmonates are Commonly Occurring Metabolites of Jasmonic Acid and Contribute to a Partial Switch-off in Jasmonate Signaling. *New. Phytol.* **2008,** *177* (1), 114–127.

5. Aubert, Y.; Widemann, E.; Miesch, L.; Pinot, F.; Heitz, T. CYP94-mediated Jasmonoyl-isoleucine Hormone Oxidation Shapes Jasmonate Profiles and Attenuates Defence Responses to *Botrytis cinerea* Infection. *J. Exp. Bot.* **2015,** *66* (13), 3879–3892.
6. Wang, Y.; Yang, W.; Zuo, Y.; Zhu, L.; Hastwell, A. H.; Chen, L.; Tian, Y., Su, C.; Ferguson, B. J.; Li, X.; GmYUC2a Mediates Auxin Biosynthesis During Root Development and Nodulation in Soybean. *J. Exp. Bot.* **2019,** *70* (12), 3165–3176.
7. Heitz, T.; Smirnova, E.; Marquis, V.; Poirier, L. Metabolic Control Within the Jasmonate Biochemical Pathway. *Plant Cell Physiol.* **2019,** *60* (12), 2621–2628.
8. Nelson, D.; Werck-Reichhart, D. A P450-centric View of Plant Evolution. *Plant J.* **2011,** *66* (1), 194–211.
9. Paschold, A.; Bonaventure, G.; Kant, M. R.; Baldwin, I. T. Jasmonate Perception Regulates Jasmonate Biosynthesis and JA-Ile Metabolism: The Case of COI1 in *Nicotiana attenuata. Plant Cell Physiol.* **2008,** *49* (8), 1165–1175.
10. Koo, A. J.; Cooke, T. F.; Howe, G. A. Cytochrome P450 CYP94B3 Mediates Catabolism and Inactivation of the Plant Hormone Jasmonoyl-L-Isoleucine. *Proc. Natl Acad. Sci. U. S. A.* **2011,** *108* (22), 9298–9303.
11. Heitz, T.; Widemann, E.; Lugan, R.; Miesch, L.; Ullmann, P.; Désaubry, L.; Désaubry, L.; Holder, E.; Grausem, B.; Kandel, S.; Miesch, M.; Pinot, F. Cytochromes P450 CYP94C1 and CYP94B3 Catalyze Two Successive Oxidation Steps of Plant Hormone Jasmonoyl-isoleucine for Catabolic Turnover. *J. Biol. Chem.* **2012,** *287* (9), 6296–6306.
12. Kitaoka, N.; Kawaide, H.; Amano, N.; Matsubara, T.; Nabeta, K.; Takahashi, K.; Matsuura, H. CYP94B3 Activity Against Jasmonic Acid Amino Acid Conjugates and the Elucidation of 12-O-β-Glucopyranosyl-jasmonoyl-l-isoleucine as an Additional Metabolite. *Phytochemistry* **2014,** *99,* 6–13.
13. Koo, A. J.; Thireault, C.; Zemelis, S.; Poudel, A. N.; Zhang, T.; Kitaoka, N.; Brandizzi, F.; Matsuura, H.; Howe, G. A. Endoplasmic Reticulum-associated Inactivation of the Hormone Jasmonoyl-L-isoleucine by Multiple Members of the Cytochrome P450 94 Family in *Arabidopsis*. *J. Biol. Chem.* **2014,** *289* (43), 29728–29738.
14. Poudel, A. N.; Zhang, T.; Kwasniewski, M.; Nakabayashi, R.; Saito, K.; Koo, A. J. Mutations in Jasmonoyl-L-isoleucine-12-hydroxylases Suppress Multiple JA-dependent Wound Responses in *Arabidopsisthaliana*. *Biochim. Biophys. Acta Mol. Cell Biol. Lipids* **2016,** *1861* (9), 1396–1408.
15. Ludwig-Müller, J.; Auxin Conjugates: Their Role for Plant Development and in the Evolution of Land Plants. *J. Exp. Bot.* **2011,** *62* (6), 1757–1773.
16. Woldemariam, M. G.; Onkokesung, N.; Baldwin, I. T.; Galis, I. Jasmonoyl-l-isoleucine Hydrolase 1 (JIH1) Regulates Jasmonoyl-l-isoleucine Levels and Attenuates Plant Defenses Against Herbivores. *Plant J.* **2012,** *72* (5), 758–767.
17. Bhosale, R.; Jewell, J. B.; Hollunder, J.; Koo, A. J.; Vuylsteke, M.; Michoel, T.; Maere, S. Predicting Gene Function from Uncontrolled Expression Variation among Individual Wild-type *Arabidopsis* Plants. *Plant Cell* **2013,** *25* (8), 2865–2877.
18. Zhang, J.; Lin, J. E.; Harris, C.; Pereira, F. C. M.; Wu, F.; Blakeslee, J. J.; Peer, W. A. DAO1 Catalyzes Temporal and Tissue-specific Oxidative Inactivation of Auxin in *Arabidopsis thaliana. Proc. Natl Acad. Sci. U. S. A.* **2016,** *113* (39), 11010–11015.
19. Widemann, E.; Miesch, L.; Lugan, R.; Holder, E.; Heinrich, C.; Aubert, Y.; Miesch, M.; Pinot, F.; Heitz, T. The Amidohydrolases IAR3 and ILL6 Contribute to

Jasmonoyl-isoleucine Hormone Turnover and Generate 12-hydroxyjasmonic Acid Upon Wounding in *Arabidopsis* Leaves. *J. Biol. Chem.* **2013,** *288* (44), 31701–31714.
20. Demole, E.; Lederer, E.; Mercier, D. Isolementetdétermination de la Structure du Jasmonate de Méthyle, Constituant Odorant Caractéristique de L'essence de Jasmin. *Helv. Chim. Acta* **1962,** *45* (2), 675–685.
21. Seo, H. S.; Song, J. T.; Cheong, J. J.; Lee, Y. H.; Lee, Y. W.; Hwang, I.; Lee, J. S.; Do Choi, Y. Jasmonic Acid Carboxyl Methyltransferase: A Key Enzyme for Jasmonate-regulated Plant Responses. *Proc. Natl Acad. Sci. U. S. A.* **2001,** *98* (8), 4788–4793.
22. Wu, Y.; Zhang, D.; Chu, J. Y.; Boyle, P.; Wang, Y.; Brindle, I. D.; De Luca, V.; Després, C. The *Arabidopsis* NPR1 Protein is a Receptor for the Plant Defense Hormone Salicylic Acid. *Cell Rep.* **2012,** *1,* 639–647.
23. Thines, B.; Katsir, L.; Melotto, M.; Niu, Y.; Mandaokar, A.; Liu, G.; Nomura, K.; He, S. Y.; Howe, G. A.; Browse, J. JAZ Repressor Proteins are Targets of the SCF COI1 Complex During Jasmonatesignalling. *Nature* **2007,** *448* (7154), 661–665.
24. Stitz, M.; Gase, K.; Baldwin, I. T.; Gaquerel, E. Ectopic Expression of AtJMT in *Nicotianaattenuata:* Creating a Metabolic Sink has Tissue-specific Consequences for the Jasmonate Metabolic Network and Silences Downstream Gene Expression. *Plant Physiol.* **2011,** *157* (1), 341–354.
25. Gidda, S. K.; Miersch, O.; Levitin, A.; Schmidt, J.; Wasternack, C.; Varin, L. Biochemical and Molecular Characterization of a Hydroxyjasmonatesulfotransferase from *Arabidopsisthaliana. J. Biol. Chem.* **2003,** *278* (20), 17895–17900.
26. Koo, A. J.; Howe, G. A. Catabolism and Deactivation of the Lipid-Derived Hormone Jasmonoyl-isoleucine. *Front Plant Sci.* **2012,** *3,* 19.
27. Wasternack, C.; Hause, B. Jasmonates: Biosynthesis, Perception, Signal Transduction and Action in Plant Stress Response, Growth and Development. An Update to the 2007 Review in Annals of Botany. *Ann. Bot.* **2013,** *111* (6), 1021–1058.
28. Heitz, T.; Smirnova, E.; Widemann, E.; Aubert, Y.; Pinot, F.; Ménard, R. The Rise and Fall of Jasmonate Biological Activities. In *Lipids in Plants and Algae Development*; Springer Cham, 2016; pp 405–426.
29. Staswick, P. E.; Tiryaki, I. The Oxylipin Signal Jasmonic Acid is Activated by an Enzyme that Conjugates it to Isoleucine in *Arabidopsis. Plant Cell* **2004,** *16* (8), 2117–2127.
30. Westfall, C. S.; Muehler, A. M.; Jez, J. M. Enzyme Action in the Regulation of Plant Hormone Responses. *J. Biol. Chem.* **2013,** *288* (27), 19304–19311.
31. Fonseca, S.; Chini, A.; Hamberg, M.; Adie, B.; Porzel, A.; Kramell, R.; Miersch, O.; Wasternack, C.; Solano, R. (+)-7-iso-Jasmonoyl-L-isoleucine is the Endogenous Bioactive Jasmonate. *Nat. Chem. Biol.* **2009,** *5* (5), 344–350.
32. Katsir, L.; Chung, H. S.; Koo, A. J.; Howe, G. A. Jasmonate Signaling: A Conserved Mechanism of Hormone Sensing. *Curr. Opin. Plant Biol.* **2008,** *11* (4), 428–435.
33. Wasternack, C.; Strnad, M. Jasmonates: News on Occurrence, Biosynthesis, Metabolism and Action of an Ancient Group of Signaling Compounds. *Int. J. Mol. Sci.* **2018,** *19* (9), 2539.
34. Sheard, L. B.; Tan, X.; Mao, H.; Withers, J.; Ben-Nissan, G.; Hinds, T. R.; Kobayashi, Y.; Hsu, F. F.; Sharon, M.; Browse, J.; Zheng, N. Jasmonate Perception by Inositol-Phosphate-Potentiated COI1–JAZ Co-receptor. *Nature* **2010,** *468* (7322), 400–405.

35. Monte, I.; Hamberg, M.; Chini, A.; Gimenez-Ibanez, S.; García-Casado, G.; Porzel, A.; Pazos, F.; Boter, M.; Solano, R. Rational Design of a Ligand-based Antagonist of Jasmonate Perception. *Nat. Chem. Biol.* **2014,** *10* (8), 671–676.
36. Koo, A. J.; Gao, X.; Daniel Jones, A.; Howe, G. A. A Rapid Wound Signal Activates the Systemic Synthesis of Bioactive Jasmonates in *Arabidopsis. Plant J.* **2009,** *59* (6), 974–986.
37. Kang, S. M.; Min, J. Y.; Kim, Y. D.; Kang, Y. M.; Park, D. J.; Jung, H. N.; Kim, S. W.; Choi, M. S.; Effects of Methyl Jasmonate and Salicylic Acid on the Production of Bilobalide and Ginkgolides in Cell Cultures of *Ginkgo biloba. In Vitro Cell Dev. Biol. Plant* **2006,** *42* (1), 44–49.
38. Meesters, C.; Moenig, T.; Oeljeklaus, J.; Krahn, D.; Westfall, C. S.; Hause, B.; Jez, J. M.; Kaiser, M.; Kombrink, E. A Chemical Inhibitor of Jasmonate Signaling Targets JAR1 in *Arabidopsis thaliana. Nat. Chem. Biol.* **2014,** *10* (10), 830–836.
39. Wasternack, C. Jasmonates: An Update on Biosynthesis, Signal Transduction and Action in Plant Stress Response, Growth and Development. *Ann. Bot.* **2007,** *100,* 681–697.
40. Yan, J.; Li, S.; Gu, M.; Yao, R.; Li, Y.; Chen, J.; Yang, M.; Tong, J.; Xiao, L.; Nan, F.; Xie, D. Endogenous Bioactive Jasmonate is Composed of a Set of (+)-7-iso-JA-amino Acid Conjugates. *Plant Physiol.* **2016,** *172* (4), 2154–2164.
41. Chini, A.; Cimmino, A.; Masi, M.; Reveglia, P.; Nocera, P.; Solano, R.; Evidente, A. The Fungal Phytotoxin Lasiojasmonate A Activates the Plant Jasmonic Acid Pathway. *J. Exp. Bot.* **2018,** *69* (12), 3095–3102.
42. Chen, I. C.; Huang, I. C.; Liu, M. J.; Wang, Z. G.; Chung, S. S.; Hsieh, H. L. Glutathione S-transferase Interacting with Far-red Insensitive 219 is Involved in Phytochrome A-mediated Signaling in *Arabidopsis. Plant Physiol.* **2007,** *143* (3), 1189–1202.
43. Kurotani, K. I.; Hayashi, K.; Hatanaka, S.; Toda, Y.; Ogawa, D.; Ichikawa, H.; Ishimaru, Y.; Tashita, R.; Suzuki, T.; Ueda, M.; Takeda, S. Elevated Levels of CYP94 Family Gene Expression Alleviate the Jasmonate Response and Enhance Salt Tolerance in Rice. *Plant Cell Physiol.* **2015,** *56* (4), 779–789.
44. Marquis, V.; Smirnova, E.; Poirier, L.; Zumsteg, J.; Schweizer, F.; Reymond, P.; Heitz, T.; Stressand Pathwayspecific Impacts of Impaired Jasmonoyl-isoleucine (JA-Ile) Catabolism on Defense Signalling and Biotic Stress Resistance. *Plant Cell Environ.* **2020,** *43* (6), 1558–1570.
45. Liu, Y.; Du, M.; Deng, L.; Shen, J.; Fang, M.; Chen, Q.;Lu, Y.; Wang, Q.; Li, C.;Zhai, Q.; MYC2 Regulates the Termination of Jasmonate Signaling via an Autoregulatory Negative Feedback Loop. *Plant Cell* **2019,** *31* (1), 106–127.
46. Mao, D.; Xin, Y.; Tan, Y.; Hu, X.; Bai, J.; Liu, Z.Y.; Yu, Y.; Li, L.; Peng, C.; Fan, T.; Chen, C.; Natural Variation in the HAN1 Gene Confers Chilling Tolerance in Rice and Allowed Adaptation to a Temperate Climate. *Proc. Natl Acad. Sci. U. S. A.* **2019,** *116* (9), 3494–3501.
47. Kimberlin, A.; Leiboff, S.; Koo, A. J.; Hake, S. Tasselseed5 Overexpresses a Wound-inducible Enzyme, ZmCYP94B1, that Affects Jasmonate Catabolism, Sex Determination, and Plant Architecture in Maize. *Commun. Biol.* **2019,** *2* (1), 1–11.
48. Hazman, M.; Sühnel, M.; Schäfer, S.; Zumsteg, J.; Lesot, A.; Beltran, F.; Marquis, V.; Herrgott, L.; Miesch, L.; Riemann, M.; Heitz, T. Characterization of

Jasmonoyl-isoleucine (JA-Ile) Hormonal Catabolic Pathways in Rice Upon Wounding and Salt Stress. *Rice* **2019,** *12* (1), 1–14.
49. Bruckhoff, V.; Haroth, S.; Feussner, K.; König, S.; Brodhun, F.; Feussner, I. Functional Characterization of CYP94-genes and Identification of a Novel Jasmonate Catabolite in Flowers. *PLoS One* **2016,** *11* (7), e0159875.
50. Broekaert, W. F.; Delauré, S. L.; De Bolle, M. F.; Cammue, B. P. The Role of Ethylene in Host-pathogen Interactions. *Annu. Rev. Phytopathol.* **2006,** *44,* 393–416.
51. Bruce, T. J.; Matthes, M. C.; Chamberlain, K.; Woodcock, C. M.; Mohib, A.; Webster, B.; Smart, L. E.; Birkett, M. A.; Pickett, J. A.; Napier, J. A. Cis-Jasmone Induces Arabidopsis Genes that Affect the Chemical Ecology of Multitrophic Interactions with Aphids and their Parasitoids. *Proc. Natl Acad. Sci. U. S. A.* **2008,** *105* (12), 4553–4558.
52. Matthes, M. C.; Bruce, T. J.; Ton, J.; Verrier, P. J.; Pickett, J. A.; Napier, J. A. The Transcriptome of Cis-jasmone-induced Resistance in *Arabidopsis thaliana* and Its Role in Indirect Defence. *Planta* **2010,** *232* (5), 1163–1180.
53. Von Dahl, C. C.; Baldwin, I. T. Methyl Jasmonate and Cis-jasmone do Not Dispose of the Herbivore-induced Jasmonate Burst in *Nicotianaattenuata. Physiol. Plant.* **2004,** *120* (3), 474–481.
54. Schulze, B.; Dąbrowska, P.; Boland, W. Rapid Enzymatic Isomerization of 12-oxophytodienoic Acid in the Gut of Lepidopteran Larvae. *Chem. Bio. Chem.* **2007,** *8* (2), 208–216.
55. Miersch, O.; Herrmann, G.; Kramell, H. M.; Sembdner, G.; Biological Activity of Jasmonic Acid Glucosyl Ester. *Biochem. Physiol. Pflanz.* **1987,** *182* (5), 425–428.
56. Andolfi, A.; Maddau, L.; Cimmino, A.; Linaldeddu, B. T.; Basso, S.; Deidda, A.; Serra, S.; Evidente, A.; Lasiojasmonates A–C, Three Jasmonic Acid Esters Produced by *Lasiodiplodia* sp., a Grapevine Pathogen. *Phytochemistry* **2014,** *103,* 145–153.
57. Ueda, M.; Yang, G.; Nukadzuka, Y.; Ishimaru, Y.; Tamura, S.; Manabe, Y. Functional Importance of the Sugar Moiety of Jasmonic Acid Glucoside for Bioactivity and Target Affinity. *Org. Biomol. Chem.* **2015,** *13,* 55–58.
58. Wakuta, S.; Hamada, S.; Ito, H.; Matsuura, H.; Nabeta, K.; Matsui, H. Identification of a β-glucosidase Hydrolyzing Tuberonic Acid Glucoside in Rice (*Oryza sativa* L.). *Phytochemistry* **2010,** *71* (11–12), 1280–1288.
59. Seo, J. S.; Joo, J.; Kim, M. J.; Kim, Y. K.; Nahm, B. H.; Song, S. I.; Cheong, J. J.; Cheong, J. J.; Lee, J. S.; Kim, J. K; Lee, J. S., Kim, J. K..; Choi, Y. D. OsbHLH148, a Basic Helix-loop-helix Protein, Interacts with OsJAZ Proteins in a Jasmonate Signaling Pathway Leading to Drought Tolerance in Rice. *Plant J.* **2011,** *65* (6), 907–921.
60. Mugford, S. G.; Yoshimoto, N.; Reichelt, M.; Wirtz, M.; Hill, L.; Mugford, S. T.; Nakazato, Y.; Noji, M.; Takahashi, H.; Kramell, R.; Gigolashvili, T. Disruption of Adenosine-5′-phosphosulfate Kinase in *Arabidopsis* Reduces Levels of Sulfated Secondary Metabolites. *Plant Cell.* **2009,** *21* (3), 910–92.
61. Hayward, A.; Moreno, M. A.; Howard, T. P.; Hague, J.; Nelson, K.; Heffelfinger, C.; Romero, S.; Kausch, A. P.; Glauser, G.; Acosta, I. F.; Mottinger, J. P.; Dellaporta, S. L. Control of Sexuality by the *sk1*-encoded UDP-glycosyltransferase of Maize. *Sci. Adv.* **2016,** *2,* e1600991.
62. Świątek, A.; Van Dongen, W.; Esmans, E. L.; Van Onckelen, H.; Metabolic Fate of Jasmonates in Tobacco Bright Yellow-2 Cells. *Plant Physiol.* **2004,** *135* (1), 161–172.
63. Jones, J. D.; Dangl, J. L. The Plant Immune System. *Nature* **2006,** *444* (7117), 323–329.

64. Pieterse, C. M.; Van der Does, D.; Zamioudis, C.; Leon-Reyes, A.; Van Wees, S. C. Hormonal Modulation of Plant Immunity. *Annu. Rev. Cell Dev. Biol.* **2012,** *28,* 489–521.
65. Robert-Seilaniantz, A.; Grant, M.; Jones, J. D. Hormone Crosstalk in Plant Disease and Defense: More than Just Jasmonate-salicylate Antagonism. *Annu. Rev. Phytopathol.* **2011,** *49,* 317–343.
66. De Vos, M.; Van Oosten, V. R.; Van Poecke, R. M.; Van Pelt, J. A.; Pozo, M. J.; Mueller, M. J.; Buchala, A. J.; Métraux, J. P.; Van Loon, L. C.; Dicke, M.; Pieterse, C. M. Signal Signature and Transcriptome Changes of *Arabidopsis* During Pathogen and Insect Attack. *Mol. Plant Microbe. Interact.* **2005,** *18* (9), 923–937.
67. Loake, G.; Grant, M. Salicylic Acid in Plant Defence—The Players and Protagonists. *Curr. Opin. Plant Biol.* **2007,** *10,* 466–472.
68. Glazebrook, J. Contrasting Mechanisms of Defense Against Biotrophic and Necrotrophic Pathogens. *Annu. Rev. Phytopathol.* **2005,** *43,* 205–227.
69. Grant, M.; Lamb, C. Systemic Immunity. *Curr. Opin. Plant Biol.* **2006,** *9* (4), 414–420.
70. Lumba, S.; Cutler, S.; Mccourt, P. Plant Nuclear Hormone Receptors: A Role for Small Molecules in Protein–Protein Interactions. *Annu. Rev. Cell Dev. Biol.* **2010,** *26,* 445–469.
71. Fu, Z. Q.; Yan, S.; Saleh, A.; Wang, W.; Ruble, J.; Oka, N.; Mohan, R.; Spoel, S. H.; Tada, Y.; Zheng, N.; Dong, X. NPR3 and NPR4 are Receptors for the Immune Signal Salicylic Acid in Plants. *Nature* **2012,** *486* (7402), 228–232.
72. Kelley, D. R.; Estelle, M. Ubiquitin-Mediated Control of Plant Hormone Signaling. *Plant Physiol.* **2012,** *160* (1), 47–55.
73. Kagale, S.; Links, M. G.; Rozwadowski, K. Genome-Wide Analysis of Ethylene-Responsive Element Binding Factor-associated Amphiphilic Repression Motif-containing Transcriptional Regulators in *Arabidopsis*. *Plant Physiol.* **2010,** *152,* 1109–1134.
74. Sun, T. P. The Molecular Mechanism and Evolution of the GA–GID1–DELLA Signaling Module in Plants. *Curr. Biol.* **2011,** *21* (9), R338–R345.
75. De Lucas, M.; Daviere, J. M.; Rodríguez-Falcón, M.; Pontin, M.; Iglesias-Pedraz, J. M.; Lorrain, S.; Fankhauser, C.; Blázquez, M. A.; Titarenko, E.; Prat, S. A Molecular Framework for Light and Gibberellin Control of Cell Elongation. *Nature* **2008,** *451* (7177), 480–484.
76. Silverstone, A. L.; Jung, H. S.; Dill, A.; Kawaide, H.; Kamiya, Y.; Sun, T. P. Repressing a Repressor: Gibberellin-induced Rapid Reduction of the RGA Protein in *Arabidopsis*. *Plant Cell.* **2001,** *13* (7), 1555–1566.
77. Tyler, L..; Thomas, S. G..; Hu, J.; Dill, A.; Alonso, J. M.; Ecker, J. R.; Sun, T. P. DELLA Proteins and Gibberellin-regulated Seed Germination and Floral Development in *Arabidopsis*. *Plant Physiol.* **2004,** *135* (2), 1008–1019.
78. Harberd, N. P.; Belfield, E.; Yasumura, Y. The Angiosperm Gibberellin-GID1-DELLA Growth Regulatory Mechanism: How an "Inhibitor of an Inhibitor" Enables Flexible Response to Fluctuating Environments. *Plant Cell* **2009,** *21* (5), 1328–1339.
79. Chen, Z.; Zheng, Z.; Huang, J.; Lai, Z.; Fan, B. Biosynthesis of Salicylic Acid in Plants. *Plant Signal Behav.* **2009,** *4* (6), 493–496.
80. Qi, T.; Huang, H.; Wu, D.; Yan, J.; Qi, Y.; Song, S.; Xie, D. *Arabidopsis* DELLA and JAZ Proteins Bind the WD-repeat/bHLH/MYB Complex to Modulate Gibberellin and Jasmonate Signaling Synergy. *Plant Cell* **2014,** *26* (3), 1118–1133.

81. Navarro, L.; Bari, R.; Achard, P.; Lisón, P.; Nemri, A.; Harberd, N. P.; Jones, J. D. G. DELLAs Control Plant Immune Responses by Modulating the Balance of Jasmonic Acid and Salicylic Acid Signaling. *Curr. Biol.* **2008,** *18,* 650–655.
82. Hou, X.; Lee, L. Y. C.; Xia, K.; Yan, Y.; Yu, H. DELLAs Modulate Jasmonate Signaling via Competitive Binding to JAZs. *Dev. Cell* **2010,** *19* (6), 884–894.
83. Wild, M.; Davière, J. M.; Cheminant, S.; Regnault, T.; Baumberger, N.; Heintz, D.; Baltz, R.; Genschik, P.; Achard, P. The *Arabidopsis* DELLA RGA-LIKE3 is a Direct Target of MYC2 and Modulates Jasmonate Signaling Responses. *Plant Cell* **2012,** *24* (8), 3307–3319.
84. Yang, D. L.; Yao, J.; Mei, C. S.; Tong, X. H.; Zeng, L. J.; Li, Q.; Xiao, L. T.; Sun, T. P.; Li, J.; Deng, X. W.; Lee, C. M. Plant Hormone Jasmonate Prioritizes Defense Over Growth by Interfering with Gibberellin Signaling Cascade. *Proc. Natl Acad. Sci. U. S. A.* **2012,** *109* (19), E1192–E1200.
85. Um, T. Y.; Lee, H. Y.; Lee, S.; Chang, S. H.; Chung, P. J.; Oh, K. B.; Kim, J. K.; Jang, G.; Choi, Y. D. Jasmonate Zim-domain Protein 9 Interacts with Slender Rice 1 to Mediate the Antagonistic Interaction Between Jasmonic and Gibberellic Acid Signals in Rice. *Front. Plant Sci.* **2018,** *9,* 1866.
86. Wang, K. L. C.; Li, H.; Ecker, J. R. Ethylene Biosynthesis and Signaling Networks. *Plant Cell* **2002,** *14* (Suppl 1), S131–S151.
87. Hua, J.; Sakai, H.; Nourizadeh, S.; Chen, Q. G.; Bleecker, A. B.; Ecker, J. R.; Meyerowitz, E. M.; EIN4 and ERS2 are Members of the Putative Ethylene Receptor Gene Family in *Arabidopsis*. *Plant Cell* **1998,** *10* (8), 1321–1332.
88. Tian, D.; Peiffer, M.; De Moraes, C. M.; Felton, G. W. Roles of Ethylene and Jasmonic Acid in Systemic Induced Defense in Tomato (*Solanumlycopersicum*) Against *Helicoverpazea*. *Planta* **2014,** *239* (3), 577–589.
89. Berrocal-Lobo, M.; Molina, A.; Solano, R. Constitutive Expression of Ethylene-Response-Factor1 in *Arabidopsis* Confers Resistance to Several Necrotrophic Fungi. *Plant J.* **2002,** *29* (1), 23–32.
90. Lorenzo, O.; Solano, R. Molecular Players Regulating the Jasmonate Signalling Network. *Curr. Opin. Plant Biol.* **2005,** *8* (5), 532–540.
91. Zhu, Z.; An, F.; Feng, Y.; Li, P.; Xue, L.; Mu, A.; Jiang, Z.; Kim, J. M.; To, T. K.; Li, W.; Zhang, X. Derepression of Ethylene-stabilized Transcription Factors (EIN3/EIL1) Mediates Jasmonate and Ethylene Signaling Synergy in *Arabidopsis*. *Proc. Natl Acad. Sci. U. S. A.* **2011,** *108* (30), 12539–12544.
92. Pré, M.; Atallah, M.; Champion, A.; De Vos, M.; Pieterse, C. M.; Memelink, J. The AP2/ERF Domain Transcription Factor ORA59 Integrates Jasmonic Acid and Ethylene Signals in Plant Defense. *Plant Physiol.* **2008,** *147* (3), 1347–1357.
93. Zarei, A.; Körbes, A. P.; Younessi, P.; Montiel, G.; Champion, A.; Memelink, J. Two GCC Boxes and AP2/ERF-domain Transcription Factor ORA59 in Jasmonate/Ethylene-mediated Activation of the PDF1. 2 Promoter in *Arabidopsis*. *Plant Mol. Biol.* **2011,** *75* (4–5), 321–331.
94. Schweighofer, A.; Meskiene, I. Regulation of Stress Hormones Jasmonates and Ethylene by MAPK Pathways in Plants. *Mol. Biosyst.* **2008,** *4* (8), 799–803.
95. Chang, K. N.; Zhong, S.; Weirauch, M. T.; Hon, G.; Pelizzola, M.; Li, H.; Huang, S. S. C.; Schmitz, R. J.; Urich, M. A.; Kuo, D.; Nery, J. R. Temporal Transcriptional

Response to Ethylene Gas Drives Growth Hormone Cross-regulation in *Arabidopsis*. *eLife* **2013,** *2,* e00675.
96. Song, S.; Qi, T.; Wasternack, C.; Xie, D. Jasmonate Signaling and Crosstalk with Gibberellin and Ethylene. *Curr. Opin. Plant Biol.* **2014,** *21,* 112–119.
97. Zhang, X.; Zhu, Z.; An, F.; Hao, D.; Li, P.; Song, J.; Yi, C.; Guo, H. Jasmonate-activated MYC2 Represses Ethylene Insensitive3 Activity to Antagonize Ethylene-promoted Apical Hook Formation in *Arabidopsis. Plant Cell* **2014,** *26* (3), 1105–1117.
98. Fernández-Calvo, P.; Chini, A.; Fernández-Barbero, G.; Chico, J. M.; Gimenez-Ibanez, S.; Geerinck, J.; Eeckhout, D.; Schweizer, F.; Godoy, M.; Franco-Zorrilla, J. M.; Pauwels, L. The Arabidopsis bHLH Transcription Factors MYC3 and MYC4 are Targets of JAZ Repressors and Act Additively with MYC2 in the Activation of Jasmonate Responses. *Plant Cell* **2011,** *23* (2), 701–715.
99. Chung, S. H.; Rosa, C.; Scully, E. D.; Peiffer, M.; Tooker, J. F.; Hoover, K.; Luthe, D. S.; Felton, G. W. Herbivore Exploits Orally Secreted Bacteria to Suppress Plant Defenses. *Proc. Natl Acad. Sci. U. S. A.* **2013,** *110* (39), 15728–15733.
100. Spoel, S. H.; Johnson, J. S.; Dong, X. Regulation of Tradeoffs Between Plant Defenses Against Pathogens with Different Lifestyles. *Proc. Natl Acad. Sci. U. S. A.* **2007,** 104, 18842–18847.
101. Uppalapati, S. R.; Ishiga, Y.; Wangdi, T.; Kunkel, B. N.; Anand, A.; Mysore, K. S.; Bender, C. L. The Phytotoxincoronatine Contributes to Pathogen Fitness and is Required for Suppression of Salicylic Acid Accumulation in Tomato Inoculated with *Pseudomonas syringae* pv.tomato DC3000. *Mol. Plant Microbe. Interact.* **2007,** *20* (8), 955–965.
102. Van der Does, D.; Leon-Reyes, A.; Koornneef, A.; Van Verk, M. C.; Rodenburg, N.; Pauwels, L.;Goossens, A.; Körbes, A. P.; Memelink, J.; Ritsema, T.; Pieterse, C. M. Salicylic Acid Suppresses Jasmonic Acid Signaling Downstream of SCFCOI1-JAZ by Targeting GCC Promoter Motifs via Transcription Factor ORA59. *Plant Cell* **2013,** *25* (2), 744–761.
103. Petersen, M.; Brodersen, P.; Naested, H.; Andreasson, E.; Lindhart, U.; Johansen, B.; Nielsen, H. B.; Lacy, M.; Austin, M. J.; Parker, J. E.; Mundy, J. *Arabidopsis* MAP Kinase 4 Negatively Regulates Systemic Acquired Resistance. *Cell* **2000,** *103* (7), 1111–1120.
104. Zander, M.; Chen, S.; Imkampe, J.; Thurow, C.; Gatz, C. Repression of the Arabidopsis thaliana Jasmonic Acid/Ethylene-induced Defense Pathway by TGA-interacting Glutaredoxins Depends on Their C-terminal ALWL Motif. *Mol. Plant* **2012,** *5* (4), 831–840.
105. Ndamukong, I.; Abdallat, A. A.; Thurow, C.; Fode, B.; Zander, M.; Weigel, R.; Gatz, C. SA-inducible *Arabidopsis* Glutaredoxin Interacts with TGA Factors and Suppresses JA-responsive PDF1. 2 Transcription. *Plant J.* **2007,** *50* (1), 128–139.
106. Dong, X. NPR1, All Things Considered. *Curr. Opin. Plant Biol.* **2004,** *7* (5), 547–552.
107. Spoel, S. H.; Koornneef, A.; Claessens, S. M. C.; Korzelius, J. P.; Van pelt, J. A.; Mueller, M. J.; Buchala, A. J.; Métraux, J. P.; Brown, R.; Kazan, K; Loon, L. C. V.; Dong, X. N.; Pieterse, C. M. J. NPR1 Modulates Cross-talk between Salicylate-and Jasmonate-dependent Defense Pathways Through a Novel Function in the Cytosol. *Plant Cell* **2003,** *15,* 760–770.

108. Laurie-Berry, N.; Joardar, V.; Street, I. H.; Kunkel, B. N. The *Arabidopsis thaliana* Jasmonate Insensitive 1 Gene is Required for Suppression of Salicylic Acid-dependent Defenses During Infection by *Pseudomonas syringae. Mol. Plant Microbe. Interact.* **2006,** *19* (7), 789–800.
109. Liu, L.; Li, H.; Zeng, H.; Cai, Q.; Zhou, X.; Yin, C. Exogenous Jasmonic Acid and Cytokinin Antagonistically Regulate Rice Flag Leaf Senescence by Mediating Chlorophyll Degradation, Membrane Deterioration, and Senescence-associated Genes Expression. *J. Plant Growth Regul.* **2016,** *35* (2), 366–376.
110. Jang, G.; Yoon, Y.; Choi, Y. D. Crosstalk with Jasmonic Acid Integrates Multiple Responses in Plant Development. *Int. J. Mol. Sci.* **2020,** *21* (1), 305.
111. Bishopp, A.; Help, H.; El-Showk, S.; Weijers, D.; Scheres, B.; Friml, J.; Benková, E.; Mähönen, A. P.; Helariutta, Y. A Mutually Inhibitory Interaction between Auxin and Cytokinin Specifies Vascular Pattern in Roots. *Curr. Biol.* **2011,** *21* (11), 917–926.
112. Miyamoto, K.; Oka, M.; Ueda, J. Update on the Possible Mode of Action of the Jasmonates: Focus on the Metabolism of Cell Wall Polysaccharides in Relation to Growth and Development. *Physiol. Plant* **1997,** *100* (3), 631–638.
113. Jang, G.; Chang, S. H.; Um, T. Y.; Lee, S.; Kim, J. K.; Do Choi, Y. Antagonistic Interaction Between Jasmonic Acid and Cytokinin in Xylem Development. *Sci. Rep.* **2017,** *7* (1), 1–13.
114. Ursache, R.; Miyashima, S.; Chen, Q.; Vatén, A.; Nakajima, K.; Carlsbecker, A.; Zhao, Y.; Helariutta, Y.; Dettmer, J. Tryptophan-dependent Auxin Biosynthesis is Required for HD-ZIP III-mediated Xylem Patterning. *Development* **2014,** *141* (6), 1250–1259.
115. Chen, Q.; Sun, J.; Zhai, Q.; Zhou, W.; Qi, L.; Xu, L.;Wang, B.; Chen, R.; Jiang, H.; Qi, J.; Li, C. The Basic Helix-loop-helix Transcription Factor MYC2 Directly Represses Plethora Expression During Jasmonate-mediated Modulation of the Root Stem Cell Niche in *Arabidopsis. Plant Cell* **2011,** *23* (9), 3335–3352.
116. Mähönen, A. P.; Ten Tusscher, K.; Siligato, R.; Smetana, O.; Díaz-Triviño, S.; Salojärvi, J.; Wachsman, G.; Tholl, D.; Chételat, A.; Haupt, I.; Kennerley, B. J.; Liu, B.; Qi, T.; Tong, J.; Xiao, L.; Xie, D.; Song, S. *Arabidopsis* MYB24 Regulates Jasmonate-mediated Stamen Development. *Front Plant Sci.* **2017,** *8,* 1525.
117. Lim, P. O.; Woo, H. R.; Nam, H. G. Molecular Genetics of Leaf Senescence in *Arabidopsis. Trends Plant Sci.* **2003,** *8* (6), 272–278.
118. Leopold, A. C. *Auxins and Plant Growth;* University of California Press, 2021.
119. Jiang, Y.; Liang, G.; Yang, S.; Yu, D. *Arabidopsis* WRKY57 Functions as a Node of Convergence for Jasmonic Acid–and Auxin-mediated Signaling in Jasmonic Acid–induced Leaf Senescence. *Plant Cell* **2014,** *26* (1), 230–245.
120. Campos, M. L.; De Almeida, M.; Rossi, M. L.; Martinelli, A. P.; Litholdo Junior, C. G.; Figueira, A.; Rampelotti-Ferreira, F. T.; Vendramim, J. D.; Benedito, V. A.; Pereira Peres, L. E. Brassinosteroids Interact Negatively with Jasmonates in the Formation of Anti-herbivory Traits in Tomato. *J. Exp. Bot.* **2009,** *60* (15), 4347–4361.
121. Li, L.; Zhao, Y. F.; McCaig, B. C.; Wingerd, B. A.; Wang, J. H.; Whalon, M. E.; Pichersky, E.; Howe G. A. The Tomato Homolog of Coronatine-Insensitive1 is Required for the Maternal Control of Seed Maturation, Jasmonate-signaled Defense Responses, and Glandular Trichome Development. *Plant Cell* **2004,** *16,* 126–143.

122. Perazza, D.; Vachon, G.; Herzog, M. Gibberellins Promote Trichome Formation by Up-regulating Glabrous1 in *Arabidopsis*. *Plant Physiol.* **1998,** *117* (2), 375–383.
123. Zarate, S. I.; Kempema, L. A.; Walling, L. L. Silver Leaf Whitefly Induces Salicylic Acid Defenses and Suppresses Effectual Jasmonic Acid Defenses. *Plant Physiol.* **2007,** *143* (2), 866–875.
124. Nakashita, H.; Yasuda, M.; Nitta, T.; Asami, T.; Fujioka, S., Arai, Y.; Sekimata, K.; Takatsuto, S.; Yamaguchi, I.; Yoshida, S. Brassinosteroid Functions in a Broad Range of Disease Resistance in Tobacco and Rice. *Plant J.* **2003,** *33* (5), 887–898.
125. Koka, C. V.; Cerny, R. E.; Gardner, R. G.; Noguchi, T.; Fujioka, S.; Takatsuto, S.; Youshida, S.; Clouse, S. D. A Putative Role for the Tomato Genes DUMPY and CURL-3 in Brassinosteroid Biosynthesis and Response. *Plant Physiol.* **2000,** *122* (1), 85–98.
126. Moreno, J. E.; Tao, Y.; Chory, J.; Ballaré, C. L. Ecological Modulation of Plant Defense via Phytochrome Control of Jasmonate Sensitivity. *Proc. Natl Acad. Sci. U. S. A.* **2009,** *106* (12), 4935–4940.

CHAPTER 4

Modulation of Jasmonate Signaling Under Various Conditions

ARIF HUSSAIN BHAT[1], SUHAIL ALTAF[2], SHEIKH MEHRAJ[3], ZAHOOR AHMAD SHAH[4], MUDASIR BHAT[5], SYED SAMI ULLAH[6], ADIL NAZIR[7], M. S. DAR[7], and SHAFAT AHMAD AHANGER[8]

[1]*Plant pathology, Department of Agriculture, BFC, Bathinda, Punjab, India*

[2]*Department of Plant Pathology, SKUAST, Kashmir, India*

[3]*Division of Fruit Science, SKUAST, Kashmir, India*

[4]*KVK-Ganderbal, SKUAST, Kashmir, India*

[5]*Department of Agriculture, Jammu & Kashmir, India*

[6]*ICAR-CITH, Srinagar, India*

[7]*Plant Pathology, MRCFC, Khudwani, SKUAST, Kashmir, India*

[8]*Plant Pathology, FoA, Wadura, Sopore, SKUAST, Kashmir, India*

ABSTRACT

Jasmonates have been reported to perform a number of diverse functions in the growth, development, and defense mechanisms of plants. A considerable amount of work in the last two decades has been undertaken to understand the mechanism of the jasmonic acid (JA) signaling pathway. The current review discusses the synthesis of JA compounds, signal initiation, conductance, and perception of JC signaling. Also, the modulation of JC signaling under different types of environmental stress factors has been described in detail. The signaling networks are critical

Jasmonates and Plant Defense: An Eco-Friendly and Sustainable Approach for Stress Tolerance.
Uzma Azeem, & Khalid Rehman Hakeem (Eds.)

in establishing communication channels between the environmental inputs and cellular responses. Jasmonates as signaling molecules, could efficiently moderate plant responses in mitigating environmental stress conditions through the induction of a series of gene expressions. The signaling pathways mediated by jasmonates and salicylic acid are majorly concerned with plant resistance, initiating plant responses to mechanical stress and pathological stress, thereby inducing resistant gene expressions.

4.1 INTRODUCTION

Plants always grow in open environments and hence are very often inflicted by many biotic and abiotic stresses.[1–2] The survival of plants in such challenging environments is dependent on their potential to adapt according to the prevailing conditions which is achieved through various signaling networks.[3] The signaling networks are critical in establishing communication channels between the environmental inputs and cellular responses.[4] Phytohormones are known for their pivotal role in the institution of signaling networks for regulating different developmental processes and stress-related responses in plants. Plant hormones, being natural in origin and non-toxic in nature could be employed as safe and eco-friendly chemical control entities in different stress management programs. Jasmonic acid (3-*oxo*-2-20-*cis*-pentenyl-cyclopentane-1-acetic acid, abbreviated as JA) is a phytohormone present especially in higher plants which is very crucial in regulating the growth and developmental processes of plants. Jasmonic acid, the methyl ester of JA (MeJA), and the isoleucine conjugate of JA (JA-Ile) are derived from a class of fatty acids and together they are called jasmonates (JAs). Originally considered as a stress-related hormone, jasmonates are now very well known to play important roles in different metabolic processes of plants.[5–6] Jasmonates as signaling molecules, could efficiently moderate plant responses in mitigating environmental stress conditions through the induction of a series of gene expressions.[7] The signaling pathways mediated by jasmonates and salicylic acid are majorly concerned with plant resistance, initiating plant responses to mechanical stress and pathological stress, thereby inducing resistant gene expressions. In the last few years of research, a number of genes and transcription factors (TFs) having regulatory functions in JA biosynthesis and the pathway leading to signal transduction have been characterized. These genes involve a number of inhibitors and activators related to signaling in response to environmental factors.[8] The current chapter deals with the initiation, synthesis, transmission, biological functions, and JA signaling pathways that regulate plant responses to abiotic stress.

4.2 JASMONIC ACID SIGNAL INITIATION

4.2.1 PERCEPTION AND TRANSDUCTION OF SIGNALS

A number of investigations targeting the mechanism of biotic and abiotic stress signal perception by plants and how JA biosynthesis is initiated have been conducted in the past few years. A compound called systemin was found to respond to mechanical injury stress in tomatoes.[9] Systemin is a polypeptide signaling molecule which consists of 18 amino acids. It is derived from its precursor prosystemin which consists of 200 amino acid units.[10] When a tomato undergoes mechanical damage, prosystemin gets hydrolyzed to systemin. The systemin is capable of transportation to other cells through apoplastic movement and interacts with the cell surface receptor SR160 to finally activate the JC signaling pathway.[11,12] Besides traumatic signals, the pathogen, and fungal elicitor-induced oligosaccharide signaling has been reported in tomatoes and this signaling is believed to finally result in the activation of the JA signaling pathway. "In addition to the traumatic signals, oligosaccharide signals induced by pathogens and fungal elicitors were also found in tomatoes, which ultimately activate the JA signaling pathway. There are speculations that the mechanism of action of oligosaccharides and systemin follows a similar pathway, however, the exact mechanism leading to their induction is not clear yet.[13] Later, a polypeptide (AtPEP1) with same function as systemin, comprising 23 amino acids units was reported in *Arabidopsis thaliana*. In a similar way as that of systemin, mechanical damage or the invasion by some pathogen induces the hydrolysis of PROPEP1 (precursor of AtPEP1 comprising 92 amino acid units) to AtPEP1 which on the plasma membrane binds to the receptor molecule PEPR1. This receptor molecule (PEPR1) finally leads to the activation of the JA signaling pathway.[14] There is a complex mechanism behind the binding of systemin and AtPEP1 to the receptors to activate the JC signaling pathway. First, phospholipase on the plasma membrane is activated, phospholipase then acts on the membrane leading to the release of linolenic acid. Linolenic acid is a precursor in the process of JA synthesis.[15] However, the mechanism behind the activation of phospholipases by systemin is not clear until now. There are a number of phospholipase molecules identified so far which could be induced by systemin and AtPEP1. The identified phospholipase molecules are PLA2 in tomato, DAD1, DGL, and PLD in *Arabidopsis*. These phospholipase molecules work through the same mechanism of action.[16]

4.2.2 SYNTHESIS AND METABOLISM OF JASMONIC ACID COMPOUNDS

Biosynthesis of JA and the compounds derived from it has been investigated in a number of monocot and dicot plants in the recent past. The major portion of this work has been carried out utilizing the most familiar model plants *A. thaliana* and *Lycopersicon esculentum* (tomato). Until now, there are different types of enzymes that have been recognized for their role in JA synthesis pathway. Also, our understanding of the relationship between the JA synthesis pathway and other metabolic pathways in plants is improving with time. Three pathways for the biosynthesis of jasmonates are reported in *Arabidopsis*. These include the octadecane pathway which starts from linolenic acid and the hexadecane pathway which starts with hexadecatrienoic acid. The reaction sites required by these three pathways are chloroplast, peroxisome, and cytoplasm. The initial compound 12-*oxo*-phytodienoic acid (12-OPDA) is first synthesized inside chloroplast from unsaturated fatty acid molecules. It is later converted to JA inside the peroxisome. The JA goes through diverse chemical reactions inside cytoplasm resulting in different metabolic products, namely, MeJA, JA-Ile, *cis*-jasmone (CJ), and 12-hydroxyjasmonic acid (12-OH-JA).[16]

4.3 JA SIGNAL TRANSMISSION

When a traumatic signal is perceived, the defense response by the plants may result in a localized defense response at the site of injury and/or a systemic acquired resistance (SAR) beyond the site of injury, and/or even induced defense response from adjacent plants. Such defense responses involve the short-distance and long-distance transmissions of JA signals.[17] Various studies focusing on the mechanism of hormone signaling networks have revealed the interaction of JA with salicylic acid, ethylene, auxin, and other plant hormones in regulating the adaptation of plants to challenging environments. The comprehension of the mechanism of complex regulatory networks and different metabolic pathways initiated after the perception of environmental signals is very scanty until now. The different modes of JA signal transmission are described below:

4.3.1 SHORT-DISTANCE SIGNAL TRANSMISSION

In plant systems, the mechanical injuries resulting from insect feeding or any other entity lead to prompt and transitory build-up of JA and JA-Ile at the damage sites inducing defense-related gene expression in the vicinity of the trauma site. The gene expression results in a localized defense response. In this local defense reaction, two types of short-distance transmission of the JA signaling are executed. First, the systemin produced as a result of injury, acting as a signal molecule is transmitted to the adjacent sites via apoplastic movement and phloem to activate the JA cascade reaction pathway. In the second transmission, JA and JA-Ile are incited by the action of systemin function as signal molecules and are transmitted to nearby sites for initiating defense reactions.[18]

4.3.2 LONG-DISTANCE SIGNAL TRANSMISSION

Long-distance transmission of JA signals can be achieved by two means, namely, vascular bundle transmission and airborne transmission.

4.3.2.1 VASCULAR BUNDLE TRANSMISSION

Earlier it was believed that systemin has a direct role in long-distance signal transmission and it is a mobile signaling molecule. However, a series of investigations involving grafting experiments exploiting tomato jasmonate-insensitive mutant (jai1), systemin-insensitive mutant (spr1), and the JA biosynthesis deletion mutants spr2 and acx1A proved the fact that systemin was not the systemically transmitted signal.[18] When the synthesis of JA is induced, the process is followed by the systemic transmission of JA and MeJA in plants.[19] The transmission of MeJA to vascular bundles has been demonstrated by radio labeling experiments.[20] Some other workers have established that jasmonates are not transported along the vascular bundles as such but undergo the process of resynthesis during their journey through the xylem and phloem.[19] In the vascular tissues of tomato plants different JA synthetases have been located in the companion cell–sieve element complex (CC-SE).[21] Further it has been confirmed that the sieve molecules inside phloem have the capability to form the JA precursor OPDA.[22] Some recent studies have made it clear that the systemic JA and JA-Ile induced by the stimulus of injury are not wholly transferred from the site of injury,[23] rather a portion of it is resynthesized and the rest is synthesized away from the site of injury by cascading cycles.[24]

4.3.2.2 AIRBORNE TRANSMISSION

There is enough experimental proof that besides transmission through vascular tissues, JA signals are also transmitted through other long-distance transmission routes. It has been confirmed that the rate of flow of "tomato phloem signal" is 1–5 cm/h,[25] but the accumulation of JA and its derivative JA-Ile can be seen throughout the plant system within 15 min of mechanical injury.[26] The "ring-cutting experiments" have made it clear that there is some other way of JA signal transmission beyond vascular bundles, because of the evidence that an instant and strong defense gene expression was observed in the distal leaves even though the vascular bundle transmission was blocked because of the cuts.[26] In comparison to JA which cannot penetrate a cell membrane without the help of a carrier, MeJA makes easy penetration of the cell membranes. Additionally, MeJA has strong volatility and hence can be transmitted by airborne transmission to distant leaves and even to adjacent plants.[27] The airborne transmission of MeJA from injured to uninjured leaves or to adjacent plants has been confirmed in a variety of plants, such as *A. thaliana*,[19] *Nicotiana tabacum*,[28] *Phaseolus lunatus*,[29] and *Artemisia kawakamii*.[30]

4.4 PERCEPTION OF THE JASMONIC ACID SIGNAL AND RESPONSE INITIATION

4.4.1 JASMONIC ACID RECEPTOR

The process of nuclear transport of jasmonates has been systematically studied through molecular genetics, molecular biology, biochemistry, and cell biology approaches. The ABC transporter AtJAT1/AtABCG16 with jasmonate transportation capability has been investigated by utilizing a yeast system.[31] The autoradiographic and radioactive studies have revealed the functioning of AtJAT1/AtABCG16 as a high-affinity transporter in regulating the subcellular dispersal of jasmonates.[31] The transporter "AtJAT1/AtABCG16" is located on the nuclear and plasma membranes of plant cells and regulates the transportation of JAs through plasma membrane. It also regulates the transmission of JA-Ile through the nuclear membrane to initiate JA responses at low concentrations. At higher concentrations of jasmonates, the activity of JA transporter increases on the cytoplasmic membrane leading to reduced intracellular concentrations of JA and JA-Ile, which in turn results in desensitization of JA signaling. The apoplastic movement of JA activates the acid signaling pathway in other cells. The transporter AtJAT1/AtABCG16 very efficiently regulates the cellular dynamics of JA/JA-Ile. This efficient regulation results in the hasty movement of JA-Ile into the nucleus whenever the plants face stressed conditions. Also, this regulation causes very rapid desensitization of the JA signal which is directed towards inhibiting the growth and development of plants under the defense response mechanism.[16] The knowledge of JA receptors has gone through a complex process over time. In one of the earliest studies Feys (1994) for the first time reported that the *Arabidopsis coronatine* insensitive1 (*COI1*) mutant lost all responses related to JA.[32] Other studies have revealed that the COI1 gene codes for an F-box protein which is a constituent of E3 ubiquitin ligase.[33] In this case, COI1 couples with two proteins SKP1 and Cullin resulting in the formation of SCF-type E3ubiquitin ligase (SCFCOI1). It acts upon repressor proteins bringing about their degradation by the process of ubiquitination.[34] The discovery of COI1 protein has proved to be of great significance in studying the JA signaling pathway. Earlier COI1 was thought to be the receptor for JA signaling till the discovery of a jasmonate Zinc finger Inflorescence Meristem (ZIM)-domain (JAZ) protein family. The discovery of this protein family completely changed our comprehension of the JA signal transduction pathway.[16] In 2007, three independently working research groups simultaneously reached the conclusion that JAZ proteins act as repressors in the JA signaling pathway.[35–37] Until now, in *Arabidopsis* 13 JAZ proteins have been found, mostly possessing two conserved domains, namely, Jas and ZIM.[38] The JAZ protein utilizes the Jas domain for its interaction with COI1, while the interaction with MYC2 is done through the ZIM domain.[39] That is the reason behind the belief of many researchers who are of the opinion that since JAZ proteins are the target of COI1 Therefore, "many researchers believe that JAZ proteins are the target protein of COI1 which causes their degradation and this degradation is a critical point in preventing the inhibition the JA pathway.[16] However, Sheard *et al.* (2010) performed an analysis of crystal structure of JAs receptors through which it was confirmed that "the COI1–JAZ complex is a high-affinity receptor for the bioactive JA-Ile. Hence COI1 and JAZ are coreceptors of the JA signaling pathway.[40] The current opinion is that the environmental stimuli perceived by the plants result in the generation of JA-Ile. JA-Ile plays a role in bringing about the interaction between COI1 and JAZ. This interaction results in the degradation of JAZ after their transfer to the 26S proteasome. At the same time, TFs are released for activating the expression of downstream genes.[41]

4.4.2 JASMONIC ACID SIGNAL-REGULATED TRANSCRIPTION FACTOR

The MYC TFs are activated by JA-Ile when they bind to JAZ and COI1, resulting in the degradation of JAZ by means of 26S proteasome pathway. It has also been reported that MYB TFs also bind to JAZ repressors and are activated when JAZ is degraded in the presence of JA-Ile. Similarly, many other TFs, namely, NAC, ERF, and WRKY have also been identified as playing a vital role in JA signaling pathways. These TFs are involved in the regulation of a number of gene expressions in relation to the growth and development of plants, in particular, the plant response and adaptation to different environmental conditions. A number of studies have shown that JA signaling can bring about the induction of many pathways such as MAP kinase,[42,] and calcium channel.[43] Further, this signaling system can also induce many processes via interaction with other signaling molecules, namely, ethylene, salicylic acid, and abscisic acid help in the regulation of plant growth and development.[16]

4.4.2.1 MYC TRANSCRIPTION FACTOR

The MYC2 TF of the basic helix–loop–helix (bHLH) is considered an important regulatory protein that is encoded by the JIN1 gene. It has been observed that almost all members of the JAZ family work through interacting with MYC2.[44] For quite a long time the general perception was that only MYC2 can have a direct interaction with JAZ protein. However, in the year 2011, this perception changed with the findings of Fernandez-Calvo et al. who observed that MYC3 and MYC4 (bHLH proteins) share high genetic similarity with MYC2. This finding led to the speculation that there must be a similarity in function between these proteins. Later the interaction of MYC3 and MYC4 with JAZ was observed both under in vitro and in vivo conditions. They also have identical DNA-binding specificities to that of MYC2, and have both synergistic and distinct interactions with MYC2.[45] Besides the above TFs another similar kind of TF MYC5 (bHLH28) has been reported to be induced by jasmonates and plays a role in the regulation of male fertility.[46] In addition to transcriptional activators, another jasmonate-associated and MYC2-like protein family (JAM—JAM1, JAM2, and JAM3) has been discovered to be functioning as transcriptional repressors. They regulate jasmonate responses by way of their protein-protein interactions with JAZ proteins.[47]

4.4.2.2 MYB TRANSCRIPTION FACTOR

R2R3-MYB is a family of TFs. Most of the jasmonate responsive MYB TFs which are extensively distributed in plant systems belong to this family. These TFs are required by plants in many vital processes. Two members of this family MYB51 and MYB34 have been reported to be involved in the regulation of tryptophan and glucosinolate synthesis and they also have been observed functioning downstream of MYC2.[45] However, the works of many researchers have revealed that MYB TFs can directly bind to JAZ proteins resulting in the activation of their target genes. For example, in *Arabidopsis* the key players in stamen and pollen maturation have been reported to be MYB21 and MYB24,[48] and MYB75 can easily regulate the anthocyanin accumulation and trichome initiation.[49] Some recent studies have revealed that a group of MYB TFs viz. MYB11-16 were identified to be acting as repressors for regulating the rutin biosynthesis in Tartary buckwheat.[50,51]

4.4.2.3 NAC TRANSCRIPTION FACTOR

The TFs, ATAF1 and ATAF2 in *Arabidopsis* NAC family, are induced by JA signaling and play a very important role in plant resistance to stress (drought, salt, etc.), *Botrytis cinerea*, and other plant pathogens.[52] Simultaneously,

ATAF1 and ATAF2 also play an important part in regulating oxidative stress, flowering, and pod development.[53] Additionally, other NAC TFs in *Arabidopsis*, ANAC019, and ANAC055, have also been found existing downstream to the MYC2 protein and regulating the germination seeds, cell division, and the synthesis of secondary cell walls.[54] These TFs (ATAF1, ATAF2, ANAC019, and ANAC055) are also involved in the crosstalk between JA and salicylic acid signaling mechanism.[52–54]

4.4.2.4 ETHYLENE-RESPONSIVE TRANSCRIPTION FACTOR (ERF)

Transcription of many ERF genes by the JA signaling pathway has been confirmed by genetic-level microarray experimentation. The link between AP2/ERF TFs and JA signaling was for the first time observed in *Catharanthus roseus*. The ORCA proteins—ORCA2 and ORCA3—are the members of AP2/ERF-domain family. They are induced by jasmonates and can regulate the gene expression of monoterpenoid indole alkaloid biosynthesis genes.[55] Observations on ORCAs have revealed that the *Arabidopsis* ERF proteins (ERF1 and ORA59) are dependent on JAs and/or ET for their functioning in defenses against *Botrytis cinerea*.[56–57] Again, instead of ERF, ORA59 performs the role of integrator between JAs and ET signals[57] and regulates the biosynthesis of hydroxycinnamic acid amides.[58] ORA47 is a jasmonate-induced protein that can bring about the activation of gene expression of jasmonate biosynthesis gene AOC2. This indicates that ORA47 acts as a prime regulator in the positive jasmonate responsive feedback loop.[59] Moreover, the repressor role of JAs-responsive AtERF3 and AtERF4 has been evidenced by their function of downregulating their target gene expression, as well as by their interference with the functioning of other activators.[60] The activity of aforementioned TFs is not reported to be repressed by JAZ proteins in a direct manner. This observation leads us to believe that some adaptors or corepressors are involved in the JA signaling pathway.

4.4.2.5 WRKY TRANSCRIPTION FACTOR

WRKY TFs in addition to playing an important role in the regulation of plant development and senescence also play a vital role in managing environmental stress in plants. In *Arabidopsis*, there are 89 member factors in the WRKY TF family and some of them have been reported to be regulated by the JA signaling pathway.[41] These TFs (WRKY TFs) are mainly involved in the regulation of plant defense mechanisms. In *Nicotiana attenuata*, two WRKYTFs (NaWRKY3 and NaWRKY6) have been reported regulating the expression of JAs biosynthesis-related genes (LOX, AOS, AOC, and OPR) resulting in increased levels of JA and JA-Ile.[61] In *Arabidopsis*, it has been observed that WRKY57 TF communicates with inhibitor-JAZ4/JAZ8 in the JA signaling pathway and inhibitor-IAA29 in the auxin signaling pathway. In this way, it regulates the interaction between JA and auxin-mediated signaling pathways and influences plant leaf senescence.[62]

4.5 JASMONIC ACID SIGNALING AND ENVIRONMENTAL STRESS

The plant signaling molecules, JA and its derivatives, are closely related to plant defenses and resistance to microbial pathogens, herbivorous insects, and different types of stress (wounding, drought, salinity, low temperature, etc.). Besides traumatic signals, oligosaccharide signals which are induced by plant pathogens and fungal elicitors were also observed in tomatoes, which eventually resulted in activating the JA signaling pathway. When a tomato undergoes physical damage, the compound prosystemin through the process of hydrolysis is converted into systemin, which can be transported to other cells through the apoplastic movement and reacts with a receptor on the cell surface SR160 (a protein rich in leucine repeat units) to finally activate JA responses. This section will discuss the role of JA signaling in regulating the plant responses to different environmental conditions.-

4.5.1 EFFECT OF LIGHT

From the recent research, it is clear that JA signaling is involved in the mediation of plant responses to light from two major aspects; photomorphogenesis and the damage inflicted by ultraviolet B (UV-B). In *Arabidopsis* and rice, the photomorphogenesis which is mediated by red/far-red light has been reported to be influenced by jasmonate signaling.[63] The role of JA signaling was also observed in blue light-mediated morphogenesis in *Arabidopsis* and tomato.[64,65] In *Nicotiana* and *Brassica generato*, an enhanced exposure to UVB has been found to induce jasmonate biosynthesis and initiation of the JA signaling pathway. This phenomenon plays a role in the chemical defense mechanism of plants.[66] Through continuous efforts over a period of time, it has been established that light plays a critical regulatory role both in the biosynthesis of JA and in its signal transduction.[67,68] The extra-floral nectar (EFN) secretion has been observed to be affected by the influence of both light and JA. The EFN secretion was found to decrease in the darkness under the influence of JA while as opposite reaction was seen when darkness was removed and light was provided that is EFN secretion was induced. JA-Ile which is a derivative of JA behaved differently than JA, it enhanced the EFN secretion during light hours but did not decrease its secretion under darkness.[69] In a study JA-free hebiba mutant of *Oryza sativa* was grown under dark and red light conditions for studying the effect of JA applications. It was observed that the application of JA or the precursor of JA 12-oxophytodienoic acid could rescue the growth and development of these plants under such conditions. Hence it can be concluded that the absence of JA has the potential phenotypic effects of mutation.[70] The mutants of *A. thaliana* which were having decreased JA biosynthesis levels and signaling demonstrated lower abilities of response to higher levels of irradiance under FR light. It has been confirmed by studies that the signaling pathways of both phyA and JA work through by degrading the protein JAZ1 and both of these are essential for plants in their battle against light stress.[41] It has been observed that exogenous application of MeJA increases the interaction between FR-insensitive 219 (FIN219) and the C terminus of cryptochrome 1 (CCT1) when subjected to blue light exposure. Other studies have shown that FIN219 and cryptochrome1 exhibit an antagonistic relationship. This suggests that the two pathways viz. JA signaling and blue light signaling have a crucial relationship among them and this relationship plays a critical role in the development of seedlings and response to stress conditions.[71] The studies of Liu et al.[72] have confirmed that exogenous JA application results in the mitigation of negative impacts of UV-B stress upon exposure of wheat seedlings. Exposure to UV-B radiations in *Nicotiana* induced overlapping biochemical changes and gene transcription which included jasmonate biosynthesis and signal transduction.[73] Also it has been concluded through various experiments that UV-B exposure: (a) initiates production of phenolics through two pathways, one is jasmonate dependent pathway and the other jasmonate independent pathway, and (b) increases the sensitivity to JAs, resulting in increased expression of the gene regulating wound-response, trypsin proteinase inhibitor.[74]

4.5.2 EFFECT OF TEMPERATURE

The crop productivity is severely affected by cold stress (chilling and freezing) because of the fact that many crops are sensitive to low temperature. The effects of cold stress in plants may cause chlorotic and necrotic symptoms, damaged cell membranes, alterations in the viscosity of cytoplasm, and altered enzyme activities[75] which could ultimately result in the death of whole plant. The plant systems are therefore left with no choice than adaptation to chilling and freezing stress, and have accordingly evolved many stress tolerance mechanisms for their survival under adverse environmental conditions.[76] The JA signaling is significantly associated with the response and adaptation of plants to low temperatures. Recent studies on banana (*Musa acuminata*) have reported that the application of MeJA to banana resulted in the expression of many TFs belonging to MYC-family. Also, the expression of many cold-response genes was induced after the banana were cold stored which results in the protection against cold stress. Further, upon cold exposure the endogenous jasmonate content of banana diminished slightly but the decrease in content was statistically insignificant.[77] The studies on *Arabis alpina* reported variations in jasmonates and many other hormones in cold stress conditions.[78] At a normal temperature of 22°C, all the plants (tolerant and non-tolerant) recorded a very high level of JA. When the plants were subjected to cold stress at 4°C, no significant change of JA was seen in the tolerant plants while non-tolerant plants recorded a 10-fold reduction. Further, when the plants were exposed to frost (below 0°C), a drastic dip of JA content was seen in tolerant plants and the jasmonate content of non-tolerant plants remained almost unchanged. The findings suggest that in *A. alpina*, JA might be the principal jasmonate mediator for cold stress. Methyl jasmonate (MJ) has been reported to be capable of reducing the adverse effects of chilling injuries in crop plants. The role of MJ is thought to be by means of alterations in tissue metabolism leading to an elevated synthesis of ABA at chilling temperature exposure. It is suggested that MJ acts as secondary messenger of plants in signal transduction and regulates the defense gene expression in response to different stress conditions.[79–81] Under cold stress conditions, the exogenous application of MeJA results in the rapid activation of MYC gene leading to resyntheis of jasmonates in huge amounts for providing protection against cold stress. Additionally, various investigations in tomato (*L. esculentum*),[82] pomegranate (*Punica granatum*),[83] loquat (*Eribotrya japonica*),[84] mango (*Mangifera indica*),[85] and guava (*Psidium guajava*)[86] have shown that exogenous MeJA treatment can induce heat shock protein family transcription, increased antioxidant synthesis, reduction in lipoxygenase activity, and increased resistance to cold stress (with 0°C upper temperature). Jasmonic acid is a key signal molecule for the formation of sesquiterpene which is induced by high temperature shock in eaglewood (*Aquilaria sinensis*).[87] Like cold stress, exposure to high temperatures also affects the normal growth and developmental processes in plants. The high temperature stress causes harmful changes in plants cells leading to alterations in cellular homeostasis.[88] The exposure to high temperature induces the synthesis of HSPs (heat shock proteins) in plants. The HSPs prevent denaturation of proteins and assist in refolding of damaged proteins.[89] In *A. thaliana* it has been reported that JAs promote the development of thermotolerance in plants. In heat stressed plants the exogenous applications of very low doses of MJ maintains the cell viability by controlling electrolyte leakages.[90] Further, the role of JA in thermotolerance has been proved by mutation studies of salicylic acid and JA signaling mutants (COI1–1, opr3, and jar1–1cpr5–1).Other studies have shown that the genes responsible for JA biosynthesis and signaling, were up regulated by drought and cold stress, but the same genes were downregulated by high temperature exposure (heat stress) in rice plants. Also, the endogenous JA content increased under cold stress, while as a decrease in JA content was seen under heat stress.[91]

4.5.3 EFFECT OF DROUGHT

Climate change is leading the world to global warming, the impact of which can be seen in the form of very severe drought events with increased frequency in a number of agriculturally important regions worldwide. The reduction in crop yield (by more than 50% in many crops) and even crop failures can be directly seen as the impact of drought stress on crop cultivation.[92] In general, the effects of drought stress include reduced plant growth,[93,94] declining photosynthetic rates,[95] and faster leaf senescence.[96,97]

Additionally, drought stress can also induce oxidation reactions, initiate accumulation of lipids on cell membranes and antioxidant enzyme expressions.[98,99] The levels of endogenous jasmonates alleviate very quickly under drought conditions but return back to actual levels if the stress periods are prolonged. There are a number of studies confirming the involvement of jasmonate signaling of plant systems in drought stress management. In *A. thaliana*[100] and citrus,[101] it was found that the increase in endogenous jasmonate content was very quick and transitional following drought stress exposure and came down to the base levels if the stressed conditions were of long duration. However, the prevention of drought stress damage in plants by exogenous applications of jasmonates has been well documented. The seedlings of Peanut (*Arachis hypogaea*) treated with MeJA showed enhanced drought resistance.[102] It was also observed that MeJA treatment can improve drought resistance in rice (*O. sativa*),[103] soybean (*Glycine max*),[104] and broccoli (*Brassica oleracea*)[105] through appropriate changes in the plant metabolism. It has been found that MeJA application in broad bean (*Vicia faba*) and barley (*Hordeum vulgare*) can influence the closure of stomata by interfering with the K^+ channel pathway, thereby increasing the ability of plants to resist drought.[106,107] It has been reported that in *A. thaliana* the JA minimizes loss of water by regulating stomatal opening and closure.[108] Stomatal closure in plants reduces the rate of transpiration resulting in retention of water inside the plant tissues under drought conditions. The interactions of ABA with other plant hormones such as JA and NO to induce stomatal closure have been reported. JA produced under drought stress helps in ABA-mediated stomatal closure by inducing the entry of Ca^{2+} from extracellular space and by means of initiating the H_2O_2/NO signaling system.[109] There is one hypothesis that the interaction of ABA and MJ takes place in guard cells of stomata and this interaction results in the generation of reactive oxygen species (ROS) and nitrous oxide (NO).[110] Many JA-associated signaling genes have a regulatory role in the management of drought stress.[111] The JAZ proteins act as regulators and more specifically repressors in the JA signaling pathway. It has been demonstrated that OsJAZ1 is involved in a negative regulation of genes related to drought stress tolerance of rice, especially for abscisic acid and JA signaling pathways.[112] Also, it has been reported that OsbHLH148, a basic helix–loop–helix protein, plays a role as a transcriptional regulator, upregulating the OsDREB1 gene responsible for drought stress management and OsJAZ gene which plays a role in JA signaling.[113] Further, the transitional accumulation of JA in a drought–tolerant genotype belonging to *Prunus armeniaca* can result in the promotion of senescence in leaves, prevention of excess loss of water and hence improving the chances of plant survival when the soils are deficient in water.[114] It has been reported that drought stress causes as high as 19 times increase in Methyl jasmonate content and a 1.4-fold increase in ABA content.[115] The regulation of plant response to water deficiency by JA has been reported.[116] The studies have shown that under drought conditions initiated by PEG 6000 in the crop *Agropyron cristatum*, the jasmonate levels were observed to increase.[117] Multiple studies have shown that the exogenous application of jasmonates could mitigate the damage caused by water-deficient stress in *P. armeniaca*. The applications of MeJA on the foliage of soybean can increase the drought stress tolerance capacity. Further, increased levels of sugar compounds, flavonoid compounds, and phenolics were also recorded.[118] A 10 min exposure to MJ caused a drastic reduction in the size of the stomatal aperture in the leaves of *Arabidopsis*.[119] The increase in the growth of shoots, roots, and bulbs of garlic (*Allium sativum* L.) was observed upon treatment with MJ under drought-stress conditions. The total chlorophyll content, anthocyanin content, and permeability of membranes also recorded significant values.[120] It has been, however, observed that a combination of IBA and MJ is more effective in the promotion of growth and development under normal conditions. While IBA alone is not effective under drought stress conditions, the application of MJ alone elevates both the growth and yield of crop plants. It is clear from the above discussion that both endogenous and exogenous jasmonates participate in the mitigation of drought stress effects in plants.

4.5.4 SALINITY STRESS

The salinity stress affects the metabolism of plants, resulting in oxidative stress, malnourishment, damage to cell membranes leading to different disorders, and toxicity at the genetic level.[121,122] It has been established by a number of studies that both the endogenous and exogenous JA could increase the salinity stress tolerance in plants.[123] The studies have confirmed that endogenous JA content increased significantly in *A. thaliana*,[124] tomato (*L. esculentum*),[125] potato (*Solanum tuberosum*),[123] and other plants after salt treatment. It was observed that the jasmonate content of plants that are sensitive to salt stress increased quickly as compared to salt-tolerant plants where the increase was not significant.[123] The endogenous JA in tomato has been reported to increase salt tolerance by means of maintaining homeostasis between the ROS.[126] However, a number of studies have made it clear that the application of jasmonates exogenously resulted in the reduction of salt stress damage to plant systems via increase in rate of photosynthesis, and proline and ABA contents,[127] and antioxidant enzyme activity,[128] or "via reductions in Na+ accumulation rates in shoots".[129] The exogenous JA application in two maize genotypes has demonstrated surface salt stress tolerance. The possible mechanism could be through improving Na+ exclusion from root and decreasing Na+ uptake.[130] An exposure of 3 days to exogenous JA resulted in significantly reduced concentrations of malondialdehyde (MDA) and H_2O_2 in wheat seedlings, leading to improved tolerance of wheat seedlings to salinity stress. Additionally, the level of transcription, SOD, peroxidase, CAT, and APX functioning also increased significantly.[131] The results suggest that JA can improve salinity stress tolerance in plants by increasing the concentrations of antioxidant compounds and increasing the antioxidant enzyme activities. A number of studies have also demonstrated that exogenous JA application improves the salinity resistance in pepper (*Capsicum annuum*)[132] and verbena (*Rupestris riparia*).[133]

4.5.5 METAL AND HEAVY METAL STRESS

It is a known fact that a variety of metallic ions constitute components of plant nutrients, however some are toxic to plants.[134–137] As we know if the metallic ions or metalloids are in excess, it may cause toxicity symptoms in plants. This toxicity may cause oxidative stress conditions which ultimately leads to some alterations in the physiology of plants.[138] Application of JA increased the accumulation of osmolytes while carotenoids resulted in enhancement of the antioxidant levels. This in return protects the plants from the toxicity of excessive metal ion concentrations.[139] In pigeon peas exposure of plants to JA resulted in the accumulation of proteins (and total chlorophyll, carotenoids) and decreased the adverse effect of Cu^{2+} on the growth of seedlings. If boron is available in the soil in excessive concentration, the plants may absorb abnormally high amounts from the soil and the toxicity symptoms in the plants can be seen in the form of burning of leaves and decreased fruit size,[140,141] besides ROS production in wheat,[142] barley,[143] and tomato.[144] It has been observed that the foliar applications of MJ decreased the symptoms considerably which may be because of the stimulation of antioxidant production that consequently reduced the lipid peroxidation.[110] Heavy metal pollution of the environment has severe implications on plant growth and development.[145] Exposure of *Suaeda glauca* and *A. thaliana* to higher concentrations of Pb, Ni, Cd, and Mn resulted in the decline of both the fresh weight of plants and photosynthetic pigment concentration. Most of these heavy metals do not have any beneficial roles in plants, and may, in fact, be detrimental to plant tissues even at very low concentrations.[146] Zhao et al.[147] in an experimental study compared the Cd stress responses in two types of plants were in one was wild-type and the other was JA deficient mutant (spr2). A considerable increase in the Cd concentrations was recorded at higher doses of $CdCl_2$, and the increase was more sgnificant in spr2 plants. From the aforementioned study, it can be inferred that a lack of endogenous JA could increase the vulnerability of tomato seedlings to Cd. "Excess cadmium produces ROS, H_2O_2, and the radicals of superoxide radicals that are the cause of oxidative damage in plant tissues.[148] However, a reduction of 30% in this oxidative damage has been observed in soybean and *A. thaliana* following MJ applications.[149] Heavy metal damages can also be managed by the activation of plants' antioxidant system. It has been observed that MJ application helps *A. thaliana* plants to combat the toxicity of copper and cadmium metals. This remarkable impact is brought about by the ability of MJ to accumulate chelating ligands. These chelating ligands engage the metal ions in complex formation, hence rendering them unavailable for the uptake by plants.[150] Another study reported that when cadmium-stressed *O. sativa* plants were treated with MJ, remarkable changes in the activity of CAT, SOD, and GR were observed besides increased glutathione pools.[151] MJ application at 0.1 μM to *Capsicum frutescens* plants which were previously treated with Cd brought about the changes as an increase in the dry weight of plants, chlorophyll content of leaves, and the activities of CAT and APX activity.[152] Also, in *Avicennia marina* it was observed that MJ applications at 1.0 and 10 μM concentration reduced the absorption of Cd and enhanced the APX activity in plants.[153] The endogenous level of JA was maintained by the exogenous application of MJ in *Kandelia obovata*. The exogenous application has other effects on the plants like control on the closure of stomata, decreased rate of transpiration, prevention of Cd uptake, and decreased photosynthetic damage.[154] The studies also suggest that jasmonates are also helpful in combating the harmful effects of Cd toxicity in the plant *V. faba*. The mechanism behind this is the prevention of accumulations of Cd, H_2O_2, and MDA in plants and an increase in the activities of proline, glycine betaine, SOD, CAT, APX, and GR which helps in the reduction of oxidative stress effects.[155] A recent study also reported alleviation of arsenic-induced oxidative stress in *Brassica napus* upon exogenous application of MJ to plants besides enhanced plant growth and photosynthesis.[156] In similar studies, it has been observed that the exogenous application of jasmonates prior to $NiCl_2$ exposure could alleviate soybean (*G. max*) seedling tolerance to Ni^{2+} stress. It is suggested that jasmonates protect vulnerable seedling plants by means of appropriate regulations in the antioxidant pathway and total protein DNA synthesis prevention.[157] Other studies revealed that exogenous application of JA alleviated the deleterious impacts of oxidative stress on plant growth, normal plant biomass production, and concentration of proteins in Ni-exposed plants because the exogenous JA application resulted in further enhancement of antioxidant enzyme activities.[158] Exogenous JC application could minimize the rate of Cd accumulation in faba bean root, shoot, and leafy tissues by increasing osmotic activity, and antioxidant activity, and also by inhibiting H_2O_2 and MDA accumulation.[159] Jasmonic acid caused the inhibition of the activities of lipid peroxidase by means of the activation of ascorbate or glutathione antioxidant machinery.[160] Additionally, the considerable increase in HO-1 antioxidant enzyme activity under heavy metal stress conditions could be regulated strictly by ROS homeostasis. It is clear from the above findings that jasmonates regulate the plant responses to heavy metal stress conditions by the regulation of the antioxidant mechanism in plants.

4.5.6 CO_2 CONCENTRATION

It has been observed that under high CO_2 concentrations, the plants of lima bean (*P. lunatus*) released higher amounts of jasmonates into the environment.[161] Another study showed that the induced defense mechanism of plants suffering from nematode infections could be influenced by increased CO_2 concentrations, and carbon dioxide induced changes in plant resistance may result in genotype-specific responses of plants to nematode infestations.[162] However, only a small number of studies have focused on the behavior of JA signaling pathway under CO_2 stress in plants until now.

4.5.7 *EFFECT OF OZONE*

In a number of studies, the effect of ozone treatment on JA signaling pathways has been studied. The endogenous JA content of wild *Arabidopsis* was observed to enhance considerably following exposure to ozone treatments. The experimental studies utilizing the ozone-sensitive mutant plants, namely, rcd1 (radical-induced cell death 1)[163] and oji1(ozone-sensitive and jasmonate-insensitive)[164] and the jasmonate signal mutant jar1[165] have exhibited that the exogenous MeJA applications can prevent the progression of programmed cell death inflicted through ozone toxicity. Suppression of the JA signaling pathway makes the plant show the reaction to ozone more intensely.[166] Applications of exogenous MeJA to hybrid poplar (*Populus maximowizii*, *P. trichocarpa*)[164] and tomato (*L. esculentum*)[165] have resulted in decreased sensitivity to ozone.

4.6 SYSTEMATIC BIOLOGY APPROACHES TO STUDY JASMONIC ACID SIGNAL CHANGES

In the present times, using systematic biology approaches (gene chip, proteomics, etc.) we have been able to study different physiological processes in plant systems at the genetic level, and thereby analyze the relationships among different metabolic pathways. RNA-sequencing and proteomics are establishing the level of involvement and role of jasmonate signaling pathway in regulating the physiological processes of plants. In one of the studies on MeJA-treated *Arabidopsis* plants 137 genes that had altered gene expressions were investigated with the help of gene chip technology. Out of these genes, 74 genes depicted upregulation, including jasmonate biosynthesis genes, different defense-related genes, genes related to oxidative stress, genes responsible for aging, genes concerned with cell wall modifications, genes related to the metabolism of hormones, and the genes responsible for storage, signal transductions, metabolism in plants. On the other hand, the 63 genes which underwent downregulation constitute photosynthesis-related genes, cold-regulating genes, drought-responsible genes, defense-related genes, genes responsible for the growth and developmental processes of plants, cell wall-related genes, etc.[164] By proteomics technology, the protein content of *A. thaliana* was compared before and after MeJA exposure and 186 differentially expressed proteins were identified. The proteins were found to be involved in photosynthesis, secondary metabolism, hormone metabolism, carbohydrate metabolism, stress and defense, product transport and gene transcription.[165]

KEYWORDS

- **jasmonates**
- **jasmonic acid**
- **signaling**
- **biotic and abiotic stress**
- **transcription factors**

REFERENCES

1. Mittler, R. Abiotic Stress, the Field Environment and Stress Combination. *Trends Plant Sci.* **2006,** *11*, 15–19.
2. Suzuki, N.; Rivero, R. M.; Shulaev, V.; Blumwald, E.; Mittler, R. Abiotic and Biotic Stress Combinations. *New Phytol.* **2014,** *203*, 32–43.
3. Mur, L. A.; Kenton, P.; Atzorn, R.; Miersch, O.; Wasternack, C. The Outcomes of Concentration-Specific Interactions Between Salicylate and Jasmonate Signaling Include Synergy, Antagonism, and Oxidative Stress Leading to Cell Death. *Plant Physiol.* **2006,** *140*, 249–262.
4. Clarke, S. M.; Cristescu, S. M.; Miersch, O.; Harren, F. J.;Wasternack, C.; Mur, L. A. Jasmonates Act with Salicylic Acid to Confer Basal Thermotolerance in *Arabidopsis thaliana*. *New Phytol.* **2009,** *182*, 175–187.
5. Wasternack, C.; Hause, B. Jasmonates: Biosynthesis, Perception, Signal Transduction and Action in Plant Stress Response, Growth and Development. An Update to the 2007 Review in Annals of Botany. *Ann. Bot.* **2013,** *111*, 1021–1058.
6. Campos, M. L.; Kang, J. H.; Howe, G. A. Jasmonate-Triggered Plant Immunity. *J. Chem. Ecol.* **2014,** *40*, 657–675.
7. Gupta, A.; Hisano, H.; Hojo, Y.; Matsuura, T.; Ikeda, Y.; Mori, I. C.; Kumar, M. S. Global Profiling of Phytohormone Dynamics During Combined Drought and Pathogen Stress in *Arabidopsis thaliana* Reveals ABA and JA as Major Regulators. *Sci. Rep.* **2017,** *7*, 4017.

8. Howe, G. A.; Major, I. T.; Koo, A. J. Modularity in Jasmonate Signaling for Multi Stress Resilience. *Annu. Rev. Entomol.* **2018,** *69*, 387–415.
9. Pearce, G.; Strydom, D.; Johnson, S.; Ryan, C. A. A Polypeptide from Tomato Leaves Induces Wound-Inducible Proteinase Inhibitor Proteins. *Science* **1991,** *253*, 895–897.
10. Ryan, C. A.; Pearce, G. Systemins: A Functionally Defined Family of Peptide Signals That Regulate Defensive Genes in *Solanaceae* Species. *Proc. Natl. Acad. Sci. USA* **2003,** *100*, 14577–14580.
11. Li, C.; Liu, G.; Xu, C.; Lee, G. I.; Bauer, P.; Ling, H. Q.; Ganal, M. W.; Howe, G. A. The Tomato Suppressor of Prosystemin-Mediated Responses2 Gene Encodes A Fatty Acid Desaturase Required for the Biosynthesis of Jasmonic Acid and the Production of a Systemic Wound Signal for Defense Gene Expression. *Plant Cell* **2003,** *15*, 1646–1661.
12. Scheer, J. M.; Ryan, C. A. The Systemin Receptor SR160 from *Lycopersicon peruvianum* Is a Member of the LRR Receptor Kinase Family. *Proc. Natl. Acad. Sci. USA* **2002,** *99*, 9585–9590.
13. Stratmann, J. W.; Ryan, C. A. Myelin Basic Protein Kinase Activity in Tomato Leaves Is Induced Systemically by Wounding and Increases in Response to Systemin and Oligosaccharide Elicitors. *Proc. Natl. Acad. Sci. USA* **1997,** *94*, 11085–11089.
14. Yamaguchi, Y.; Hu_aker, A.; Bryan, A. C.; Tax, F. E.; Ryan, C. A. PEPR2 Is a Second Receptor for the Pep1 and Pep2 Peptides and Contributes to Defense Responses in *Arabidopsis*. *Plant Cell* **2010,** *22*, 508–522.
15. Hind, S.; Malinowski, R.; Yalamanchili, R.; Stratmann, J. W. Tissue-Type Specific Systemin Perception and the Elusive Systemin Receptor. *Plant Signal. Behav.* **2010,** *5*, 42–44.
16. Ruan, J.; Zhou, Y.; Zhou, M.; Yan, J.; Khurshid, M.; Weng, W.; Cheng, J.; Zhang, K. Jasmonic Acid Signaling Pathway in Plants. *Int. J. Mol. Sci.* **2019,** *20*, 2479.
17. Ryan, C. A.; Moura, D. S. Systemic Wound Signaling in Plants: A New Perception. *Proc. Natl. Acad. Sci. USA* **2002,** *99*, 6519–6520.
18. Truman, W.; Bennett, M. H.; Kubigsteltig, I.; Turnbull, C.; Grant, M. *Arabidopsis* Systemic Immunity Uses Conserved Defense Signaling Pathways and Is Mediated by Jasmonates. *Proc. Natl. Acad. Sci. USA* **2007,** *104*, 1075–1080.
19. Heil, M.; Ton, J. Long-Distance Signaling in Plant Defense. *Trends Plant Sci.* **2008,** *13*, 264–272.
20. Thorpe, M. R.; Ferrieri, A. P.; Herth, M. M.; Ferrieri, R. A. 11Cimaging: Methyl Jasmonate Moves in Both Phloem and Xylem, Promotes Transport of Jasmonate and of Photoassimilate Even After Proton Transport Is Decoupled. *Planta* **2007,** *226*, 541–551.
21. Hause, B.; Stenzel, I.; Miersch, O.; Maucher, H.; Kramell, R.; Ziegler, J.;Wasternack, C. Tissue-Specific Oxylipin Signature of Tomato Flowers: Allene Oxide Cyclase Is Highly Expressed in Distinct Flower Organs and Vascular Bundles. *Plant J.* **2000,** *24*, 113–126.
22. Hause, B.; Hause, G.; Kutter, C.; Miersch, O.; Wasternack, C. Enzymes of Jasmonate Biosynthesis Occur in Tomato Sieve Elements. *Plant Cell Physiol.* **2003,** *44*, 643–648.
23. Koo, A. J; Gao, X.; Jones, A. D.; Howe, G. A. A Rapid Wound Signal Activates Systemic Synthesis of Bioactive Jasmonates in *Arabidopsis*. *Plant J.* **2009,** *59*, 974–986.

24. Larrieu, A.; Vernoux, T. Q&A: How Does Jasmonate Signaling Enable Plants to Adapt and Survive? *BMC Boil.* **2016,** *14*, 79.
25. Nelson, C. E.; Walker-simmons, M.; Makus, D.; Zuroske, G.; Graham, J.; Ryan, C. A. Regulation of Synthesis and Accumulation of Proteinase-Inhibitors in Leaves of Wounded Tomato Plants. *ACS Symp. Ser.* **1983,** *208*, 103–122.
26. Malone, M. Rapid, Long-Distance Signal Transmission in Higher Plants. *Adv. Bot. Res.* **1996,** *22*, 163–228.
27. Farmer, E. E.; Ryan, C. A. Interplant Communication: Airborne Methyl Jasmonate Induces Synthesis of Proteinase Inhibitors in Plant Leaves. *Proc. Natl. Acad. Sci. USA* **1990,** *87*, 7713–7716.
28. Park, S. W.; Kaimoyo, E.; Kumar, D.; Mosher, S.; Klessig, D. F. Methyl Salicylate Is a Critical Mobile Signal for Plant Systemic Acquired Resistance. *Science* **2007,** *318*, 113–116.
29. Kost, C.; Heil, M. The Defensive Role of Volatile Emission and Extra-Floral Nectar Secretion for Lima Bean in Nature. *J. Chem. Ecol.* **2008,** *34*, 2–13.
30. Karban, R.; Baldwin, I. T.; Baxter, K. J.; Laue, G.; Felton, G. W. Communication Between Plants: Induced Resistance in Wild Tobacco Plants Following Clipping of Neighboring Sagebrush. *Oecologia* **2000,** *125*, 66–71.
31. Li, Q.; Zheng, J.; Li, S.; Huang, G.; Skilling, S. J.; Wang, L.; Li, L.; Li, M.; Yuan, L.; Liu, P. Transporter-Mediated Nuclear Entry of Jasmonoyl-Isoleucine Is Essential for Jasmonate Signaling. *Mol. Plant.* **2017,** *10*, 695–708.
32. Feys, B.; Benedetti, C. E.; Penford, C. N.; Turner, J. G. *Arabidopsis* Mutants Selected for Resistance to the Phytotoxin Coronatine Are Male Sterile, Insensitive to Methyl Jasmonate, and Resistant to a Bacterial Pathogen. *Plant Cell* **1994,** *6*, 751–759.
33. Xie, D. X.; Feys, B. F.; James, S.; Nieto-Rostro, M.; Turner, J. G. COI1: An *Arabidopsis* Gene Required for Jasmonate-Regulated Defense and Fertility. *Science* **1998,** *280*, 1091–1094.
34. Zhai, Q.; Zhang, X.; Wu, F.; Feng, H.; Deng, L.; Xu, L.; Zhang, M.; Wang, Q.; Li, C. Transcriptional Mechanism of Jasmonate Receptor COI1-Mediated Delay of Flowering Time in *Arabidopsis*. *Plant Cell* **2015,** *27*, 2814–2828.
35. Chini, A.; Fonseca, S.; Fernandez, G.; Adie, B.; Chico, J.; Lorenzo, O.; Garcia-Casado, G.; Lopez-Vidriero, I.; Lozano, F.; Ponce, M. The JAZ Family of Repressors Is the Missing Link in Jasmonate Signalling. *Nature* **2007,** *448*, 666–671.
36. Thines, B.; Katsir, L.; Melotto, M.; Niu, Y.; Mandaokar, A.; Liu, G.; Nomura, K.; He, S. Y.; Howe, G. A.; Browse, J. JAZ Repressor Proteins Are Targets of the SCFCOI1 Complex During Jasmonate Signaling. *Nature* **2007,** *448*, 661–665.
37. Yan, Y.; Stolz, S.; Chételat, A.; Reymond, P.; Pagni, M.; Dubugnon, L.; Farmer, E. E. A Downstream Mediator in the Growth Repression Limb of the Jasmonate Pathway. *Plant Cell* **2007,** *19*, 2470–2483.
38. Melotto, M.; Mecey, C.; Niu, Y.; Chung, H. S.; Katsir, L.; Yao, J.; Zeng, W.; Thines, B.; Staswick, P.; Browse, J. et al. A Critical Role of Two Positively Charged Amino Acids in the Jas Motif of *Arabidopsis* JAZ Proteins in Mediating Coronatine- and Jasmonoyl Isoleucine-Dependent Interactions with the COI1 F-Box Protein. *Plant J.* **2008,** *55*, 979–988.

39. Chini, A.; Fonseca, S.; Chico, J. M.; Fernández-Calvo, P.; Solano, R. The ZIM Domain Mediates Homo- and Heteromeric Interactions Between *Arabidopsis* JAZ Proteins. *Plant J.* **2009,** *59*, 77–87.
40. Sheard, L. B.; Tan, X.; Mao, H.; Withers, J.; Ben-Nissan, G.; Hinds, T. R.; Kobayashi, Y.; Hsu, F. F.; Sharon, M.; Browse, J. et al. Jasmonate Perception by Inositolphosphate-Potentiated COI1-JAZ Co-Receptor. *Nature* **2010,** *468*, 400–405.
41. Wang, J.; Song, L.; Gong, X.; Xu, J.; Li, M. Functions of Jasmonic Acid in Plant Regulation and Response to Abiotic Stress. *Int. J. Mol. Sci.* **2020,** *21*, 1446.
42. Li, Y.; Qin, L.; Zhao, J.; Muhammad, T.; Cao, H.; Li, H.; Zhang, Y.; Liang, Y. SlMAPK3 Enhances Tolerance to Tomato Yellow Leaf Curl Virus (TYLCV) by Regulating Salicylic Acid and Jasmonic Acid Signaling in Tomato (*Solanum lycopersicum*). *PLoS ONE* **2017,** *12* (2), e0172466.
43. Kenton, P.; Mur, L. A. J.; Draper, J. A Requirement for Calcium and Protein Phosphatase in the Jasmonate-Induced Increase in Tobacco Leaf Acid Phosphatase Specific Activity. *J. Exp. Bot.* **1999,** *50*, 1331–1341.
44. Fernandez-Calvo, P.; Chini, A.; Fernandez-Barbero, G.; Chico, J. M.; Gimenez-Ibanez, S.; Geerinck, J.; Eeckhout, D.; Schweizerd, F.; Godoy, M.; Franco-Zorrillae, J. M. et al. The *Arabidopsis* bHLH Transcription Factors MYC3 and MYC4 Are Targets of JAZ Repressors and Act Additively with MYC2 in the Activation of Jasmonate Responses. *Plant Cell* **2011,** *23*, 701–715.
45. Schmiesing, A.; Emonet, A.; Gouhier-Darimont, C.; Reymond, P. *Arabidopsis* MYC Transcription Factors Are the Target of Hormonal Salicylic Acid/Jasmonic Acid Cross Talk in Response to *Pieris brassicae* Egg Extract. *Plant Physiol.* **2016,** *170*, 2432–2443.
46. Figueroa, P.; Browse, J. Male Sterility in *Arabidopsis* Induced by Overexpression of a MYC5-SRDX Chimeric Repressor. *Plant J.* **2015,** *81*, 849–860.
47. Sasaki-Sekimoto, Y.; Jikumaru, Y.; Obayashi, T.; Saito, H.; Masuda, S.; Kamiya, Y.; Ohta, H.; Shirasu, K. Basic Helix-Loop-Helix Transcription Factors Jasmonate-Associated MYC2-LIKE1 (JAM1), JAM2, and JAM3 Are Negative Regulators of Jasmonate Responses in *Arabidopsis*. *Plant Physiol.* **2013,** *163*, 291–304.
48. Song, S.; Qi, T.; Huang, H.; Ren, Q.; Wu, D.; Chang, C.; Peng, W.; Liu, Y.; Peng, J.; Xie, D. The Jasmonate-ZIM Domain Proteins Interact with the R2R3-MYB Transcription Factors MYB21 and MYB24 to Affect Jasmonate-Regulated Stamen Development in *Arabidopsis*. *Plant Cell* **2011,** *23*, 1000–1013.
49. Qi, T.; Song, S.; Ren, Q.; Wu, D.; Huang, H.; Chen, Y.; Fan, M.; Peng, W.; Ren, C.; Xie, D. The Jasmonate-ZIM-Domain Proteins Interact with the WD-Repeat/bHLH/MYB Complexes to Regulate Jasmonate-Mediated Anthocyanin Accumulation and Trichome Initiation in *Arabidopsis thaliana*. *Plant Cell* **2011,** *23*, 1795–1814.
50. Zhou, M.; Sun, Z.; Ding, M.; Logacheva, M. D.; Kreft, I.; Wang, D.; Yan, M.; Shao, J.; Tang, Y.; Wu, Y. et al. FtSAD2 and FtJAZ1 Regulate the Activity of FtMYB11 Transcription Repressor in Phenylpropanoid Pathway of *Fagopyrum tataricum*. *New Phytol.* **2017,** *216*, 814–828.
51. Zhang, K.; Logacheva, M. D.; Meng, Y.; Hu, J.; Wan, D.; Li, L.; Dagmar, J.; Wang, Z.; Georgiev, M. I.; Yu, Z. et al. Jasmonate-Responsive MYB Factors Spatially Repress Rutin Biosynthesis in *Fagopyrum tataricum*. *J. Exp. Bot.* **2018,** *69*, 1955–1966.

52. Delessert, C.; Kazan, K.; Wilson, I. W.; van der Straeten, D.; Manners, J.; Dennis, E. S.; Dolferus, R. The Transcription Factor ATAF2 Represses the Expression of Pathogenesis-Related Genes in *Arabidopsis*. *Plant J.* **2005,** *43*, 745–757.
53. Nuruzzaman, M.; Sharoni, A. M.; Kikuchi, S. Roles of NAC Transcription Factors in the Regulation of Biotic and Abiotic Stress Responses in Plants. *Front. Microbiol.* **2013,** *4*, 248.
54. Bu, Q.; Jiang, H.; Li, C. B.; Zhai, J.; Wu, X.; Sun, J.; Xie, Q.; Li, C. Role of the *Arabidopsis thaliana* NAC Transcription Factors ANAC019 and ANAC055 in Regulating Jasmonic Acid-Signaled Defense Responses. *Cell Res.* **2008,** *18*, 756–767.
55. Van der Fits, L.; Memelink, J. ORCA3, a Jasmonate-Responsive Transcriptional Regulator of Plant Primary and Secondary Metabolism. *Science* **2001,** *289*, 295–297.
56. Lorenzo, O.; Piqueras, R.; Sánchez-Serrano, J. J.; Solano, R. ETHYLENE RESPONSE FACTOR1 Integrates Signals from Ethylene and Jasmonate Pathways in Plant Defense. *Plant Cell* **2003,** *15*, 165–178.
57. Pré, M.; Atallah, M.; Champion, A.; de Vos, M.; Pieterse, C. M. J.; Memelink, J. The AP2/ERF Domain Transcription Factor ORA59 Integrates Jasmonic Acid and Ethylene Signals in Plant Defense. *Plant Physiol.* **2008,** *147*, 1347–1357.
58. Li, J.; Zhang, K.; Meng, Y.; Hu, J.; Ding, M.; Bian, J.; Yan, M.; Han, J.; Zhou, M. Jasmonic Acid/Ethylene Signaling Coordinates Hydroxylcinnamic Acid Amides Biosynthesis Through ORA59 Transcription Factor. *Plant J.* **2018,** *95*, 444–457.
59. Chen, H. Y.; Hsieh, E. J.; Cheng, M. C.; Chen, C. Y.; Hwang, S. Y.; Lin, T. P. ORA47 (Octade-Canoid-Responsive AP2/ERF-Domain Transcription Factor 47) Regulates Jasmonic Acid and Abscisic Acid Biosynthesis and Signaling Through Binding to a Novel Cis-Element. *New Phytol.* **2016,** *211*, 599–613.
60. Fujimoto, S. Y.; Ohta, M.; Usui, A.; Shinshi, H.; Ohme-Takagi, M. *Arabidopsis* Ethylene-Responsive Element Binding Factors Act as Transcriptional Activators or Repressors of GCC Box–Mediated Gene Expression. *Plant Cell* **2000,** *12*, 393–404.
61. Skibbe, M.; Qu, N.; Galis, I.; Balawin, I. T. Induced Plant Defenses in the Natural Environment: *Nicotiana attenuala* WRKY3 and WRKY6 Coordinate Responses to Herbivory. *Plant Cell* **2008,** *20*, 1984–2000.
62. Jiang, Y.; Liang, G.; Yang, S.; Yu, D. *Arabidopsis* WRKY57 Functions as a Node of Convergence for Jasmonic Acid–and Auxin-Mediated Signaling in Jasmonic Acid–Induced Leaf Senescence. *Plant Cell* **2014,** *26*, 230–245.
63. Mewis, I.; Schreiner, M.; Nguyen, C. N.; Krumbein, A.; Ulrichs, C.; Lohse, M.; Zrenner, R. UV-B Irradiation Changes Specifically the Secondary Metabolite Profile in Broccoli Sprouts: Induced Signaling Overlaps with Defense Response to Biotic Stressors. *Plant Cell Physiol.* **2012,** *53*, 1546–1560.
64. Cerrudo, I.; Keller, M. M.; Cargnel, M. D.; Demkura, P. V.;Wit, M.; Patitucci, M. S.; Pierik, R.; Pieterse, C. M. J.; Ballare, C. L. Low Red/Far-Red Ratios Reduce *Arabidopsis* Resistance to *Botrytis cinerea* and Jasmonate Responses via a COI1-JAZ10-Dependent, Salicylic Acid Independent Mechanism. *Plant Physiol.* **2012,** *158*, 2042–2052.
65. Gupta, N.; Prasad, V. B. R.; Chattopadhyay, S. LeMYC2 Acts as a Negative Regulator of Blue Light Mediated Photomorphogenic Growth, and Promotes the Growth of Adult Tomato Plants. *BMC Plant Boil.* **2014,** *14*, 38.

66. Svyatyna, K.; Riemann, M. Light-Dependent Regulation of the Jasmonate Pathway. *Protoplasma* **2012,** *249*, 137–145.
67. Kazan, K.; Manners, J. M. The Interplay Between Light and Jasmonate Signalling During Defence and Development. *J. Exp. Bot.* **2011,** *62*, 4087–4100.
68. Izaguirre, M. M.; Mazza, C. A.; Svatos, A.; Baldwin, I. T.; Ballaré, C. L. Solar Ultraviolet-B radiation and Insect Herbivory Trigger Partially Overlapping Phenolic Responses in *Nicotiana attenuata* and *Nicotiana longiflora*. *Ann. Bot.* **2007,** *99*, 103–109.
69. Radhika, V.; Kost, C.; Mithofer, A.; Boland, W. Regulation of Extrafloral Nectar Secretion by Jasmonates in Lima Bean Is Light Dependent. *Proc. Natl. Acad. Sci. USA* **2010,** *107*, 17228–17233.
70. Riemann, M.; Axel, M.; Korte, A.; Furuya, M.; Weiler, E. W.; Nick, P. Impaired Induction of the Jasmonate Pathway in the Rice Mutant Hebiba. *Plant Physiol.* **2003,** *133*, 1820–1830.
71. Chen, H. J.; Fu, T. Y.; Yang, S. L.; Hsien, H. L. FIN219/JAR1 and Cryptochrome1 Antagonize Each Other to Modulate Photomorphogenesis Under Blue Light in *Arabidopsis*. *PLoS Genet.* **2018,** *14* (3), e1007248.
72. Liu, X.; Chi, H.; Yue, M.; Zhang, X. F.; Li, W. J.; Jia, E. P. The Regulation of Exogenous Jasmonic Acid on UV-B Stress Tolerance in Wheat. *Plant Growth Regul.* **2012,** *31*, 436–447.
73. Ahmad, P.; Alyemeni, M. N.; Vijaya, L.; Alam, P.; Ahanger, M. A.; Alamri, S. A. Jasmonic Acid Alleviates Negative Impacts of Cadmium Stress by Modifying Osmolytes and Antioxidants in Faba Bean (*Vicia faba* L.). *Arch. Agron. Soil Sci.* **2017,** *63*, 1–11.
74. Demkura, P. V.; Abdala, G.; Baldwin, I. T.; Ballaré, C. L. Jasmonate-Dependent and -Independent Pathways Mediate Specific Effects of Solar Ultraviolet B Radiation on Leaf Phenolics and Antiherbivore Defense. *Plant Physiol.* **2010,** *152*, 1084–1095.
75. Ruelland, E. and Zachowski, A. How Plants Sense Temperature. *Environ. Exp. Bot.* **2010,** *69*, 225–232.
76. Shinozaki, K.; Yamaguchi-Shinozaki, K.; Seki, M. Regulatory Network of Gene Expression in the Drought and Cold Stress Responses. *Curr. Opin. Plant Biol.* **2003,** *6*, 410–417.
77. Zhao, M. L.;Wang, J. N.; Shan,W.; Fan, J. G.; Kuang, J. F.;Wu, K. Q.; Li, X. P.; Chen,W. X.; He, F. Y.; Chen, J. Y. et al. Induction of Jasmonate Signaling Regulators MaMYC2s and Their Physical Interactions with MaICE1 in Methyl Jasmonate-Induced Chilling Tolerance in Banana Fruit. *Plant Cell Environ.* **2012,** *36*, 30–51.
78. Kolaksazov, M.; Laporte, F.; Ananieva, K.; Dobrev, P.; Herzog, M.; Ananiev, E. D. Effect of Chilling and Freezing Stresses on Jasmonate Content in *Arabis alpina*. *Bulgarian J. Agric. Sci.* **2013,** *19*, 15–17.
79. Farmer, E. E.; Johnson, R. R.; Ryan, C. A. Regulation of Expression of Proteinase Inhibitor Genes by Methyl Jasmonate and Jasmonic Acid. *Plant Physiol* **1992,** *98*, 995–1002.
80. Gundlach, H.; Mtller, M.; Kutchan, T. M.; Zenk M. H. Jasmonic Acid Is a Signal Transducer in Elicitor-Induced Plant Cell Culture. *Proc. Natl. Acad. Sci. USA* **1992,** *89*,2389–2393.

81. Reinbothe, S.; Reinbothe, C.; Parthier, B. Methyl Jasmonate Represses Translation Initiation of a Specific Set of mRNAs in Barley. *Plant J* **1993,** *4*, 459–467.
82. Li, S.; Yang, Y.; Zhang, Q.; Liu, N.; Xu, Q.; Hu, L. Differential Physiological and Metabolic Response to Low Temperature in Two *zoysia* Grass Genotypes Native to High and Low Latitude. *PLoS One* **2018,** *13*.
83. Sayyari, M.; Babalar, M.; Kalantari, S.; Martínez-Romero, D.; Guillen, F.; Serrano, M.; Valero, D. Vapour Treatments with Methyl Salicylate or Methyl Jasmonate Alleviated Chilling Injury and Enhanced Antioxidant Potential During Postharvest Storage of Pomegranates. *Food Chem.* **2011,** *124*, 964–970.
84. Cao, S. F.; Zheng, Y. H.; Wang, K. T.; Jin, P.; Rui, H. J. Methyl Jasmonate Reduces Chilling Injury and Enhances Antioxidant Enzyme Activity in Postharvest Loquat Fruit. *Food Chem.* **2009,** *115*, 1458–1463.
85. González-Aguilar, G. A.; Fortiz, J.; Cruz, R.; Baez, R.; Wang, C. Y. Methyl Jasmonate Reduces Chilling Injury and Maintains Postharvest Quality of Mango Fruit. *J. Agric. Food Chem.* **2000,** *48*, 515–519.
86. González-Aguilar, G. A.; Tiznado-Hernández, M. E.; Zavaleta-Gatica, R.; Martínez-Téllez, M. A. Methyl Jasmonate Treatments Reduce Chilling Injury and Activate the Defense Response of Guava Fruits. *Biochem. Bioph. Res. Commun.* **2003,** *313*, 694–701.
87. Xu, Y. H.; Liao, Y. C.; Zhang, Z.; Liu, J.; Sun, P. W.; Gao, Z. H.; Sui, C.; Wei, J. H. Jasmonic Acid Is a Crucial Signal Transducer in Heat Shock Induced Sesquiterpene Formation in *Aquilaria sinensis*. *Sci. Rep.* **2016,** *6*, 21843.
88. Bokszczanin, K. L. Solanaceae Pollen Thermotolerance Initial Training Network (SPOT-ITN) Consortium, Fragkostefanakis S. Perspectives on Deciphering Mechanisms Underlying Plant Heat Stress Response and Thermotolerance. *Front Plant. Sci.* **2013,** *4*, 315.
89. Boston, R. S.; Viitanen, P. V.; Vierling, E. Molecular Chaperones and Protein Folding in Plants. *Plant Mol. Biol.* **1996,** *32*, 191–222.
90. Clarke, S. M.; Cristescu, S. M.; Miersch, O.; Harren, F. J.; Wasternack, C.; Mur, L. A. Jasmonates Act with Salicylic Acid to Confer Basal Thermotolerance in *Arabidopsis thaliana*. *New Phytol.* **2009,** *182*, 175–187.
91. Du, H.; Liu, H.; Xiong, L. Endogenous Auxin and Jasmonic Acid Levels Are Differentially Modulated by Abiotic Stresses in Rice. *Front. Plant Sci.* **2013,** *4*, 397.
92. Abdullah, A. S.; Aziz, M. M.; Siddique, K. H. M.; Flower, K. C. Film Antitranspirants Increase Yield in Drought Stressed Wheat Plants by Maintaining High Grain Number. *Agric. Water Manag.* **2015,** *159*, 11–18.
93. Sun, X. G.; Shi, J.; Ding, G. J. Combined Effects of *Arbuscular mycorrhiza* and Drought Stress on Plant Growth and Mortality of Forage Sorghum. *Appl. Soil Ecol.* **2017,** *119*, 384–391.
94. Javed, N.; Ashraf, M.; Akram, N. A.; Ai-Qurainy, F. Alleviation of Adverse Effects of Drought Stress on Growth and Some Potential Physiological Attributes in Maize (*Zea mays* L.) by Seed Electromagnetic Treatment. *Photochem. Photobiol.* **2011,** *87*, 1354–1362.
95. Huang, J.; Chen, C.; Zhang,W. X.; Ding, C. J.; Su, X. H.; Huang, Q. J. Effects of Drought Stress on Anatomical Structure and Photosynthetic Characteristics of

Transgenic JERF36 *Populus alba, P. berolinensis* Seedling Leaves. *Sci. Silvae Sin.* **2017,** *119*, 384–391.
96. Munne-bosch, S.; Alegre, L. Die and Let Live: Leaf Senescence Contributes to Plant Survival Under Drought Stress. *Funct. Plant Biol.* **2004,** *31*, 8808–8818.
97. Ma, X. Q.; Zhang, J.; Burgess, P.; Rosso, S.; Huang, B. Interactive Effects of Melatonin and Cytokinin on Alleviating Drought-Induced Leaf Senescence in Creeping Bentgrass (*Agrostis stolonifera*). *Environ. Exp. Bot.* **2018,** *145*, 1–11.
98. Lei, Y. Physiological Responses of *Populus przewalskii* to Oxidative Burst Caused by Drought Stress. *Russ. J. Plant Physiol.* **2008,** *55*, 857–864.
99. Wang, L.; He, J.; Ding, H. D.; Liu, H.; Lv, B.; Liang, J. S.; Wang, L.; He, J.; Ding, H. D.; Liu, H. et al. Overexpression of AT14A Confers Tolerance to Drought Stress-Induced Oxidative Damage in Suspension Cultured Cells of *Arabidopsis thaliana. Protoplasma* **2014,** *252*, 1111–1120.
100. Balbi, V.; Devoto, A. Jasmonate Signalling Network in *Arabidopsis thaliana*: Crucial Regulatory Nodes and New Physiological Scenarios. *New Phytologist.* **2008,** *177*, 301–318.
101. Ollas, C.; Hernando, B.; Arbona, V.; Gómez-Cadenas, A. Jasmonic Acid Transient Accumulation Is Needed for Abscisic Acid Increase in Citrus Roots Under Drought Stress Conditions. *Physiol. Plantarum.* **2013,** *147*, 296–306.
102. Todaka, D.; Shinozaki, K.; Yamaguchi-Shinozaki, K. Recent Advances in the Dissection of Drought-Stress Regulatory Networks and Strategies for Development of Drought–Tolerant Transgenic Rice Plants. *Front. Plant Sci.* **2015,** *6*, 84.
103. Fu, J.; Wu, H.; Ma, S.; Xiang, D.; Liu, R.; Xiong, L. OsJAZ1 Attenuates Drought Resistance by Regulating JA and ABA Signaling in Rice. *Front. Plant Sci.* **2017,** *8*, 2108.
104. Mohamed, H. I.; Latif, H. H. Improvement of Drought Tolerance of Soybean Plants by Using Methyl Jasmonate. *Physiol. Mol. Biol.* Plants **2017,** *23*, 545–556.
105. Wu, H. L.;Wu, X. L.; Li, Z. H.; Duan, L. S.; Zhang, M. C. Physiological Evaluation of Drought Stress Tolerance and Recovery in Cauliflower (*Brassica oleracea* L.) Seedlings Treated with Methyl Jasmonate and Coronatine. *J. Plant Growth Regul.* **2012,** *31*, 113–123.
106. Evans, N. H. Modulation of Guard Cell Plasma Membrane Potassium Currents by Methyl Jasmonate. *Plant Physiol.* **2003,** *131*, 8–11.
107. Horton, R. F. Methyl Jasmonate and Transpiration in Barley. *Plant Physiol.* **1991,** *96*, 1376–1378.
108. Savchenko, T.; Kolla, V. A.; Wang, C. Q.; Nasafi, Z.; Hicks, D. R.; Phadungchob, B.; Chehab, W. E.; Brandizzi, F.; Froehlich, J.; Dehesh, K. Functional Convergence of Oxylipin and Abscisic Acid Pathways Controls Stomatal Closure in Response to Drought. *Plant Physiol.* **2014,** *164*, 1151–1160.
109. Harrison, M. A. Cross-Talk Between Phytohormone Signaling Pathways Under Both Optimal and Stressful Environmental Conditions. In: *Phytohormones and Abiotic Stress Tolerance In Plants*; Khan, N. A., Nazar, R., Iqbal, N., Anjum, N. A., Eds.; Springer-Verlag: Berlin Heidelberg, 2012; pp 49–76.
110. Siddiqi, K. S.and Husen, A. Plant Response to Jasmonates: Current Developments and Their Role in Changing Environment. *Bull. Natl. Res. Centre* **2019,** *43*, 153.

111. Huang, D.; Wu, W.; Abrams, S. R.; Adrian, J.; Cutler, A. J. The Relationship of Drought-Related Gene Expression in *Arabidopsis thaliana* to Hormonal and Environmental Factors. *J. Exp. Bot.* **2008,** *59*, 2991–3007.
112. Fu, J.; Wu, H.; Ma, S. Q.; Xiang, D. H.; Liu, R. Y.; Xiong, L. Z. OsJAZ1 Attenuates Drought Resistance by Regulating JA and ABA Signaling in Rice. *Front. Plant Sci.* **2017,** *8*, 1–13.
113. Seo, J. S.; Joo, J.; Kim, M. J.; Kim, Y. K.; Nahm, B. H.; Song, S. I.; Cheong, J. J.; Lee, J. S.; Kim, J. K.; Choi, Y. D. OsbHLH148, a Basic Helix-Loop-Helix Protein, Interacts with OsJAZ Proteins in a Jasmonate Signaling Pathway Leading to Drought Tolerance in Rice. *Plant J.* **2011,** *65*, 907–921.
114. Ge, Y. X.; Zhang, L. J.; Li, F. H.; Chen, Z. B.; Wang, C.; Yao, Y. C.; Han, Z. H.; Zhang, J.; Shi, Z. S. Relationship Between Jasmonic Acid Accumulation and Senescence in Drought-Stress. *Afr. J. Agric. Res.* **2010,** *5*, 1978–1983.
115. Kim, E. H.; Kim, Y. S.; Park, S. H.; Koo, Y. J.; Choi, Y. D.; Chung, Y. Y.; Lee, I. J.; Kim, J. K. Methyl Jasmonates Reduces Grain Yield by Mediating Stress Signals to Alter Spikelet Development in Rice. *Plant Physiol.* **2009,** *149*, 1751–1760.
116. Liu, X.; Shi, W. L.; Zhang, S. Q.; Lou, C. H. Nitric Oxide Participates in the Signal Transduction of Stomatal Closure Induced by Jasmonic Acid in *Vicia faba* L. *Chin Sci Bull* **2005,** *50*, 453–458.
117. Shan, C.; Liang, Z. Jasmonic Acid Regulates Ascorbate and Glutathione Metabolism in *Agropyron cristatum* Leaves Under Water Stress. *Plant Sci.* **2010,** *178*, 130–139.
118. Mohamed, H. I.; Latif, H. H. Improvement of Drought Tolerance of Soybean Plants by Using Methyl Jasmonate. *Physiol. Mol. Biol.* Plants **2017,** *23*, 545–556.
119. Munemasa, S.; Oda, K.; Watanabe-Sugimoto, M.; Nakamura, Y.; Shimoishi, Y.; Murata, Y. The Coronatine-Insensitive 1 Mutation Reveals the Hormonal Signaling Interaction Between Abscisic Acid and Methyl Jasmonate in *Arabidopsis* Guard Cells: Specific Impairment of Ion Channel Activation and Second Messenger Production. *Plant Physiol.* **2007,** *143*, 1398–1407.
120. Bideshki, A.; Arvin, M. J. Interactive Effects of Methyl Jasmonate (MJ) and Indole-3 Butyric Acid (IBA) on Growth and Bio Chemical Parameters, Bulb and Allicin Yield of Garlic (*Allium sativum* L.) Under Drought Stress in Iran. *Int. J. Agric. Res. Rev.* **2013,** *3*, 349–360.
121. Syeed, S.; Anjum, N. A.; Nazar, R.; Iqbal, N.; Masood, A.; Khan, N. A. Salicylic Acid-Mediated Changes in Photosynthesis, Nutrients Content and Antioxidant Metabolism in Two Mustard (*Brassica juncea* L.) Cultivars Differing in Salt Tolerance. *Acta Physiol. Plant.* **2011,** *33*, 877–886.
122. Kadri, K.; Abdellaoui, R.; Mhamed, H. C.; Da Silva, J. A. T.; Naceur, M. B. Analysis of Salt-Induced mRNA Transcripts in Tunisian Local Barley (*Hordeum vulgare*) Leaves Identified by Differential Display RT-PCR. *Biochem. Genet.* **2014,** *52*, 106–115.
123. Domenico, S. D.; Taurino, M.; Gallo, A.; Poltronieri, P.; Pastor, V.; Flors, V.; Santino, A. Oxylipin Dynamics in *Medicago truncatula* in Response to Salt and Wounding Stresses. *Physiol. Plant.* **2019,** *165*, 198–208.
124. Ellouzi, H.; Hamed, K. B.; Cela, J.; Müller, M.; Abdelly, C.; Munné-Bosch, S. Increased Sensitivity to Salt Stress in Tocopherol-Deficient *Arabidopsis* Mutants Growing in a Hydroponic System. *Plant Signal Behav.* **2013,** *8*, e23136.

125. Pedranzani, H.; Racagni, G.; Alemano, S.; Miersch, O.; Ramírez, I.; Pea-Cortes, H.; Taleisnik, E.; Machado-Domenech, E.; Abdala, G. Salt Tolerant Tomato Plants Show Increased Levels of Jasmonic Acid. *J. Plant Growth Regul.* **2003,** *41*, 149–158.
126. Abouelsaad, I.; Renault, S. Enhanced Oxidative Stress in the Jasmonic Acid-Deficient Tomato Mutant Def-1 Exposed to NaCl Stress. *J. Plant Physiol.* **2018,** *226*, 136–144.
127. Bandurska, H.; Stroinski, A.; Kubis, J. The Effect of Jasmonic Acid on the Accumulation of ABA, Proline and Spermidine and Its Influence on Membrane Injury Under Water Deficit in Two Barley Genotypes. *Acta Physiol. Plant.* **2003,** *25*, 279–285.
128. Walia, H.;Wilson, C.; Condamine, P.; Liu, X.; Ismail, A. M.; Close, T. J. Large- Scale Expression Profiling and Physiological Characterization of Jasmonic Acid-Mediated Adaptation of Barley to Salinity Stress. *Plant Cell Environ.* **2007,** *30*, 410–421.
129. Khan, M. I. R.; Syeed, S.; Nazar, R.; Anjum, N. A. An Insight into the Role of Salicylic Acid and Jasmonic Acid in Salt Stress Tolerance. In: *Phytohormones and Abiotic Stress Tolerance in Plants*; Springer: Berlin/Heidelberg, Germany, 2012; pp 277–300.
130. Shahzad, A. N.; Pitann, B.; Ali, H.; Qayyum, M. F.; Fatima, A.; Bakhat, H. F. Maize Genotypes Differing in Salt Resistance Vary in Jasmonic Acid Accumulation During the First Phase of Salt Stress. *J. Agron. Crop Sci.* **2015,** *201*, 443–451.
131. Qiu, Z. B.; Guo, J. L.; Zhu, A. J.; Zhang, L.; Zhang, M. M. Exogenous Jasmonic Acid Can Enhance Tolerance of Wheat Seedlings to Salt Stress. *Ecotoxicol. Environ. Saf.* **2014,** *104*, 202–208.
132. Molina, C.; Zaman-Allah, M.; Khan, F.; Fatnassi, N.; Horres, R.; Rotter, B.; Steinhauer, D.; Amenc, L.; Drevon, J. J.; Winter, P. et al. The Salt-Responsive Transcriptome of Chickpea Roots and Nodules via DeepSuperSAGE. *BMC Plant Biol.* **2011,** *11*, 31.
133. Ismail, A.; Riemann, M.; Nick, P. The Jasmonate Pathway Mediates Salt Tolerance in Grapevines. *J. Exp. Bot.* **2012,** *63*, 2127–2139.
134. Hou, W.; Chen, X.; Guanling, S.; Wang, Q.; Chang, C. C. Effects of Copper and Cadmium on Heavy Metal Polluted Water Body Restoration by Duck Weed (*Lemna minor*). *Plant Physiol. Biochem.* **2007,** *45*, 62–69.
135. Asgher, M.; Khan, M. I. R.; Anjum, N. A.; Khan, N. A. Minimising Toxicity of Cadmium in Plants-Role of Plant Growth Regulators. *Protoplasma* **2015,** *252*, 399–413.
136. Iqbal, M.; Ahmad, A.; Ansari, M. K. A.; Qureshi, M. I.; Aref, I. M.; Khan, P. R.; Hegazy, S. S.; El-Atta, H.; Husen, A.; Hakeem, K. R. Improving the Phytoextraction Capacity of Plants to Scavenge Metal (Loid)-Contaminated Sites. *Environ. Rev.* **2015,** *23*, 44–65.
137. Dhankar, R.; Solanki, R. Effect of Copper and Zinc Toxicity on Physiological and Biochemical Parameters in *Vigna mungo* (L.). Hepper. *Int. J. Pharm. Bio. Sci.* **2011,** *2*, 553–565.
138. Poonam, S.; Kaur, H.; Geetika, S. Effect of Jasmonic Acid on Photosynthetic Pigments and Stress Markers in *Cajanus cajan* (L.) Mill sp. Seedlings Under Copper Stress. *Am. J. Plant Sci.* **2013,** *4*, 817–823.
139. Paull, J. G.; Nable, R. O.; Lake, A. W. H.; Materne, M. A.; Rathjen, A. J. Response of Annual Medics (*Medicago* spp.) and Field Peas (*Pisum sativum*) to High Concentrations of Boron: Genetic Variation and Mechanism of Tolerance. *Aust. J. Agric. Res.* **1992,** *43*, 203–213.
140. Nable, R. O.; Bañuelos, G. S.; Paull, J. G. Boron Toxicity. *Plant Soil* **1997,** *193*, 181–198.

141. Gunes, A.; Inal, A.; Bagci, E. G.; Coban, S.; Sahin, O. Silicon Increases B Tolerance and Reduces Oxidative Damage of Wheat Grown in Soil with Excess B. *Biol Plant* **2007,** *51*, 571–574.
142. Inal, A.; Pilbeam, D. J.; Gunes, A. Silicon Increases Tolerance to B Toxicity and Reduces Oxidative Damage in Barley. *J Plant Nutr* **2009,** *32*, 112–128.
143. Cervilla, L. M.; Blasco, B.; Rios, J. J.; Romero, L.; Ruiz, J. M. Oxidative Stress and Antioxidants in Tomato (*Solanum lycopersicum*) Plants Subjected to Boron Toxicity. *Ann. Bot.* **2007,** *100*, 747–756.
144. Zhang, X.; Li, M.; Yang, H. H.; Li, X. X.; Cui, Z. J. Physiological Responses of *Suaeda glauca* and *Arabidopsis thaliana* in Phytoremediation of Heavy Metals. *J. Environ. Manag.* **2018,** *223*, 132–139.
145. Ali, I.; Ali, R.; Alothman, Z. A.; Ali, J.; Habila, M. A. Assessment of Toxic Metals in Wheat Crops Grown on Selected Soils, Irrigated by Different Water Sources. *Arab. J. Chem.* **2012,** *9*, 1555–1562.
146. Zhao, S. Y.; Ma, Q. F.; Xu, X.; Li, G. Z.; Hao, L. Tomato Jasmonic Acid-Deficient Mutant spr2 Seedling Response to Cadmium Stress. *J. Plant Growth Regul.* **2016,** *35*, 603–610.
147. Gallego, S. M.; Pena, L. B.; Bracia, R. A.; Azpilicueta, C. E.; Lannone, M. F.; Rosales, E. P.; Zawoznik, M. S.; Groppa, M. D.; Benavides, M. P. Unravelling Cadmium Toxicity and Tolerance in Plants: Insight into Regulatory Mechanisms. *Environ. Exp. Bot.* **2012,** *83*, 33–46.
148. Maksymiec, W.; Wianowska, D.; Dawidowicz, A. L.; Radkiewicz, S.; Mardarowicz, M.; Krupa, Z. The Level of Jasmonic Acid in *Arabidopsis thaliana* and *Phaseolus coccineus* Plants Under Heavy Metal Stress. *J. Plant Physiol.* **2005,** *162*, 1338–1346.
149. Yan, Z.; Chen, J.; Li, X. Methyl Jasmonate as Modulator of CD Toxicity in *Capsicum frutescens* var. *fasciculatum* Seedlings. *Ecotoxicol. Environ. Saf.* **2013,** *98*, 203–209.
150. Singh, I.; Shah, K. Exogenous Application of Methyl Jasmonate Lowers the Effect of Cadmium-Induced Oxidative Injury in Rice Seedlings. *Phytochemistry* **2014,** *108*, 57–66.
151. Yan, Z.; Li, X.; Chen, J.; Tam, N. F. Combined Toxicity of Cadmium and Copper in *Avicennia marina* Seedlings and the Regulation of Exogenous Jasmonic Acid. *Ecotoxicol. Environ. Saf.* **2015,** *113*, 124–132.
152. Chen, J.; Yan, Z.; Li, X. Effect of Methyl Jasmonate on Cadmium Uptake and Antioxidative Capacity in *Kandelia obovata* Seedlings Under Cadmium Stress. *Ecotoxicol. Environ. Saf.* **2014,** *104*, 349–356.
153. Ahmad, P.; Alyemeni, M. N.; Wijaya, L.; Alam, P.; Ahanger, M. A.; Alamri, S. A. Jasmonic Acid Alleviates negative Impacts of Cadmium Stress by Modifying Osmolytes and Antioxidants in Faba Bean (*Vicia faba* L.). *Arch. Agro. Soil Sci.* **2017,** *63*, 1889–1899.
154. Farooq, M. A.; Zhang, K.; Islam, F.; Wang, J.; Athar, H. U. R.; Nawaz, A.; UllahZafar, Z.; Xu, J.; Zhou, W. Physiological and iTRAQ-Based Quantitative Proteomics Analysis of Methyl Jasmonate-Induced Tolerance in *Brassica napus* Under Arsenic Stress. *Proteomics* **2018,** *8*, e1700290.
155. Sirhindi, G.; Mir, M. A.; Sharma, P.; Gill, S. S.; Kaur, H.; Mushtaq, R. Modulatory Role of Jasmonic Acid on Photosynthetic Pigments, Antioxidants and Stress Markers of *Glycine max* L. Under Nickel Stress. *Physiol. Mol. Biol. Plants* **2015,** *21*, 559–565.

156. Azeem, U. Ameliorating Nickel Stress by Jasmonic Acid Treatment in *Zea mays* L. *Russ. Agric. Sci.* **2018,** *44*, 209–215.
157. Noriega, G.; Cruz, D. S.; Batlle, A.; Tomaro, M.; Balestrasse, K. Hemeoxygenase Is Involved in the Protection Exerted by Jasmonic Acid Against Cadmium Stress in Soybean Roots. *J. Plant Growth Regul.* **2012,** *31*, 79–89.
158. Carvalho, R. F.; Campos, M. L.; Azevedo, R. A. The Role of Phytochromes in Stress Tolerance. In: *Salt Stress Plants*; Springer: Berlin/Heidelberg, Germany, 2013; pp 283–299.
159. Ballhorn, D. J.; Reisdor_, C.; Pfanz, H. Quantitative Effects of Enhanced CO_2 on Jasmonic Acid Induced Plant Volatiles of Lima Bean (*Phaseolus lunatus* L.). *J. Appl. Bot. Food Qual.* **2011,** *84*, 65–71.
160. Sun, Y.; Yin, J.; Cao, H.; Li, C.; Kang, L.; Ge, F. Elevated CO_2 Influences Nematode-Induced Defense Responses of Tomato Genotypes Differing in the JA Pathway. *PLoS ONE* **2011,** *6*, e19751.
161. Overmyer, K.; Tuominen, H.; Kettunen, R.; Betz, C.; Langebartels, C.; Sandermann, H. J.; Kangasjarvi, J. Ozone-Sensitive *Arabidopsis* RCD1 Mutant Reveals Opposite Roles for Ethylene and Jasmonate Signaling Pathways in Regulating Superoxide-Dependent Cell Death. *Plant Cell* **2000,** *12*, 1849–1862.
162. Kanna, M.; Tamaoki, M.; Kubo, A.; Nakajima, N.; Rakwal, R.; Agrawal, G. K.; Tamogami, S.; Ioki, M.; Ogawa, D.; Saji, H. et al. Isolation of an Ozone-Sensitive and Jasmonate-Semi-Insensitive *Arabidopsis* Mutant (OJI1). *Plant Cell Physiol.* **2003,** *44*, 1301–1310.
163. Rao, M. V.; Lee, H.; Creelman, R. A.; Mullet, J. E.; Davis, K. R. Jasmonic Acid Signaling Modulates Ozone-Induced Hypersensitive Cell Death. *Plant Cell* **2000,** *12*, 1633–1646.
164. Koch, J. R.; Creelman, R. A.; Eshita, S. M.; Seskar, M.; Mullet, J. E.; Davis, K. R. Ozone Sensitivity in Hybrid Poplar Correlates with Insensitivity to Both Salicylic Acid and Jasmonic Acid: The Role of Programmed Cell Death in Lesion Formation. *Plant Physiol.* **2000,** *123*, 487–496.
165. Devoto, A.; Ellis, C.; Magusin, A.; Chang, H. S.; Chilcott, C.; Zhu, T.; Turnter, J. G. Expression Profiling Reveals COI1 to Be a Key Regulator of Genes Involved in Wound- and Methyl Jasmonate Induced Secondary Metabolism, Defense and Hormone Interactions. *Plant Mol. Biol.* **2005,** *58*, 497–513.
166. Chen Chen, Y. Z.; Pang, Q. Y.; Dai, S. J.; Wang, Y.; Chen, S. X.; Yan, X. F. Proteomic Identification of Differentially Expressed Proteins in *Arabidopsis* in Response to Methyl Jasmonate. *Plant Physiol.* **2011,** *168*, 995–1008.

CHAPTER 5

Jasmonates-Mediated Defense Responses to Mechanical Injury and Herbivorous Insects

ISHFAQ UL REHMAN, SUMMIA REHMAN, AANIQA QAYOOM, ISHFAQ AHMAD SHEERGOJRI, IRFAN RASHID, and NAFEESA FAROOQ KHAN

Department of Botany, University of Kashmir, Srinagar, J&K, India

ABSTRACT

Plant reactions to wounding and herbivore attacks are coordinated by intricate signaling pathways that connect the manufacture of chemical and physical indications at the wound position to initiate gene expression and other cellular processes. The universal nature of several wound-induced responses offers a striking opportunity to study signaling pathways between cells that function over long distances inside the plant. Jasmonate (JA) performs an ultimate part in wound-induced adaptive reactions via diverse cell-independent and dependent ways. Here, in this script, we will evaluate recent advancements in understanding the molecular mechanism of jasmonates action in plants as defense against insects.

5.1 INTRODUCTION

Plants on land encompass various ecological conditions having the potential to cause damage or destroy whole plant tissues whether above or below

Jasmonates and Plant Defense: An Eco-Friendly and Sustainable Approach for Stress Tolerance.
Uzma Azeem, & Khalid Rehman Hakeem (Eds.)

the soil. Insect herbivores and pathogens cause maximum severe threats to plant existence. About, 45% of the 792,000 insect species feed on plants.[1] About hundreds of millions of years ago the battle between plants and these plant-eating insects has continued. With the passage of time, plants have developed sophisticated protection systems to counter attacks from these herbivores. These defenses are comprised of mobile compounds that directly or indirectly function as defenses. Direct defenses having toxic, repellant, or antidigestive or repellant effects on herbivores include glucosinolates, alkaloids, cyanogenic glycosides, phenolics, proteinase inhibitors (PIs), etc.[2] Physical barriers like thorns, cuticles, trichomes are also included as direct defenses for herbivores. Attacked plants also emit volatile compounds or produce extra floral nectar that attracts the predators of insect herbivores to gain an indirect means of protection.[3] Two main theories explain how plants differentiate insect herbivory from automatic damage. First is that plants identify complexes in insect oral exudations. This belief is braced by the documentation of numerous insect-derived factors that provoke resistance responses when applied to simulated wounds. Plants can also discriminate mechanical injury from herbivory over the use of as until now unidentified mechanisms that estimate the amount and value of tissue injury. Caterpillar feeding, which takes out equal-sized parts of leaf tissue in an extremely choreographed and predictable way includes the action of particular mandibles. Almost all studies show that mechanical injury and herbivory (or simulated herbivory) stimulate different responses and have relied on wound treatments that do not approximate tissue injury caused by insect grazing.[4] Soft tissue damage inflicted by herbivore does not go overlooked; greater selection pressure imposed by biotic pressure has driven the progression of advanced mechanisms through which plants recognize and react to tissue damage.[5] Phytohormones play a vital part in modifying these extremely active and adaptive reactions.[6,7] Wiesner in the 19th century developed the notion that plants vigorously react to tissue injury.[8] Hormone traumatin, which shows mitogenic properties associated with wound healing was developed and purified through different Bioassays.[9] Traumatin formed by the hydroperoxylyase branch of the lipoxygenase pathway is a responsive fatty acid derivative (12-oxo-trans-10-dodecenoic acid).[10] Restricted activity of traumatic to specific bean variety minimized interest for the opinion of a wound hormone, which helped to provoke interest in the biochemical investigation on lipoxygenase-dependent pathways for metabolism of polyunsaturated fatty acids oxidatively.[11] It has been extensively believed that oxylipins (oxygenated fatty acid derivatives) formed by these pathways execute a variety of crucial functions in the plant life cycle.[12] Signaling molecules, called jasmonates (JAs), are arranged locally and systemically and help plants to respond to many stresses both biotic as well as abiotic. Presently, we assess the recent advances in the signaling properties of different JAs, signal transduction of JA via an E3 ubiquitin ligase and JA biosynthesis regulation. The term jasmonate comprises the biologically dynamic intermediates in the path for jasmonic acid biosynthesis, besides the biologically active derivatives of jasmonic acid. These compounds distributed widely in plants affect a number of processes[13] which include the ripening of fruits, production of fertile pollens, growth of roots, coiling of tendrils, plant response to injury and abiotic strain, and insect and pathogen resistance. Farmer and Ryan suggested the role of JAs in defense[14] by providing proof for a causal relation among wounding (as caused by insect herbivores), formation of jasmonates, and induction of PIs genes that discourage insect feeding. Specifically, they suggested that wounding triggered the release of linolenic acid (LA), a presumed precursor of Jasmonates, from membrane lipids. Novel indications indicate that JA signaling in plants is generally more complex than proposed by Farmer and Ryan. Recent evidences indicate that intermediates in JA biosynthesis have diverse biological activity, the JA signaling path interacts with further defense signaling pathways and an E3 ubiquitin ligase perhaps regulates maximum JA reactions in Arabidopsis.

5.2 OVERVIEW OF THE JA PATHWAY

Two main pathways, octadecanoid and hexadecanoid pathways occur in chloroplast to yield Jasmonate. In the octadecanoid pathway chloroplast membrane containing linoleic acid (18:2) is converted into linoleic acid (18:3) by ω-3 fatty acid desaturase (FAD)[15] enzyme and phospholipase A2 (PLA2).[16] In the case of the hexadecanoid pathway hexadecatrionic acids (16:3) are freed from the chloroplast membrane. The linoleic acid (18:3) and hexadecatrionic acid (16:3) are converted into 13-(*S*)-hydroxyperoxy-octadecadi(tri)enoic acid (13-HPOT) and 11-(*S*)-hydroperoxy-hexadeca(tri) enoic acid (11-HPHT) through oxygenation property of lipoxygenase (13-LOX).[17] Enzyme, which over six pathways was transformed into

JA.[18] The for the octadecanoid pathway(13-HPOT) acted as a precursor and (11-HPHT) performed the role of precursor for the hexadecanoid pathway, these two routes then run equivalent to each other.[19] In the first step, unstable 12,13-(*S*)-epoxy-octadecatrienoic acid (12, 13-EOT) and 10,11(*S*)-epoxy-hexadeca(tri)enoic acid (10,11-EHT) are formed by enzyme allene-oxide synthase (AOS).[20] The enzyme alleneoxidecyclase (AOC)[21] yields cis-OPDA in the octadecanoid pathway and donor OPDA (dnOPDA) in the hexadecanoid pathway. The mechanism of how OPDAs move to the peroxisome is unclear but it is well-known that in Arabidopsis the ATP-binding cassette (ABC) transporter COMATOSE (CTS/PXA1/PED3) had been revealed to catalyze the ATP-dependent entry of fatty acids inside peroxisomes that represented as an agent for β-oxidation.[22] It is challenging for scientists to hunt for new pathways for the entry of OPDA into the peroxisome. As soon as the respective OPDAs go into the peroxisome it is degraded to respective oxophytodienoic acid reductase (OPR)[23] that then leads toward the three steps of β-oxidation process for the making of JA.[24] The 9S, 13*S*-OPDA, also named *cis*-OPDA, is degraded by an enzyme 12-oxophytodienoate reductase (OPR3) for the formation of 3-*oxo*-2-(2′(*Z*)-pentenyl) cyclopentane-1-octanoic acid (OPC-8:0), and dnOPDA was also degraded respectively to its corresponding hexanoic acid derivative (OPC-6:0). The initiation of OPC-8:0 and OPC-6:0 then takes place via CoA esterification of carboxylic moiety and OPC-8:0 CoA ligase 1 (OPCL1) but particular ligase is still not known for OPC-6:0. The hexanoic and octanoic cross chains such as OPC-8:0 and OPC-6:0, respectively, is reduced by two or more steps of β-oxidation. Three enzymes; acyl-CoA oxidase (ACX), 3-ketoacyl-CoA thiolase (KAT), and a diverse functional protein (MFP, comprising of enoyl-CoA hydratase and β-hydroxy-acyl-CoA dehydrogenase activities) are involved in the process of β-oxidation forming JA-CoA. The final last step occurs with the assistance of an enzyme acyl-thioesterase (ACH) that results in the formation of (+)-7-*epi*-jasmonic acid which is then changed into the more stable (−)-7-*epi*-jasmonic acid.[25] After releasing into the cytoplasm, it was further changed into methyl-(+)-7-*epi*-jasmonate (Me-7-epi-JA) with the aid of JA methyl transferase and into (+)-7-*epi*-jasmonyl-Lisoleucine(+)-7-*epi*-JA-L-Ile) catalyzed by a JA amidosynthetase or other derivatives. Thus, the production of jasmonic acid derivatives results in the regulation of targeted gene expression.[26] The byproducts produced by gene regulation in consecutive manner, turn as defensive protein counter to the external attacking entities.

5.3 PLANT PERCEPTION OF INSECT HERBIVORY AND MECHANICAL INJURY

The presence of chemical compounds in the oral discharges of feeding herbivores plays a significant part in the detection of herbivory and then stimulation of herbivore-induced defenses in plants. This recognition may well be measured more thoroughly analogous to the credit exhibited by the human immune system rather than to the old human senses that (sometimes, though by no means always) incite cognitive perception of the incentives being practiced. However, particular alertness of sensory contributions is perhaps not a pertinent dissimilarity for plants (or most other organisms); certainly, the immune system itself could be successfully regarded as a sensory organ.[27] In any case, a survey of the biochemical and molecular indications that plants use to identify and reply to insect feeding is a key attention of investigation on plant–insect interactions. Subsequently, an abundant agreement of the latest work emphasizes on explaining so-called herbivore-associated molecular patterns (or HAMPs), language equivalent to that used in work talking about plant reactions to pathogens and other microbes (PAMPs and MAMPs, respectively).[28] As usually well-defined, these patterns comprise both molecular factors derived from the pathogen or herbivore and plant-derived elements that are consistently linked with the existence of the antagonists (e.g., molecules by now present in the unspoiled plant that is then discharged from the sections in which they generally reside and/or go through chemical alteration or degradation during insect feeding).[29] From a sensual view, the latter might possibly be regarded as portion of the plant's perceptual system, while the previous comprises the environmental stimuli being distinguished, though this is generally a semantic distinction. To further make difficult problems, plant-associated microbial communities associated with plants can affect in turn be affected by plant reactions to herbivory[28] and plants might also identify the existence of herbivores using signals originating from insect-associated microbes.[30] In both cases of insect and pathogen-derived factors, it is believed that plant opinion and reaction areprobably key to evolutionarily preserved molecules that have vital physiological roles to play for the antagonist and thus cannot be easily curbed or changed to evade detection.[31] This is reliable with the extensive taxonomic distribution of elicitors like fatty acid amides[32] as well as volicitin[33] certainly, these molecules appear to play a vital role in insect nitrogen metabolism.[34]

Caeliferinsdenotea new class of fatty-acid-derived elicitors, these are called so that they are common in grasshoppers from suborder Caelifera[35]; however, β-glucosidase is an enzymatic elicitor exists in the regurgitate of *Pierisbrassicae* caterpillars.[36] Inceptins which are peptide elicitors are formed by the degradation of the plant chloroplastic ATP synthase during insect feeding.[29] Plant recognition of volicitin had been shown to comprise the binding of this molecule by cell membrane protein in *Zea mays*.[37] However, our mutual understanding of the molecular processes underlying plant sensitivity of HAMPs and thus the subsequent signaling cascades that recruits remains scanty, while it is obvious that there is a considerable connection in the processes of signal transduction started by herbivore- and microbe-derived elicitors.[38] Besides, HAMP-induced signaling has remained implicated with the regulation of changes in gene expression and further cellular processes, along with the regulation of plant hormones signaling pathways that facilitate an extensive array of indirect and direct plant defenses. Additional details regarding the uniqueness and method of action of HAMPs facilitating plant perception and reaction to insects could be found in many reviews.[30]

5.4 SIGNALING OF JASMONIC ACID

The nuclear transportation mechanism of JAs was thoroughly investigated by means of molecular genetics, molecular biology, cell biology, and biochemistry. The ABC transporter AtJAT1/AtABCG16 having JAs transportation capability remained screened by a yeast system. Autoradiography and radioactive isotope uptake trials showed that AtJAT1/AtABCG16 acts as a high-affinity transporter to control the subcellular dissemination of Jas.[39] AtJAT1/AtABCG16 is restricted on the plasma and nuclear membranes of plant cells and facilitates the transportation of JAs over the plasma membrane and the bioactive JA-Ile over the innermost membrane of the nuclear membrane to arouse JA responses at slight concentration. At high concentrations of JAs, the function of the JA carrier on the cytoplasmic membrane is leading, which decreases the JA and JA-Ile concentrations within the cells to diminish the JA signal. In other cells, the JAs signaling path is stimulated by carrying JA to the apoplast. AtJAT1/AtABCG16 can quickly control the dynamics of JA/JA-Ile within cells leads to the rapid transportation of JA-Ile into the nucleus in stressed plants, in addition to the quick desensitization of the JA signal to evade the inhibition of plant growth and development by the protection response (Figure 5.1). Considering JA receptors practiced a complex process, it was first found that the Arabidopsis coronatine insensitive1 (coi1) mutant has lost all reactions to JA[40] and additional studies have shown that the COI1 gene codes an F-box protein which is a constituent of E3 ubiquitin ligase.[41] In this situation, COI1 links with the SKP1 protein and then links to Cullin protein to make SCF-type E3 ubiquitin ligase, which is referred to as SCFCOI1, that aims the repressor proteins for degradation through ubiquitination (Xie et al and Zhai, et al.). The finding of COI1 protein is of immense importance for studying JA signaling pathway. It was once believed that COI1 acted as the jasmonic acid signaling receptor in cells, before the detection of a jasmonate Zinc finger Inflorescence Meristem (ZIM)-domain (JAZ) protein family, that gave a novel understanding of the jasmonic acid signal transduction path. In 2007, three investigating groups simultaneously found that JAZ proteins act as repressors in the JA signaling pathway.[26,42,43]. To date, in Arabidopsis near about 13 JAZ proteins have been found, the majority having two conserved domains, Jas and ZIM.[44] The JAZ protein works together with COI1 through the Jas domain and then interacts with MYC2 through the ZIM domain[45]. Thus, numerous researchers are of the opinion that JAZ proteins are the targeting proteins of COI1 and degradation of JAZ proteins is a significant step to let go of the inhibition of the JAs pathway. On the contrary, in 2010,[46] suggested different opinions on JAs receptors from the analysis of crystal structure and made it clear that the COI1–JAZ complex could be a high-affinity receptor for the bioactive JA-Ile; that is, COI1 and JAZ are coreceptors of JA signaling.[46] It is presently thought that plants perceive stimuli from the outside environment to produce JA-Ile, which stimulates the communication between COI1 and JAZ proteins. Consequently, JAZ proteins are degraded when being relocated to the 26S proteasome, and at the same time, transcription factors (TFs) become free to initiate the expression of downstream genes (Figure 5.1). JA-Ile boosts the MYC TFs by straightly binding to JAZ and COI1 resulting in the degradation of JAZ via the 26S proteasome pathway. Current studies have revealed that the MYB TFs too link with JAZ repressors and might be stimulated by the degradation of JAZ in the presence of JA-Ile. Furthermore, some additional TFs, such as NAC, ERF, and WRKY, also play a role in the JA signaling. These JA-responsive TFs control the expression of numerous genes that have a role in the overall growth and development of plants, and particularly the reactions and variations of plants to the exterior environment.

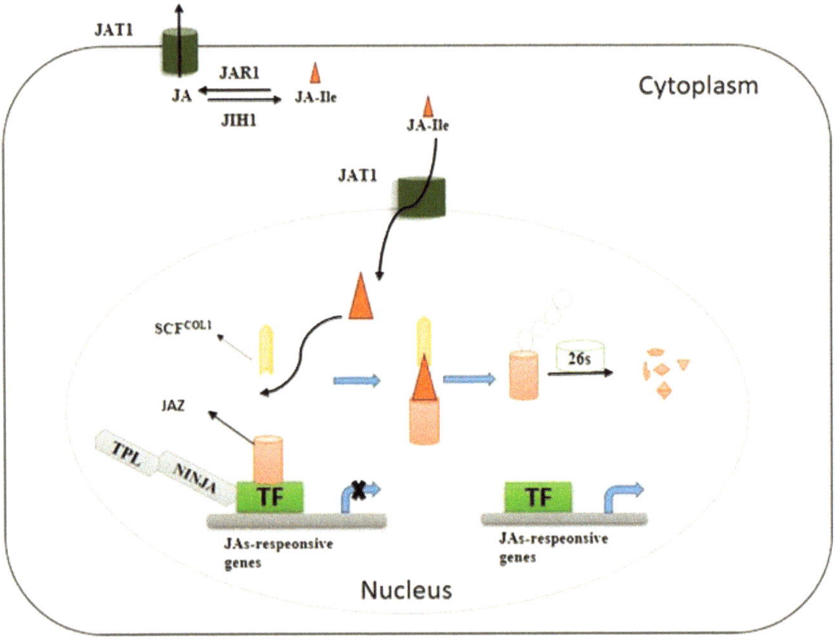

FIGURE 5.1 The working model of jasmonic acid transport and signaling pathway. JAT1: jasmonic acid transfer protein1; SCF: Skp1, Cullin and F-box proteins; COI1: coronatine insensitive1; JAZ: jasmonate ZIM-domain protein; TF: transcription factor; TPL: TOPLESS protein; NINJA: NOVEL INTERACTOR OF JAZ; 26S: 26S proteasome. Modified version of Ruan et al. (2019).

5.5 HOW JA REGULATES PLANT DEFENSE AGAINST INSECTS AND WOUND INJURY

Insect herbivores take nutrients from plants and impose mechanical injury to plant tissues. Upon pest grazing and cutting during herbivory, JA production was promptly activated, both nearby at the injured site and systemically in untouched leaves throughout the plants, to curb or check the feeding, growth, or reproduction of arthropod herbivores by stimulating expression of defense-related genes, manufacture of poisonous secondary chemicals, creation of physical barrier (such as trichome) and release of volatile organic complexes (VOCs) that attract predators or parasitic opponents of pests.[47] For instance, the *fad3/fad7/fad8* mutant displays exceptional sensitivity to the feeding of *Bradysia flies* larvae.[48] Likewise, the *opr3* and *coi1* mutants also display the most susceptiblity to the feeding of *Pierisrapae* caterpillars and aphid *Myzuspersicae* herbivory.[49] The *aos* and *myc2/myc3/myc4* mutants cause increased susceptibility to *Spider mite* herbivory.[50] The tomato mutant *def1* has compromised resistance to *Manducasexta*, *Spodopteraexigua*, *Tetranychusurticae*, and *Frankliniellaoccidentalis*.[51] Such evidences reveal that JA plays a vital and conserved role in regulating plant defense against a broad spectrum of insect herbivores.

KEYWORDS

- **jasmonates**
- **defense**
- **insects**
- **signaling**

REFERENCES

1. Strong, D. R.; Lawton, J. H.; Southwood, S. R. *Insects on Plants: Community Patterns and Mechanisms*; Blackwell Scientific Publicatons 1984.
2. Bennett, R. N.; Wallsgrove, R. M. Secondary Metabolites in Plant Defence Mechanisms. *New Phytol.* **1994,** *127*, 617–633.
3. Heil, M. Indirect Defence via Tritrophic Interactions. *New Phytol.* **2008,** *178*, 41–61.
4. Mithöfer, A.; Wanner, G.; Boland, W. Effects of Feeding Spodopteralittoralis on Lima Bean Leaves. II. Continuous Mechanical Wounding Resembling Insect Feeding Is Sufficient to Elicit Herbivory-Related Volatile Emission. *J. Plant Physiol.* **2005,** *137*, 1160–1168.
5. Howe, G. A.; Schilmiller, A. L. Oxylipin Metabolism in Response to Stress. *Curr. Opin. Plant Biol.* **2002,** *5*, 230–236.
6. Pieterse, C. M.; Leon-Reyes, A.; Van der Ent, S.; Van Wees, S. C. Networking by Small-Molecule Hormones in Plant Immunity. *Nat. Chem. Biol.* **2009,** *5*, 308–316.
7. Santner, A.; Calderon-Villalobos, L. I. A.; Estelle, M. Plant Hormones Are Versatile Chemical Regulators of Plant Growth. *Nat. Chem. Biol.* **2009,** *5*, 301–307.

8. Wiesner, J. *Die Elementarstructur und das Wachstum der lebendenSubstanz*. A. Hölder, 1892.
9. Abraham, J. K. O. O; Howe, G. A. The Wound Hormone Jasmonate. *Phytochemistry* **2009,** *70*, 1571–1580.
10. Zimmerman, D. C.; Coudron, C. A. Identification of traumatin, a Wound Hormone, as 12-Oxo-Trans-10-Dodecenoic Acid. *J. Plant Physiol.* **1979,** *63*, 536–541.
11. Matsui, K. Green Leaf Volatiles: Hydroperoxidelyase Pathway of Oxylipin Metabolism. *Curr. Opin. Plant Biol.* **2006,** *9*, 274–280.
12. Browse, J. Jasmonate Passes Muster: A Receptor and Targets for the Defense Hormone. *Annu. Rev. Plant Biol.* **2009,** *60*, 183–205.
13. Creelman, R. A.; Mullet, J. E. Biosynthesis and Action of Jasmonates in Plants. *Annu. Rev. Plant Biol.* **1997,** *48*, 355–381.
14. Farmer, E. E.; Ryan, C. A. Octadecanoid Precursors of Jasmonic Acid Activate the Synthesis of Wound-Inducible Proteinase Inhibitors. *Plant Cell* **1992,** 129–134.
15. Kodama, H.; Nishiuchi, T.; Seo, S.; Ohashi, Y.; Iba, K. Possible Involvement of Protein Phosphorylation in the Wound-Responsive Expression of Arabidopsis Plastid ω-3 Fatty Acid Desaturase Gene. *J. Plant Sci.* **2000,** *155*, 153–160.
16. Engelberth, J. Secondary Metabolites and Plant Defense. *Plant Physiol.* **2006,** *4*, 315–344.
17. Bell, E.; Creelman, R. A.; Mullet, J. E. A Chloroplast Lipoxygenase Is Required for Wound-Induced Jasmonic Acid Accumulation in Arabidopsis. *PNAS* **1995,** *92*, 8675–8679.
18. Feussner, I.; Wasternack, C. The lipoxygenase pathway. *Annu. Rev. Plant Biol.* **2002,** *53*, 275–297.
19. Weber, H.; Vick, B. A.; Farmer, E. E. Dinor-Oxo-Phytodienoic Acid: A New Hexadecanoid Signal in the Jasmonate Family. *PNAS* **1997,** *94*, 10473–10478.
20. Laudert, D.; Weiler, E. W. Allene oxide synthase: a major control point in *Arabidopsis thaliana* octadecanoid signalling. *Plant J.* **1998,** *15*, 675–684.
21. Ziegler, J.; Hamberg, M.; Miersch, O.; Parthier, B. Purification and Characterization of Allene Oxide Cyclase from Dry Corn Seeds. *Plant Physiol.* **1997,** *114*, 565–573.
22. Theodoulou, F. L.; Job, K.; Slocombe, S. P.; Footitt, S.; Holdsworth, M.; Baker, A.; Larson, T. R.; Graham, I. A. Jasmonic Acid Levels Are Reduced in COMATOSE ATP-Binding Cassette Transporter Mutants: Implications for Transport of Jasmonate Precursors into Peroxisomes. *Plant Physiol.* **2005,** *137*, 835–840.
23. Stintzi, A. The *Arabidopsis* Male-Sterile Mutant, opr3, Lacks the 12-Oxophytodienoic Acid Reductase Required for Jasmonate Synthesis. *PNAS* **2000,** *97*, 10625–10630.
24. Schilmiller, A. L.; Howe, G. A. Systemic Signaling in the Wound Response. *Curr. Opin. Plant Biol.* **2005,** *8*, 369–377.
25. Schaller, A.; Stintzi, A. Jasmonate Biosynthesis and Signaling for Induced Plant Defense Against Herbivory. In: *Induced Plant Resistance to herbivory*; Springer: Dordrecht, 2008; pp 349–366.
26. Chini, A.; Fonseca, S. G. D. C.; Fernandez, G.; Adie, B.; Chico, J. M.; Lorenzo, O.; García-Casado, G.; López-Vidriero, I.; Lozano, F. M.; Ponce, M. R.; Micol, J. L. The JAZ Family of Repressors Is the Missing Link in Jasmonate Signalling. *Nature* **2007,** *448*, 666–671.
27. Blalock, J. E. The Immune System as a Sensory Organ. *J. Immunol.* **1984** *132*, 1067–1070.

28. Felton, G. W.; Tumlinson, J. H. Plant–Insect Dialogs: Complex Interactions at the Plant–Insect Interface. *Curr. Opin. Plant Biol.* **2008,** *11*, 457–463.
29. Schmelz, E. A.; Carroll, M. J.; LeClere, S.; Phipps, S. M.; Meredith, J.; Chourey, P. S.; Alborn, H. T. Teal, P. E. Fragments of ATP Synthase Mediate Plant Perception of Insect Attack. *PNAS* **2006,** *103*, 8894–8899
30. Bonaventure, G. Perception of Insect Feeding by Plants. *J. Plant Biol.***2012,** *14*, 872–880.
31. Bent, A. F.; Mackey, D. Elicitors, Effectors, and R Genes: The New Paradigm and a Lifetime Supply of Questions. *Annu. Rev. Phytopathol.* **2007,** *45*, 399–436.
32. Yoshinaga, N.; Aboshi, T.; Ishikawa, C.; Fukui, M.; Shimoda, M.; Nishida, R.; Lait, C. G.; Tumlinson, J. H.; Mori, N. Fatty Acid Amides, Previously Identified in Caterpillars, Found in the Cricket Teleogryllustaiwanemma and Fruit Fly Drosophila Melanogaster Larvae. *J. Chem. Ecol.* **2007,** *33*, 1376–1381.
33. Alborn, H. T.; Turlings, T. C. J.; Jones, T.; Stenhagen, G.; Loughrin, J. H.; Tumlinson, J. H. An Elicitor of Plant Volatiles from Beet Armyworm Oral Secretion. *Science* **1997,** *276*, 945–949.
34. Yoshinaga, N.; Aboshi, T.; Abe, H.; Nishida, R.; Alborn, H. T.; Tumlinson, J. H.; Mori, N. Active Role of Fatty Acid Amino Acid Conjugates in Nitrogen Metabolism in *Spodopteralitura larvae*. *PNAS* **2008,** *105*, 18058–18063.
35. Alborn, H. T.; Hansen, T. V.; Jones, T. H.; Bennett, D. C.; Tumlinson, J. H.; Schmelz, E. A.; Teal, P. E. Disulfooxy Fatty Acids from the American Bird Grasshopper *Schistocercaamericana*, Elicitors of Plant Volatiles. *PNAS* **2007,** *104*, 12976–12981.
36. Mattiacci, L.; Dicke, M.; Posthumus, M, A. β-Glucosidase: An Elicitor of Herbivore-Induced Plant Odor That Attracts Host-Searching Parasitic Wasps. *PNAS* **1995,** *92*, 2036–2040.
37. Truitt, C. L.; Wei, H. X.; Pareé, P. W. A Plasma Membrane Protein from *Zea mays* Binds with the Herbivore Elicitor Volicitin. *Plant Cell* **2004,** *16*, 523–532.
38. Bonaventure, G.; VanDoorn, A.; Baldwin, I. T. Herbivore-Associated Elicitors: FAC Signaling and Metabolism. *Trends Plant Sci.* **2011,** *16*, 294–299.
39. Li, Q.; Zheng, J.; Li, S.; Huang, G.; Skilling, S. J.; Wang, L.; Li, L.; Li, M.; Yuan, L.; Liu, P. Transporter-Mediated Nuclear Entry of Jasmonoyl-Isoleucine Is Essential for Jasmonate Signaling. *Mol. Plant***2017,** *10*, 695–708.
40. Feys, B. J.; Benedetti, C. E.; Penfold, C. N.; Turner, J. G. Arabidopsis Mutants Selected for Resistance to the Phytotoxincoronatine Are Male Sterile, Insensitive to Methyl Jasmonate, and Resistant to a Bacterial Pathogen. *Plant Cell* **1994** *6*, 751–759.
41. Xie, D. X.; Feys, B. F.; James, S.; Nieto-Rostro, M.; Turner, J. G. COI1: An Arabidopsis Gene Required for Jasmonate-Regulated Defense and Fertility. *Science* **1998,** *280*, 1091–1094.
42. Thines, B.; Katsir, L.; Melotto, M.; Niu, Y.; Mandaokar, A.; Liu, G.; Nomura, K.; He, S. Y.; Howe, G. A.; Browse, J. JAZ Repressor Proteins Are Targets of the SCF COI1 Complex During Jasmonate Signalling. *Nature* **2007,** *448*, 661–665.
43. Yan, Y.; Stolz, S.; Chételat, A.; Reymond, P.; Pagni, M.; Dubugnon, L.; Farmer, E. E. A Downstream Mediator in the Growth Repression Limb of the Jasmonate Pathway. *Plant Cell* **2007,** *19*, 2470–2483.
44. Melotto, M.; Mecey, C.; Niu, Y.; Chung, H. S.; Katsir, L.; Yao, J.; Zeng, W.; Thines, B.; Staswick, P.; Browse, J.; Howe, G. A. A Critical Role of Two Positively

Charged Amino Acids in the Jas Motif of Arabidopsis JAZ Proteins in Mediating Coronatine-and Jasmonoyl Isoleucine-Dependent Interactions with the COI1 F-Box Protein. *Plant J.* **2008,** *55*, 979–988.
45. Chini, A.; Fonseca, S.; Chico, J. M.; Fernández-Calvo, P.; Solano, R. The ZIM Domain Mediates Homo-and Heteromeric Interactions Between Arabidopsis JAZ Proteins. *Plant J.* **2009,** *59*, 77–87.
46. Sheard, L. B.; Tan, X.; Mao, H.; Withers, J.; Ben-Nissan, G.; Hinds, T. R.; Kobayashi, Y.; Hsu, F. F.; Sharon, M.; Browse, J.; He, S. Y. Jasmonate Perception by Inositol-Phosphate-Potentiated COI1–JAZ Co-Receptor. *Nature* **2010,** *468*, 400–405.
47. Sun, J. Q.; Jiang, H. L.; Li, C. Y. Systemin/Jasmonate-Mediated Systemic Defense Signaling in Tomato. *Mol. Plant* **2011,** *4*, 607–615.
48. McConn, M.; Creelman, R. A.; Bell, E.; Mullet, J. E. Jasmonate Is Essential for Insect Defense in Arabidopsis. *Proc. Natl. Acad. Sci. USA* **1997,** *94*, 5473–5477.
49. Reymond, P.; Bodenhausen, N.; Van Poecke, R. M.; Krishnamurthy, V.; Dicke, M.; Farmer, E. E. A Conserved Transcript Pattern in Response to a Specialist and a Generalist Herbivore. *Plant Cell* **2004,** *16*, 3132–3147.
50. Zhurov, V.; Navarro, M.; Bruinsma, K. A.; Arbona, V.; Santamaria, M. E.; Cazaux, M.; Wybouw, N.; Osborne, E. J.; Ens, C.; Rioja, C.; Vermeirssen, V. Reciprocal Responses in the Interaction Between Arabidopsis and the Cell-Content-Feeding Chelicerate Herbivore Spider Mite. *Plant Physiol.* **2014,** *164*, 384–399.
51. Thaler, J. S.; Owen, B.; Higgins, V. J. The Role of the Jasmonate Response in Plant Susceptibility to Diverse Pathogens with a Range of Lifestyles. *J. Plant Physiol.* **2004,** *135*, 530–538.

CHAPTER 6

Jasmonates and Oxidative Stress

UZMA AZEEM[1], GURPAUL SINGH DHINGRA[2], and RICHA SHRI[3]

[1]*Government College Malerkotla, Malerkotla, Punjab, India*

[2]*Department of Botany, Punjabi University, Patiala, India*

[3]*Department of Pharmaceutical Sciences and Drug Research, Punjabi University, Patiala, India*

ABSTRACT

Oxidative stress occurs due to the overproduction of reactive oxygen species. The latter are an outcome of metabolism but under various abiotic and biotic stress conditions, their overproduction occurs. To avoid damage by reactive oxygen species, plants are equipped with various kinds of defenses. Antioxidants play a crucial role in defending plants from various biotic and abiotic stresses. In addition to various enzymatic and non-enzymatic antioxidants, many plant hormones, such as jasmonates, regulate plant cell metabolism and stress tolerance. Jasmonates modulate the activities and levels of many enzymatic antioxidants (such as superoxide dismutase, peroxidases, catalase, etc.) and non-enzymatic antioxidants like phenolic compounds, osmolytes, etc. The interplay between antioxidants and jasmonates helps in the alleviation of reactive oxygen species-induced damage which in turn boosts plant growth and development. This chapter deals with the cross talk between jasmonates and antioxidants in defending the plants from stress which ultimately improves plant growth and development.

Jasmonates and Plant Defense: An Eco-Friendly and Sustainable Approach for Stress Tolerance.
Uzma Azeem, & Khalid Rehman Hakeem (Eds.)

6.1 INTRODUCTION

Plants are sedantry and face different kinds of stresses in their habitat. Under stress conditions, the excess of reactive oxygen species causes (ROS) oxidative damage. To overcome oxidative stress, plants are consecrated with various types of defenses. Among the different types of defense mechanisms, plants possess antioxidant defense system. Antioxidants are the biochemicals (enzymatic and non-enzymatic) which protect the plants from the toxic action of ROS. These ROS are produced during cell metabolism and under stress conditions, their overproduction causes cellular damage. Therefore, their removal is essential to protect the cell and cellular components from damage.[1] In addition to antioxidants, many plant hormones like brassinosteroids, jasmonates, etc., play a defensive role under various kinds of abiotic and biotic stresses in various vital and mechanized ways.[2] Under stress conditions, the endogenous level of jasmonates increases in plants. Their exogenous applcation has also been reported to ameliorate the adverse impacts of stress on plant growth and development. Jasmonates (jasmonic acid and its derivatives: MeJA, isoleucine conjugate (JA-Ile)), are the lipid-derived plant hormones which act as critical mediators of the plant defense mechanisms and thereby regulate plant growth and development.[3] Currently, jasmonates occur in almost all higher plants and their concentration is higher in flowers and reproductive parts.[4] Generally, jasmonates affect the development of embryo, seed germination, growth and development of roots and trichomes, seedling development, formation of tubers, leaf movement, fruit ripening, leaf senescence, and accumulation of phenolic antioxidants. They modulate gene expression and help plants to overcome the insect-induced injury, pathogen attack, and environmental stresses like low temperature salinity and drought.[5] Jasmonates defend the plants from various stresses through different mechanisms. One of which is through crosstalk with antioxidants to get rid of oxidative stress-induced damages through their scavenging action on the ROS.[6]

6.2 JASMONATES AND ANTIOXIDANTS

Jasmonates modulate the levels and activities of the antioxidant enzymes such as SOD, CAT, APOX, GPOX, etc., and help in amelioration of oxidative stress. This ultimately improves the plant growth and development.[7]

The increase in the contents of non-enzymatic antioxidants such as ascorbic acid, glutathione, phenolic compounds, flavonoids, catkin, carotenoid, α–tocopherols, etc., with jasmonates also exerts an alleviating impact on plants facing stress.[8]

6.2.1 JA-ANTIOXIDANTS CROSSTALK UNDER ABIOTIC STRESS

Environmental factors such as extremes of temperature, light, water, etc., cause adverse effects on plant growth and development. The stress induced increase in the endogenous levels of jasmonates and their exogenous application exhibit ameliorative action against a variety of stresses. However, they are known to act in a concentration-dependent manner. Generally, concentrations of micromolar to nanomolar jasmonates are found more effective. The application of jasmonic acid has been reported to ameliorate the negative impacts of Ni toxicity on *Glycine max* L. by decreasing the accumulation of H_2O_2, malondialdehyde (MDA), and activity of NADPH oxidase through the enhancment of the activities of SOD, peroxidase (POD), CAT, and APX. The contents of proline, glycine betaine, and ascorbic acid also increased in Ni-stressed soybean plants. The elevated levels of antixidants caused an increase in shoot and root length, chlorophyll content, and dry weight of the treated plants.[9] Ni toxicity has also been alleviated in Ni-stressed *Zea mays* L. plants on exogenous application of jasmonic acid because of increase in the activities of the antioxidant enzymes.[10] Non-halophytes such as strawberry plants are unable to tolerate high salt concentration, which limits their growth and develeopment. Excess of NaCl in the soil exerts both osmotic and ionic stresses, leading to nutritent deficiencies and metabolic perturbations. This in turn results in poor crop production. Production of ROS is one of the major consequences of salt stress.[11] Exogenous application of MeJA alleviated the detrimental impacts of excess of ROS produced in salinity stress by enhancing the activities of SOD, APX, and POD.[12] Water deficit is one of the major abiotic stresses reducing the crop yield due to overproduction of ROS, which leads to cellular damage. The foliar application of MeJA on soybean seedlings subjected to drought has been found to increase the activities of enzymatic antioxidants (SOD, POD, and CAT), proline content, relative water content, and to decrease the membrane lipid peroxidation which in turn enhanced the grain yield and harvest index.[13] In addition to the increase in antioxidant enzyme activities, the application of jasmonates also enhances the amount of the non-enzymatic antioxidants in plants. In wound-induced broccoli, MeJA application caused accumulation of phenolic compounds like rutin, caffeic acid, and cinnamic acid and increase in the activities of antioxidant enzymes.[14] As the problem of global warming intensifies, the consequent high temperatures detereorate plant growth and development and further cause loss of yield. Application of MeJA provides an effective solution to make plants capable of combating the high temperature. The exogenous spraying of MeJA (0.1 mmol/L) at the booting stage in rice alleviated the negative effects of high temperature on rice by increasing the stomatal conductance, transpiration rate, and reducing the organ temperature and thereby decreasing the damage to the actual photochemical efficiency. A decrease in the MDA level, and increase in the carotenoid content and antioxidant capacity were also reported and the yield factor which in turn resulted in the greater number of grains per panicle and higher seed-setting rate.[15] The postharvest application of jasmonates also enhances the shelf life of crop produce and their marketability. The postharvest treatment of the various fruits such as mango, sweet pepper, and tomato fruits with methyl jasmonate (MeJA) significantly reduced the development of chilling injury symptoms and also improved their marketability.[16–18] MeJA treatment decreases in O^{2-} and H_2O_2 production by causing increase in the activities of SOD, CAT, APX, and fall in the activity of lipoxygenase in stored fruits.[19,20]

6.2.2 JA–ANTIOXIDANT CROSSTALK UNDER BIOTIC STRESS

Jasmonates play role in plant defense against biotic stress mainly under insect attack and pathogens. Insects act as vectors of plant viruses and their attack elicites a rapid increase in the concentration of jasmonates. Jasmonates are themselves non-toxic to insects but they induce the synthesis of secondary metabolites and systemic resistance in plants.[21,22] Jasmonic acid and its derivatives are known to alleviate the negaitive effects of biotic stress in a number of crops.[23–28] In *Zea mays* L. plants infected with the insect *Trigonotylus caelestialium*, the activities of the antioxidant enzymes (SOD, CAT, APX, and GPX) decreased with MeJA and MeJA+Cu treatment in leaves but increased in the roots. This happened due to the scavenging effect of MeJA on the ROS produced under insect influence.[7] The adverse effects of aphid feeding on cucumber (*Cucumis sativus* L.)

were alleviated by the treatment of aphid-attacked plants by the application of jasmonic acid and MeJA. This has been attributed to the increased activities of SOD, POD, CAT, APX, phenylalanine ammonia–lyase (PAL), and laccase, and enhanced contents of total phenols, flavonoids, tannins, lignins, and various osmolytes.[29] Jasmonates also protect the plants from many postharvest pathogens. *Monilinia laxa* (Aderh. & Ruhl.) honey is a serious fungal pathogen of apricot fruits (*Prunus armeniaca* L.) causing brown rot disease. Postharvest treatment with MeJA treatment enhances the defense against *Monilinia laxa* by decreasing the fungal growth. An increase in total soluble phenol content, total antioxidant activity, and enhanced activities of SOD, POD, and PAL are responsible for the improved defense and shelf life of the friut.[30]

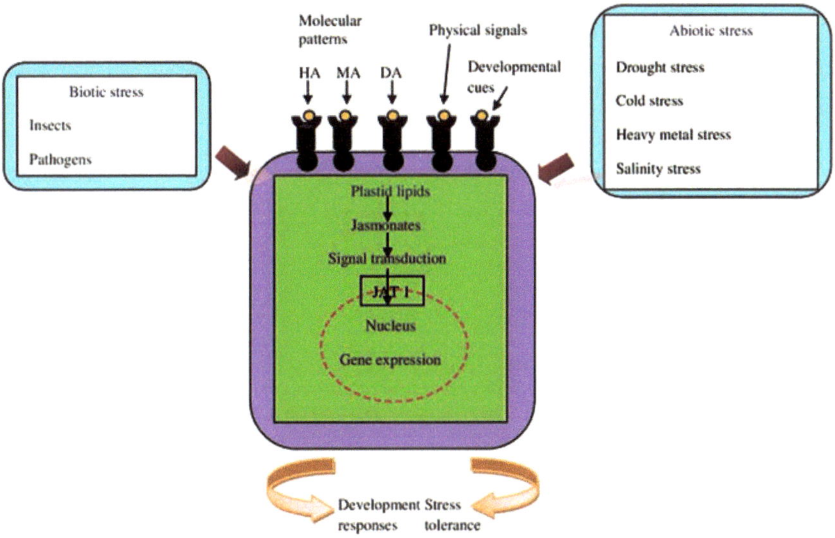

FIGURE 6.1 Stress tolerance by jasmonates in plants. JAT1, jasmonate transporter 1; HA, herbivore associated; MA, microbe associated; DA, damage-associated molecular patterns.

6.3 MECHANISM OF ACTION OF JASMONATES

Under abiotic and biotic stress, the endogenous levels of jasmonates increase in response to the increase in the ROS levels. A signal transduction cascade starts involving crosstalk between jasmonates, plant hormones such as IAA, GA, ET, ABA, SA, etc., and various enzymatic, and nonenzymatic components. At molecular level, the expression pattern of genes involved in various metabolic pathways changes. The outcome of all these events is stress tolerance in plants.

6.4 CONCLUSION

Under abiotic and biotic stress, the level of ROS increases. The endogenus levels of jasmonates also increase in response to the enhanced production of ROS. The increase in the concentrations of endogenous jasmonates and their exogenous application helps in the amelioration of oxidative stress-induced damage which in turn improves plant growth and development. Jasmonates alleviate oxidative stress by inducing the activities of the antioxidant enzymes such as SOD, POD, CAT, APX, GPX, etc. Additionally, the activities of glucose 6-phosphate dehydrogenase, phenylalanine ammonia lyase, substrate-specific peroxidases (caffeic acid peroxidase, quercetin peroxidase, and ferulic acid peroxidase) enhace on exposure to jasmonates. The contents of non-enzymatic antioxidants such as total phenolic, flavonoid, ascorbic acid, non-protein thiol (NPSH), and cysteine, and the radical reducing activity of 1,1–diphenyl–2–picrylhydrazyl (DPPH) also increase under jasmonates treatment.[31]

KEYWORDS

- **Antioxidants**
- **jasmonates**
- **reactive oxygen species**
- **oxidative stress**

REFERENCES

1. Mansoor, S.; Ali Wani, O.; Lone, J. K.; Manhas, S.; Kour, N.; Alam, P.; Ahmad, A.; Ahmad, P. Reactive Oxygen Species in Plants: From Source to Sink. *Antioxidants* **2022,** *11* (2), 225.

2. Waadt, R.; Seller, C. A.; Hsu, P. K.; Takahashi, Y.; Munemasa, S.; Schroeder, J. I. Plant Hormone Regulation of Abiotic Stress Responses. *Nat. Rev. Mol. Cell Biol.* **2022,** 1–15.
3. Li, C.; Xu, M.; Cai, X.; Han, Z.; Si, J.; Chen, D. Jasmonate Signaling Pathway Modulates Plant Defense, Growth, and Their Trade-Offs. *Int. J. Mol. Sci.* **2022,** *23* (7), 3945.
4. Dar, T. A.; Uddin, M.; Khan, M. A.; Hakeem, K. R.; Jaleel, H. Jasmonates Counter Plant Stress: A Review. *Environ. Exp. Bot.* **2015,** *115*, 49–57.
5. Yu, X.; Zhang, W.; Zhang, Y.; Zhang, X.; Lang, D.; Zhang, X. The Roles of Methyl Jasmonate to Stress in Plants. *Funct. Plant Biol.* **2018,** *46* (3), 197–212.
6. Aftab, T.; Khan, M.; Idrees, M.; Naeem, M.; Hashmi, N. Methyl Jasmonate Counteracts Boron Toxicity by Preventing Oxidative Stress and Regulating Antioxidant Enzyme Activities and Artemisinin Biosynthesis in *Artemisia annua* L. *Protoplasma* **2011,** *248* (3), 601–612.
7. Hanaka, A.; Lechowski, L.; Mroczek-Zdyrska, M.; Strubińska, J. Oxidative Enzymes Activity During Abiotic and Biotic Stresses in *Zea mays* Leaves and Roots Exposed to Cu, Methyl Jasmonate and *Trigonotylus caelestialium. Physiol. Mol. Biol. Pla.* **2018,** *24* (1), 1–5.
8. Asghari, M.; Aghdam, M. S. Impact of Salicylic Acid on Post-Harvest Physiology of Horticultural Crops. *Trends Food Sci. Tech.* **2010,** *21* (10), 502–509.
9. Sirhindi, G.; Mir, M. A.; Abd-Allah, E. F.; Ahmad, P.; Gucel, S. Jasmonic Acid Modulates the Physio-Biochemical Attributes, Antioxidant Enzyme Activity, and Gene Expression in *Glycine max* Under Nickel Toxicity. *Front. Plant Sci.* **2016,** *7*, 591.
10. Azeem, U. Ameliorating Nickel Stress by Jasmonic Acid Treatment in *Zea mays* L. *Russ. Agric. Sci.* **2018,** *44* (3), 209–215.
11. Eraslan, F.; Inal, A.; Gunes, A.; Alpaslan, M. Impact of Exogenous Salicylic Acid on the Growth, Antioxidant Activity and Physiology of Carrot Plants Subjected to Combined Salinity and Boron Toxicity. *Sci. Hort.* **2007,** *113*, 120–128.
12. Faghih, S.; Ghobadi, C.; Zarei, A.; Response of Strawberry Plant cv. 'Camarosa' to Salicylic Acid and Methyl Jasmonate Application Under Salt Stress Condition. *J. Plant Growth Regul.* **2017,** *36* (3), 651–659.
13. Anjum, S. A.; Wang, L.; Farooq, M.; Khan, I.; Xue, L. Methyl Jasmonate-Induced Alteration in Lipid Peroxidation, Antioxidative Defence System and Yield in Soybean Under Drought. *J. Agron. Crop Sci.* **2011,** *197* (4), 296–301.
14. Guan, Y.; Hu, W.; Jiang, A.; Xu, Y.; Sa, R.; Feng, K.; Zhao, M.; Yu, J.; Ji, Y.; Hou, M. and Yang, X.; Effect of Methyl Jasmonate on Phenolic Accumulation in Wounded Broccoli. *Molecules* **2019,** *24* (19), 3537.
15. Tang, S.; Zhao, Y.; Ran, X.; Guo, H.; Yin, T.; Shen, Y.; Liu, W.; Ding, Y. Exogenous Application of Methyl Jasmonate at the Booting Stage Improves Rice's Heat Tolerance by Enhancing Antioxidant and Photosynthetic Activities. *Agronomy* **2022,** *12* (7), 1573.
16. González-Aguilar, G. A.; Fortiz, J.; Wang, C. Y. Methyl Jasmonate Reduces Chilling Injury and Maintains Postharvest Quality of Mango Fruit. *J. Agric. Food Chem.* **2000,** *48*, 515–519.
17. Ding, C. K.; Wang, C. Y.; Gross, K. C.; Smith, D. L. Reduction of Chilling Injury and Transcript Accumulation of Heat Shock Protein Genes in Tomatoes by MeJA and MeSA. *Plant Sci.* **2001,** *161*, 1153–1159.

18. Fung, R. W. M.; Wang, C. Y.; Smith, D. L.; Gross, K. C.; Tian, M. S.; MeSA and MeJA Increase Steady-State Transcript Levels of Alternative Oxidase and Resistance Against Chilling Injury in Sweet Peppers (*Capsicum annuum* L.). *Plant Sci.* **2004,** *166*, 711–719.
19. Cao, S.; Zheng, Y.; Wang, K.; Jin, P.; Rui, H.; Methyl Jasmonate Reduces Chilling Injury and Enhances Antioxidant Enzyme Activity in Postharvest Loquat Fruit. *Food Chem.* **2009,** *115* (4), 1458–1463.
20. Asghari, M.; Hasanlooe, A. R. Methyl Jasmonate Effectively Enhanced Some Defense Enzymes Activity and Total Antioxidant Content in Harvested "Sabrosa" Strawberry Fruit. *Food Sci. Nutr.* **2016,** *4* (3), 377–383.
21. Wu, X.; Ye, J. Manipulation of Jasmonate Signaling by Plant Viruses and Their Insect Vectors. *Viruses* **2020,** *12* (2), 148.
22. Wang, Y.; Mostafa, S.; Zeng, W.; Jin, B. Function and Mechanism of Jasmonic Acid in Plant Responses to Abiotic and Biotic Stresses. *Int. J. Mol. Sci.* **2021,** *22* (16), 8568.
23. Cooper, W. R.; Goggin, F. L. Effects of Jasmonate-Induced Defenses in Tomato on the Potato Aphid, *Macrosiphum euphorbiae*. *Entomol. Exp. Appl.* **2005,** *115*, 107–115.
24. Senthil, N. S.; Kalaivani, K. C. M. Y.; Paik, C. H. Effects of Jasmonic Acid-Induced Resistance in Rice on the Plant Brownhopper, *Nilaparvata lugens* Stål (Homoptera: Delphacidae). *Pest. Biochem. Page 20/27 Physiol.* **2009,** *95* (2), 77–84.
25. Wakeil, N. E.; Volkmar, C.; Sallam, A. A. Jasmonic Acid Induces Resistance to Economically Important Insect Pests in Winter Wheat. *Pest. Manag. Sci.* **2010,** *66* (5), 549–554.
26. Warabieda, W.; Olszak, R. Effect of Exogenous Methyl Jasmonate on Numerical Growth of the Population of the Two-Spotted Spider Mite (*Tetranychus urticae* koch.) on Strawberry Plants and Young Apple Trees. *J. Plant Prot. Res.* **2010,** *50* (4).
27. Bravo, R.; Alba, J. M.; Pons, C.; Granell, A.; Kant, M. R.; Moriones, E.; Fernández, M. R. A Jasmonate-Inducible Defense Trait Transferred from Wild Into Cultivated Tomato Establishes Increased Whitefly Resistance and Reduced Viral Disease Incidence. *Front. Plant Sci.* **2016,** *7*.
28. Ellis, C.; Karafyllidis, I.; Turner, J. G. Constitutive Activation of Jasmonate Signaling in an *Arabidopsis* Mutant Correlates with Enhanced Resistance to *Erysiphe cichoracearum*, *Pseudomonas syringae*, and *Myzus persicae*. *Mol. Plant Microbe Interact.* **2002,** *15* (10), 1025–1030.
29. Wang, X.; Cui, J.; Li, X.; Wang, G.; Zheng, G.; Li, Y.; Li, X.; Ma, L.; Wang, L. Comparative Study on Growth, Physiological and Biotic Stress Resistance of Cucumber (*Cucumis sativus* L.) by Exogenous Application of Jasmonic Acid and Methyl Jasmonate Under Greenhouse Conditions. *Res Sq.* **2022,** *2022*, 1–27.
30. Ezzat, A.; Szabó, S.; Szabó, Z.; Hegedűs, A.; Berényi, D.; Holb, I. J. Temporal Patterns and Inter-Correlations Among Physical and Antioxidant Attributes And Enzyme Activities of Apricot Fruit Inoculated with *Monilinia laxa* Under Salicylic Acid and Methyl Jasmonate Treatments Under Shelf-Life Conditions. *J. Fungi* **2021,** *7* (5), 341.
31. Ali, M. B.; Hahn, E. J.; Paek, K. Y. Methyl Jasmonate and Salicylic Acid Induced Oxidative Stress and Accumulation of Phenolics in *Panax ginseng* Bioreactor Root Suspension Cultures. *Molecules* **2007,** *12* (3), 607–621.

CHAPTER 7

Jasmonates in Plant Defense Under Abiotic Stress

MUHAMMAD ASHAR AYUB[1], MUHAMMAD ZIA UR REHMAN[1], NAQSHE ZUHRA[1], MAIRA NAVEED[1], WAJID UMAR[2], MUHAMMAD NADEEM[1], and ZAHOOR AHMAD[3]

[1]*Institute of Soil and Environmental Sciences, University of Agriculture Faisalabad, Punjab Pakistan*

[2]*Doctoral School of Environmental Science, Hungarian University of Agricultural and Life Sciences, Gödöllő, Hungary*

[3]*Department of Botany, University of Central Punjab, Bahawalpur Campus, Punjab Group of Colleges, Pakistan*

ABSTRACT

Abiotic stress harms the yearly production of essential food crops, reducing its yield up to 70%. Phytohormones are chemical messengers that help plants recognize and react to unfavorable environmental circumstances. They are responsible for assisting plants to recognize and respond to abiotic stresses under stressful circumstances, phytochrome reprograms their development by working with phytochromes to improve their performance. The jasmonates and their derivatives are an important component of defense machinery and play a vital role in a plant's stress management. Plant development (root growth, gravitropism, trichome creation, embryo development (for example, in maize), fertility, seedling formation, tuber development, leaves, and fruit ripening, as well as leaf senescence, among other things) is stimulated by this hormone. The endogenous and

Jasmonates and Plant Defense: An Eco-Friendly and Sustainable Approach for Stress Tolerance.
Uzma Azeem, & Khalid Rehman Hakeem (Eds.)

exogenous applications of jasmonates are discussed in detail in this chapter to elaborate on its useful function for the plant.

7.1 INTRODUCTION

Plants are the sessile organs that depend on the external environment for their growth and production. However, the outer environment in which the plants live is characterized by various kinds of stresses, whether it is biotic as well as abiotic.[1] Abiotic stress has a notable impact on the molecular, biological, biochemical, and physiological processes, resulting in the loss of economy and yield. The annual yield of staple food crops is reduced by about 70% due to abiotic stress conditions.[2,3] The common abiotic stress factors that have a severe impact on the productivity and growth of plants include temperature stress (heat and cold), water stress (drought), salinity, organic pollutants, ultraviolet (UV) rays, and inorganic pollutants (heavy metal stress). The continuing climatic changes are also a constant threat to crop yield.; therefore, researchers are exploring new ways to recognize them better and activate the protective response against them. The plant's established complex signaling system sustains normal plant growth under stress conditions.[4] Chemical messengers that assist the plants in recognizing the adverse ecological condition and responding to them and help them control growth and development are known as phytohormones. Plants can reprogram their growth with the help of these phytochromes during abiotic stress conditions[5] and plants' growth and development and their response to external stress are dependent on phytochromes. With the progress of research in the plant biology field, it is evident that phytohormones can alleviate the harmful effects caused by abiotic stresses[6,7] as the hormones signaling and biosynthesis have a vital role in plant survival.[8] The plant hormones, for example, salicylic acid (SA),ethylene (ET), Jasmonic acid (JA), and to a lesser extent absicic acid (ABA) are the leading players in the regulation of signaling pathways.[9] The term "jasmonates" includes JA and its derivatives. It stimulates root growth, gravitropism, trichome formation, development of plant embryo, determination of sex (e.g., maize), fertility, seedling formation, tuber development, the emergence of leaves, fruit ripening, and the leaf senescence, etc. in plants.[10] Jasmonates are growing in plants during abiotic stress and pathogen attack as a defense mechanism.[8] JA was first isolated from the cultures of fungi named *Lasiodiplodiatheobromae*. Methyl jasmonate (MeJA) originated from the essential oil of species *Rosmarinus officinalis L.* and *Jasminum grandiforum.*[1] JA regulates root morphogenesis and leaf morphogenesis in the soybean plant.[11] In almost all the higher plants, jasmonates are present in low concentration in mature leaves and roots, more in flowers and other reproductive parts. There are many derivatives of JA, especially the hydroxylated forms such as curcubic acid, tuberonic acid, and some amino acid conjugates.[12] In potatoes, tuberonic acid (12-hydroxy-JA) and its glucoside) helps in the tuber formation.[13] Furthermore, tuber formation control by the cumulative action of temperature, light, and gibberellic acid (GA)[14] along with jasmonates crosstalk with GA.[15] In recent times, researchers are discovering the roles of these growth regulators (jasmonate) in the mitigation of abiotic stresses. The reports have recently shown that the endogenous and exogenous application of jasmonates has enhanced abiotic stress tolerance in plants. Moreover, the signaling mechanism between these hormones is also helpful in facilitating plants with abiotic stress tolerance.[8] In this chapter, we will see the role of endogenous production and exogenous application of jasmonates in response to the significant abiotic stresses (salinity, water stress, temperature, and pollutants) in plants.

7.2 TYPES OF ABIOTIC STRESS IN PLANTS

7.2.1 SALINITY

Salt stress is known for the adverse effect of sodium ion (Na^+) and chloride ion (Cl^-) on plant growth and productivity.[16] Soil salinity has become a universal issue affecting almost 20% of irrigated lands, thus reducing the overall crop yield.[17] Salt stress existed before even mankind and agriculture, and this problem is going to be worse with irrigation and agricultural practices.[18] It has become the most limiting known factor for plant growth and yield.[19] Salinity is affecting plant health *via* cytotoxicity (excess uptake of Na^+ and Cl^- ions), water imbalance, nutritional stress, and oxidative stress (generation of reactive oxygen species (ROS).[20] There are mainly two salinity types: primary salinity (natural) and secondary salinity (human-induced). Primary salinity results from the accumulation of salts through various natural processes happening within the soil or groundwater with gradual deposition. Mainly, it occurs due to the two natural processes, that is, weathering of parent material and sediment of oceanic salts. The secondary salinity originated from the negative effect of human activities on the hydrologic balance of soil between the water applied to the soil by irrigation or rainfall and the water used by the crop plants.[21] Plants respond to the salinity in two steps. In the first step, an ion-dependent reduction of growth occurs, which takes a few minutes to a few days resulting in the stomatal closure and inhibition of cell expansion, especially into the shoots. In the second phase, the concentration of ions builds up to the cytotoxic level, mainly in the older leaves, which results in premature senescence, a slow rate of metabolic processes, and even cell death.[22] Salinity affects different functions of plants such as seed germination, photosynthesis, plant growth and productivity, oxidative stress, nutrient imbalance, and plant–water relationship.[19]

7.2.2 WATER STRESS

Almost 80 to 95% of non-woody plant parts are water, which plays a critical role in plant growth, productivity, metabolism, etc.[23] Water is essential for plant survival because it helps in the transport of nutrients within the plant. Therefore, water deficiency leads to water stress or drought in the plants resulting in reduced plant growth.[24] Globally water stress is considered the most prevalent kind of stress in several areas, mainly the arid and semiarid regions.[25] The stress resulting due to water deficiency is called water stress. Drought is another form of water stress. Water stress also occurs when the soil has decreased quantity of water due to some environmental conditions causing loss of water, such as evaporation and transpiration.[26] Few other reasons for water stress include salinity, light intensity, low or high temperature, etc.[27] The elevated salt stress decreases the soil water potential due to salts' smaller osmotic potential, making it difficult for the plant roots to absorb water. The high temperature also leads to water stress through evaporation and transpiration. Moreover, the low temperature can also lead to water stress as the freezing temperature decreases ice crystals, which reduces the water potential and leads to intracellular water efflux.[26] Contrary to this, there are many conditions in which plants did not uptake water, but enough water is present in the soil. This kind of water stress is known as physiological drought or pseudo-drought.[22–27] So, various water stress causes generally lead to cellular water efflux, plasmolysis, or even cell death. Water stress can inhibit photosynthesis in plants by affecting the thylakoid membranes. It also causes an increase in the toxic ion concentration in the plants. Therefore, water stress is complex environmental stress that reduces plant growth, decreasing overall plant output.[26] It is multidimensional stress that affects plants' morphological, physiological, biochemical, ecological, and molecular characteristics. The plant response to water stress depends upon the severity and duration of the water deficit, age of the plant, plant species, and the developmental stage of the plant.[25]

7.2.3 TEMPERATURE

The temperature of the earth's surface is increasing day by day mainly due to the increasing levels of carbon dioxide and other greenhouse gases in the atmosphere. Therefore, the temperature is considered a detrimental abiotic stress factor. Climate change and global warming are increasing temperatures around the world. This increase in temperature decreases agricultural productivity by causing changes in plants' physiological, morphological, anatomical, and biochemical processes, ultimately reducing crop growth and yield.[29] This increase in temperature is becoming a major cause of temperature stress in plants. It depends upon the duration, intensity, and quality of light faced by plants. Hence, heat stress is the most critical limiting factor in agricultural production.[30] Due to the changing geographical conditions, numerous species and crops will be affected.[30,31] Heat stress affects the plant's membrane fluidity, and reproductive and vegetative tissues and inhibits protein folding.[33] Temperature stress also changes genes' expression involved in the formation of regulatory proteins, osmoprotectants, transporters, and detoxifying plants.[34] Due to high temperatures, plants are more susceptible to pathogens. When the temperature increases, the infection capacity of viruses also increases, such as tomato wilt virus, tobacco mosaic virus, and some other viral diseases in pepper (*Capsicum annuum*) and tobacco (*Nicotiana tabacum*), etc.[35] With the increase in the nighttime temperature, the sensitivity of wheat genotypes increased toward the *Cochliobolus sativus*, causing spot blotch.[36] Contrary to heat stress, the decrease in temperature causes chilling stress to the plants, thus affecting the plant physiology. Chilling stress results in the death of plants by ROS generation, inhibition of photosynthesis and several enzymatic activities, nucleic acid stabilization, and rigidity of biological membrane.[37] Decrease in induced temperature is a phenomenon known as vernalization (flowering) in the plants.[38] Due to the low temperature, the composition of the membrane alters, accumulation of solutes occurs, and the formation of dehydrin-like proteins is enhanced.[39]

Plants exhibit short-term and long-term responses under temperature stress, either its cold stress or heat stress. Phenological or morphological adaptations are long-term responses, and changes in the leaf orientation and membrane composition are short-term responses.[40]

7.2.4 INORGANIC POLLUTANTS

Ali and Khan[41] reported that the most significant environmental disasters and large-scale pollution are primarily inorganic pollutions. This inorganic pollution releases harmful inorganic chemicals into the environment that are destructive to the flora and fauna of the ecosystem. Inorganic pollutants come from the inorganic by-products that arise from heat, light, radiant energy or noise, etc. They release into the environment by mining or smelting industrial waste products, transporting fuel and smoke, etc. Inorganic pollutants are affecting the atmosphere at both intracellular and extracellular levels. Among these inorganic pollutants, heavy metals play the leading role and have a disastrous effect on the environment. The most common toxic heavy metals include mercury (Hg), copper (Cu), zinc (Zn), cadmium (Cd), nitrate (NO_3^-), and arsenic (As). Most of these metal pollutants are potent as they cannot degrade quickly. They enter the soil system via industrial effluents, household wastewater, unauthorized dumping of waste material, landfill seepage, poor agricultural management practices, wet deposition, mining operations, burning of fossil fuels, emission of industrial smoke, and inorganic fertilizers. Improper disposal of waste and its decomposition in the environment causes inorganic pollution (HMs). These toxic metals have a high binding affinity for the soil constituents compared to the alkali metals and their salts. However, their mobility is very low in the soil, but they can disrupt the biochemical processes and the physiological activities of organisms in the soil. Even if some metals are crucial for plants' health, their higher concentration can be phytotoxic. But other metals like Hg, Cd, chromium (Cr), lead (Pb) etc. are potentially harmful to plants beyond a certain threshold level as they do not play any good role in plant physiology.[42] Production of ROS, reactive nitrogen species (RNS), replacement of transcription factors and enzymes, cellular redox imbalance, deoxyribonucleic acid (DNA) damage, ionic transport imbalance, and protein oxidation are phytotoxic indicators.[43] Inhibited growth with or without chlorotic patches are signs of Cd toxicity in plants. Cr toxicity affects plants with necrotic lesions, chlorosis, tiny purplish, or brownish-red leaves, limiting growth and photosynthesis. High chromium concentrations stifle root development significantly. Inhibited growth with or without chlorotic patches are signs of cadmium toxicity in plants. Cu had the most considerable detrimental effect on seed germination and early growth of crop seedlings, followed by Pb and Zn.[44] Pb positioning also disrupts Ca metabolism, activates different enzymes, reduces carbon dioxide (CO_2) absorption, and suppresses photosynthesis in plants.[42] In white mustard (*Sinapis alba* L.), the root development and photosynthetic pigment production were affected by heavy metals, and the order of toxicity follows as Zn<Cd <Se<Cu <Pb.[45] An experiment was done in acidic sandy loam soil on spinach (*Spinacia oleracea L.*), the order of heavy metal toxicity development was Zn<Cu<Cr< Ni< Cd with soil metal levels corresponding to 80% relative yields of Zn (852 mg kg^{-1}), Cu (409 mg kg^{-1}), Cr (176.4 mg kg^{-1}), Cd (17.8 mg kg^{-1}), and Ni (153 mg kg^{-1}).[46] There are many ways in which these metal contaminants affect our environment, for example, (i) acid rain (resulting from the emission of nitrous oxide and sulfur during energy emission in the environment), (ii) contamination of water with mercury and methyl-mercury,[47] (iii) emission of alkyl lead from gasoline impacting human health,[45,46] (iv) antifouling paints (made with inorganic compounds), for example, tributyltin, treatment of wood with Cu and As, etc.,[50] (v) the addition of inorganic metal contaminants into the environment by the mining operations (such as As, Hg, and Pb) and coal burning (selenium (Se), Cd, Hg),[48,49] and (vi) the release of nanoparticles in the environment, for example, Hg, Cd, and Se, etc.[53] However, inorganic pollution has a massive legacy concerning its historical release and long-term fate in our environment.[54] The chemical nature (oxidation state, reactivity, and speciation) of these inorganic compounds and the chemical form of pollutants (colloidal, dissolved, or particulate nature) are the main determinants of their fate after their accumulation in the organism. A detailed study of chemistry and inorganic pollutants (metals, metalloids, and nutrients) is fundamental to understanding their fate in the environment and human health.[55]

7.2.5 ORGANIC POLLUTANTS

In this modern era, it is challenging to think of a life without chemicals due to urbanization and industrialization. Among the various anthropogenic sources of pollution, organic pollutants have the most damaging effect on our environment.[56] These are the carbon-based organic chemicals occurring either in a vapor state or adsorbed on the surface of other atmospheric particles. The organic pollutants show less solubility in water but are highly soluble in lipids; therefore, they mostly accumulate in the fatty tissues. So, these compounds are persistent. That is why they are commonly known as persistent organic pollutants (POPs). They can disperse longer distances in our atmosphere, causing adverse effects on human health and the environment.[57] They are deposited in the fatty tissues in the human body, posing severe health risks to human health and the environment.[58] Generally, these POPs are classified further as intentional POPs or unintentional POPs. In the intentional POPs, the compounds formed by chemical reactions known as the organochlorines (OCs) contain chlorine atoms. These compounds are highly neurotoxic as well as lipophilic. These compounds have two categories, that is, industrial chemicals and organochlorine pesticides (OCPs).[59] OCPs kill insects and pests like fungicides, insecticides, and herbicides, for example, DDT, Dieldrin, chlordane, endrin, aldrin, heptachlor, hexachlorobenzene etc.[60] They also include some synthetic classes of compounds like rodenticides, acaricides, and molluscicides. However, industrial chemicals contain polychlorinated biphenyls (PCBs), and they were discovered as pollutants in 1966 for the first time. Later on, they investigated fish tissues, birds, water and sediments, etc.[61] The organic pollutants include polyaromatic hydrocarbons (PAHs), PCBs and OCPs, dioxins, furans, and dibenzofurans.[62] All these compounds are photolytic, and they resist chemical and biological degradation. Plants absorb most organic pollutants either from soil or air. Organic contaminants such as PCBs, Polychlorinated dibenzo-p-dioxins (PCDDs), and polychlorinated dibenzofurans (PCDFs) can be absorbed by plants from the air through leaves or from the soil *via* roots. Similarly, herbicides can also be taken up by plants through soil or air. Plants can only take up the pollutants like antibiotics and BPA (Bisphenol A) via roots.[63] Organic pollutants may be hazardous to plant cells causing damage to their ultrastructure, biosynthesis, membrane stability, and DNA. Organic contaminants (i.e., diphenyl ether and bipyridylium) frequently break cell membranes, causing damage to plant tissues.[64] The assembly of α- and β-tubulins and microtubules (MT) associated proteins generate microtubes that can also be harmed by organic contaminants.[65] Likewise, the fungicides, that is, Griseofulvin (GF) and Tiabendazole (TBZ) at a concentration of 100 g mL^{-1}, cause microtubular damage in the meristematic root cells of *Allium cepa.* Moreover, GF can cause abnormality in the mitotic spindles, pre-prophasic bands, typical tripolar spindles, and phragmoplasts in a significant percentage of cells. Similarly, the BPA can cause microtubular array disruption in *pisum sativum L.* (pea) root meristematic cells. It also loosens the microtubular structures in root meristematic cells, causing an increase in their size up to 320.14 nm (BPA-treated cells) as compared to the untreated cells (230.70 nm).[66] PCBs, antibiotics, and pesticides impede photosynthesis in plant cells. Malondialdehyde (MDA) is an analytical marker that indicates membrane lipid peroxidation. The more MDA is present, the less stable the membranes become. An increase in MDA concentration has a toxic effect and oxidative stress generated by PCDD/Fs.[67] In some algal species, PCBs alter cell division rates and impair cell division in Porphyrahaitanenesis, with division rates decreasing as PCB concentrations increase.[68] TBZ can also disrupt mitotic cell division.[69]

7.3 BIOSYNTHESIS AND ROLE OF JASMONATES IN PLANTS

Jasmonates are the plant compounds that play their role during stress conditions.[70] Jasmonates consist of a pent acyclic ring structure derived from the oxygenated fatty acids *via* the octadecanoid pathway.[71] Current investigations on jasmonates tell us about their evolution whereas the other oxylipins indicate the conserved biosynthesis of jasmonates.[70] The biosynthesis of jasmonates involves a series of different enzymes. [15–72] The precursor of jasmonic acid formation is an unsaturated fatty acid, that is, α-linolic acid (18:3) in most plants. In few plants, the precursor is a fatty acid with 16 carbon atoms. At the initial level, the process of biosynthesis takes place in the plastid, where the conversion of 18-carbon precursors into 13-HODE is in the presence of the specific enzyme 13-LOX. The 13-HODE convert into (+)-12-OPDA by the catalysis of (allene oxide cyclase) AOC and AOS (allene oxide synthase). The process of biosynthesis starts with a 16-C fatty acid compound. It converts into dinor-OPDA after undergoing a specific set of reactions in the plastid from where the dnOPDA shifts into the peroxisome with the help of ABC transporter (COMATOSE, PXA1, or PED3).[73] The dnOPDA is then reduced into OPR3, activating OPCL1. In the final step, the carbon atoms are removed from the carboxyl chain to produce a 12-carbon compound called jasmonic acid. This last step is named β-oxidation, carried out through different enzymes, for example, ACX, KAT, L3HCD, and 2TECH.[74,75] The jasmonic acid formed during this process shifts to the cytoplasm, where it undergoes certain modifications to produce JA. It gets conjugated with amino acids using JAR1 to create the biologically active form of jasmonates, that is, JA-Ile.[70] JA formation and its metabolic conversion occur in chloroplasts, peroxisomes, and cytosol, respectively.[76] Jasmonates have importance in a wide variety of plant species for their growth and development.[77] Jasmonates also play a very crucial role in the developmental and physiological functions of plants such as organs development, embryo development, floral development, seedling formation, seed germination, plant sex determination, trichome formation, growth inhibition, fruit ripening, tuber formation in potato, coiling of tendrils, and senescence induction, etc.[10] MYC2, MYC3, and MYC4 mostly do the root growth inhibition in plants after their distribution in different layers of the primary root apex.[78] It was reviewed by Huang et al.[77] that this inhibitory effect of root growth can suppress the overexpression of JAZ and NINJA proteins after some mutation or variation in them, for example, JAZ3ΔC, JAZ1Δ3A, JAZ8, JAZ10.4, JAZ13, etc. A recent report stated that the interaction of JAZ and YABBY1 or YABBY3 could promote chlorophyll degradation.[79] Moreover, an increase in carotenoid and chlorophyll content in plants supplemented with jasmonate. There are so many reports on the relationship between jasmonates and photosynthesis.[80] There is a close relationship between the quantum efficiency of photosystem II and the gene expression involved in photosynthesis and the growth of plants due to the presence of jasmonates during photosynthesis.[81] In Arabidopsis, the opening and closing of stomata occur by the interaction of jasmonates (COI1) and the ABA pathway inplants.[82] An enhanced activity of ribulose-1,5-bisphosphate carboxylase-oxygenase (RUBISCO) (limiting photorespiration and CO_2 assimilation) is seen in MeJA-treated jin1-9/myc2 plants.[83] Jasmonates can control the emergence of lateral root formation by the modulation of AtAOC3 and AtAOC4.[84] The JA concentration increased during leaf senescence.[85] The leaf senescence due to jasmonates mostly depends on COI1.[86] Therefore, jasmonates play several vital functions in plants.

7.4 JASMONATES (ENDOGENOUS AND EXOGENOUSLY APPLIED) IN MITIGATION OF ABIOTIC STRESS

Jasmonates are the proven endogenous regulators of plant growth, productivity, and response to stress tolerance.[87] Abiotic stress leads to

JA-Ile formation in the plant cytosol that further enters into the nucleus. Moreover, JA-Ile is an active ligand in *A. thaliana* confirmed by high-performance liquid chromatography (HPLC) and gas chromatography-mass spectrometry.[88] The low levels of jasmonates have reportedly taken part in the photo-tolerance of mutants with the OXI1 gene encoding the AGC kinase.[89] The higher levels of jasmonates in *Saxifraga longifolia* helped them adapt to the lower and intermediate altitudes.[90] In different parts of plants, the foliar application of jasmonates can also be helpful for abiotic stress tolerance.[91] The jasmonates are capable of significant abiotic stress tolerance,[83,84] such as water stress,[85,86] salinity,[10–86] hot and cold temperatures,[96] pollutants such as heavy metals.[1]

7.4.1 SALINITY

There has been a growing interest in the mitigation of salt stress by using jasmonates. The endogenous jasmonic acid content has reportedly increased in tomato (*Lycopersicon esculentum*),[97] *A. thaliana*,[98] and potato (Solanum tuberosum)[99] after treating them with salt. In an experiment on sweet potato, the concentration of JA increased under the salt stress condition.[100] Moreover, the concentration of jasmonic acid has increased persistently in salt-stressed plants than in salt-tolerant ones.[99] In the salt-stressed seedlings of soybean, the foliar application of jasmonic acid has mitigated salt toxicity.[101] In another experiment, foliar spray of JAs has enhanced the physiological condition of salt-stressed safflower by improving the relative water content, chlorophyll content, grain yield, plant biomass, and the maximum quantum yield of photosystem II (Fy/Fm).[102] The foliar spray of JA (2 mM) has improved the salt tolerance in *Triticum aestivum* due to some antioxidant enzymes (i.e., superoxide dismutase (SOD), glutathione (GSH), catalase (CAT), and chlorophyll (Chl).[103] The exogenous form of jasmonic acid on salt-stressed plants has increased the potassium content and antioxidant enzyme activity while reducing lipid peroxidation.[104,105] The exogenous application of MeJA has increased the activity of APX and SOD in the black locust tree (*Robiniapseudoacacia*), thereby enhancing its salt tolerance.[106] It was similar to the finding of Faghih et al.,[104] who investigated that MeJA has enhanced the performance of SOD, Ascorbate peroxidase (APX), and POD. After these findings, it is evident that jasmonates can mitigate salt stress by improving antioxidant enzymes and endogenous hormones.[107] Moreover, the application of jasmonic acid on soybean seedlings has decreased its protein percentage by inhabiting nitrogen absorption in roots and increasing the sulfur content in the plant leaves.[108] The treatment of plants with jasmonic acid (JA) has improved the content of aromatic amino acids (tyrosine and phenylalanine) and sulfur-containing amino acids (i.e., methionine) in the plant seeds, after stimulation of the shikimate pathway (a pathway in plants).[109] In soybean seeds, the JA application has improved linolenic and linoleic acid production and the percentage of total oils by increasing lipoxygenase activity. Thus, it has improved the content of the unsaturated fat in soybean.[110] There are few reports on the molecular insights of the JA's role in salt tolerance in the literature. In *Hordeum vulgare*, the application of JA (12 μM) has regulated various genes of enzymes (i.e., apoplastic invertase, Rubisco activase, and arginine decarboxylase) which are involved in salinity tolerance.[111] A gene isolated from *T. aestivum* named Allene oxide cyclase has enhanced the salinity tolerance in *Triticum aestivum* and *Arabidopsis thaliana* through JA signaling.[112] The accumulation of jasmonic acid has also been reported in the roots of a recent cultivar *Solanum lycopersicum* mutant res (restored cell structure).[113] The foliar application of jasmonic acid has stimulated the activity of H+-ATPase in root tonoplast and decreased the uptake of Na salt in soybean plants.[114] Priming of barley with jasmonic acid has enhanced the salt stress tolerance in the plant.[111] Pretreatment of soybean seeds with jasmonic acid has improved the salt stress tolerance in the plant by increasing the accumulation of osmoprotectants.[115] The priming of Hassawi okra seeds with JA (50 μm) has enhanced the salt stress tolerance in the plants.[116] It was reported by Harpreet et al.[117] that priming of rapeseed with JA has improved the salinity tolerance by more accumulation of sugars. Exogenous application of jasmonates (JA and MJ) improves the activity of antioxidants in plants under water stress.[118] In tomatoes, the exogenous application of NO and JA ultimately controlled the antioxidant metabolism, osmolyte synthesis, and metabolite accumulation in plants under salt stress.[119] The priming of maize seedlings with JA has alleviated the toxicity caused by excess accumulation of Na_2CO_3 in photosynthesis and the growth of plants.[120]

7.4.2 WATER STRESS

In a few studies, the role of JA in abiotic stress such as water stress has been compared to biotic stresses.[91] Some of the previous studies have shown that the concentration of jasmonic acid increased in the plant leaves and roots under water stress.[121] Buchner et al.[122] investigated that the heat acclimatization capacity of alpine plants caused changes in the JA content, which is an indicator of drought stress. Under water-deficit conditions, jasmonates regulate the plant response to water stress.[123] JA can also control the potential of water in plant cells, helping them fight against the drought stress condition.[124] In this mechanism, JA upregulates the genes that play a crucial role in synthesizing secondary metabolites, forming the cell walls, and encoding solutes and proteins, that is, proline, thereby helping the plant adapt against water stress.[125] In the leaves of Agropyroncristatum under water stress, jasmonic acid has regulated the metabolism of ascorbate and glutathione.[126] Jasmonates involve the closing of stomata to reduce water loss due to transpiration.[127] Another plant hormone, that is, ABA, increases during the water stress as a stress signal. This plant hormone also interrelates with other hormones of the plant, such as NO and JA, to excite stomatal closure. The production of ABA hormone and JA mediates the stomatal closure by promoting the extracellular ion influx of Ca^{2+} and H_2O_2 or NO signaling.[128] In an experiment, the treatment of *A. thaliana* with JA-precursor 12-OPDA (2 μM) has reduced the stomatal aperture in response to the water stress. The water stress has stopped the conversion of 12-OPDA into JA. The 12-OPDA acts alone or in the conjugation of plant hormone ABA to arouse the stomatal closure in response to drought stress.[129] Drought-stressed *Allium sativum L.* received MJ treatment, which resulted in the increased growth of root, shoot, and bulb in plants.[130] In comparison, the combination of MJ (methyl jasmonate) and IBA (indole 3-butyric acid) acts as a plant growth promoter. Hence, it proved that MJ has successfully enhanced the growth and yield of plants, whereas the IBA alone has proved ineffective for growth.[87] JA also enhances the production of growth regulators.[91] Jasmonic acid signaling also synergistically affects other growth regulators like polyamine.[91] Many studies have reported that JA signaling alleviates water stress.[131] Water stress causes the overexpression of AtJMT (jasmonic acid carboxyl methyltransferase gene). It is a clear indication of JA production.[132] These findings revealed that JA signaling and synthesis are essential mechanisms for plants to respond to drought stress. There was a rapid increase in the endogenous content of JA in *A. thaliana*[133] and citrus (*Citrus paradisi Poncirus trifoliata)*[134] during the drought stress conditions. The adverse effects of drought stress on plants will reduce by exogenous application of plant growth regulators such as JAs. Several previous reports of researchers have shown that the exogenous application of JA can enhance plant production under water stress by stimulating the antioxidant enzyme activities. Moreover, the exogenous JA application has improved the antioxidant enzyme content and reduced the oxidative stress in the maize crop under water stress conditions.[125] Jasmonic acid has enhanced osmotic changes in plant cells through the deposition of other Osmo-regulators such as soluble carbohydrates and proteins.[125] In *Agropyroncristatum*, the amount of JA, the transcript level, and activities of GR, DHAR, APX, total ascorbate, total glutathione, monodehydro ascorbate reductase, and y-glutamylcysteine synthetase were increased when the water scarcity was induced by PEG-6000 (polyethylene glycol).[126] Under water stress conditions, the foliar application of JA has inhibited ROS generation and enhanced the activity of certain enzymes (i.e., ascorbate peroxidase, glutathione peroxidase, and catalase) in brassica species.[135] The JA application affects the enzymatic activity *via* translation, transcription, and post-transcriptional modifications in plants.[91] Under deficiency and normal conditions, foliar sprays of JA have decreased oxidative stress and many other plant stresses, whereas leaf expansion, chlorophyll synthesis, and root development inhibit in some plants. The priming of Arabidopsis plants with JA has increased the drought tolerance in the plants by enhancing the expression of water stress-sensitive genes. However, the epinastic effect of JA reduces the leaf expansion in the plants.[91] MeJA is the best water stress-resistant treatment in broccoli (*Brassica oleracea*) plants,[136] peanut (*Arachis hypogaea*),[137] soybean (Glycine max),[138] and rice (*Oryza sativa*).[139] Ref. [140] discovered that MeJA had increased the hydraulic conductivity of roots in *Arabidopsis thaliana, Phaseolus vulgaris,* and *S. Lycopersicum,* thus allowing the plants to extract increased levels of water from the soil with insufficient moisture. In maize plants (*Zea mays*), exogenous application of MeJA increased the total starch, soluble sugar content, polysaccharides, total proline content, free amino acids, protein concentration, and catalase CAT, POD, and SOD activities.[125] Moreover, MeJA has activated the enzymatic such as SOD, POD, CAT, APX, and GR and nonenzymatic (i.e., proline and soluble sugar) antioxidative systems in cauliflower (*B. oleracea*) to improve the water stress tolerance.[136] As a result, MeJA increases tolerance in plants by organic osmoprotectants and antioxidative enzymes.[103]

7.4.3 TEMPERATURE

The plant's growth and productivity are greatly affected due to a severe environmental perturbation, that is, cold or hot temperature.[131] Sharma and Laxmi[96] reviewed that the role of JAs under temperature stress response was extensively researched by many scientists. Finally, endogenous JA levels were higher under drought or cold stress, but JA levels were lower under heat stress.[87] Plants undergo biochemical, physiological, and molecular changes when the ambient temperature falls below the optimum level.[92] Cold stress or the temperature below optimum level results in plant cell dehydration and ice crystal formation.[131] Many phytohormones in plants tolerate low-temperature stress, that is, ethylene, abscisic acid, and jasmonates.[10] It is heavily reliant on the expression of JA signaling to respond to cold stress. In chilling and freezing temperatures, Arabis alpine (Brassicaceae) exhibited significant changes in levels of JA and other hormones. Wheat plants subjected to cold stress had a greater jasmonic acid concentration than control plants.[145] In addition, the increase in water pressure and temperature stress on Pinus pinaster plants triggered a rise in the JA content in the leaves. The JA and ABA in foliar sprays activate the C-repeat binding factor (CBF) pathway downstream of ABA in response to light quality, increasing cold stress tolerance in *Solanum Lycopersicum*. It comprises oxygen radical scavengers, such as antioxidants and inhibitors of protein synthesis, and cryo-protective chemicals, such as cryo-protectants. Antioxidants and defense chemicals were all shown to be produced in larger quantities by pomegranate (*Punica granatum*), tomato (*Lycopersicon esculentum*), loquat (*Eriobotrya japonica*), cowpea (*Vigna sinensis*), and peach (*Prunus persica*) (such as heat shock proteins and phenolic compounds). Agricultural plants treated with MJ treatment recover frost and freezing damage. In *A. thaliana*, JAs induce CBF/DREB1 cascade, thereby improving the cold stress tolerance.[141] These findings indicate that JAs may protect against cold injury by promoting active protection compounds and the antioxidant system.[131] Just like heat stress, cold stress also affects plant growth and development.[142] The synthesis of heat shock proteins is a crucial mechanism of heat stress tolerance in crop plants. It is a molecular defense mechanism found in cells from various plant species in response to high temperatures.[92] Clark et al.[143] discovered the JA's function signaling in heat stress tolerance in

Arabidopsis.[143] The increased jasmonic acid response to heat stress is a powerful defense mechanism in plants.[29] On the other hand, the foliar spray of JA increases the abscisic acid content of plant cells, which results in the closing of stomata, thus reducing water loss during water stress conditions.[144] The leaf electro leakage assays confirmed that foliar spray of the least quantity of JA maintained cell viability in plants under heat stress.[95] Furthermore, exogenous jasmonic acid application during environmental stress like heat stress adjusts the water capacity of plant cells, maintaining water content. The rate of transpiration of the leaves is increasing as the temperature rises.[145] A recent study found that lipoxygenase contributes in high-temperature stress and exogenous MJ or SA stimuli.[146] Exogenous SA or MJ, on the other hand, promotes the accumulation of endogenous phytohormones in *Gracilariopsislemaneiformis* (red alga).[87] More research is required to understand the mechanisms of jasmonic acid in reducing the harmful effects of heat stress on plants.

7.4.4 INORGANIC POLLUTANTS

The incorporation of JAs in plants by metal/metalloids is widely discussed and documented.[147,148] The critical components of signaling pathways launched in plants after stress are jasmonic acid (JA), methyl ester, and methyl jasmonate.[149] The effect of JAs on metal stress varies depending on the plant species, and the amount of JA used.[147] The treatment of plants with JA has enhanced osmolyte buildup, and carotenoids raised antioxidant enzyme concentrations, protecting the plants from being harmed by excessive metal ions.[80] A Pigeon pea seedling's exposure to JA increased the level of proteins (along with total chlorophyll and carotenoids) and inhibited copper (II) development. Although exposure to cadmium (Cd) may impair the development of *V. faba*, applying JA results in increased levels of proline, glycine betaine, CAT, SOD, CAT, APX, and GR. All of these and H_2O_2, MDA, and Cd counteract the effects of oxidative stress.[161] Exogenous JA also altered the tolerance to heavy metals and aggregate formation. Pre-planting priming of *glycine max L.* also helps to increase the crop's resistance to nickel (Ni). Jasmonate increased photosynthesis and chlorophyll fluorescence in cells exposed to arsenic. *Cajanus cajan (L.) Mill* sp. seedlings may have higher antioxidant capabilities and chlorophyll production improved by JA.[82] Exogenous MeJA increased the actions of

APX, SOD, and CAT in *A. thaliana*,[150] soybean (*G. max*),[80] mangrove (*Kandeliaobovata*) plants,[151] European black nightshade (*Solanum nigrum*),[146] chili pepper (*Capsicum frutescens*),[152] alleviating cadmium-induced damage. In *Cassia tora* and *Brugmansia candida*, foliar spray of MeJA (methyl jasmonate) has greatly increased the plant's resistance to aluminum (Al) toxicity by promoting lignin aggregation in the cell wall, enhancing the peroxidase activity and reducing oxidative stress.[153] MeJA has reduced the oxidative stress in *B. napus* by inducing the activation of genes that code for antioxidants and secondary metabolites.[147] Aftab et al.[151] found that supplementing *Artemisia annua* plants with methyl jasmonate significantly decreased boron toxicity by increasing antioxidant activity. MeJA (1.0 and 10 M) significantly reduced Cd uptake and increased APX activity in *Avicennia marina*.[154] Furthermore, in Cd-treated plants, the expression of type-2 metallothionein gene (*KoMT2*) was raised in the leaves.; however, the exogenous application of MeJA has substantially restored *KoMT2* expression.[155] Therefore, the exogenous application of MeJA has efficiently alleviated heavy metal injury in plants by enhancing the activity of antioxidant enzymes and the number of secondary metabolites.[131] Excess cadmium creates ROS, H_2O_2, and superoxide radicals, which cause oxidative damage in plants.[156] In soybean and *A. thaliana*, however, MJ application decreases this harm by around 30%.[149] MJ increased resistance to copper and cadmium stress in *A. thaliana* plants by accumulating chelating ligands, which create complexes of metal ions and inhibit their availability to the plant. The foliar spray of MJ stimulated antioxidant activity and thus reduced lipid peroxidation.[152] In *Brassica napus*, exogenous MJ application has alleviated the arsenic-induced oxidative stress and enhanced the photosynthesis and growth of plants.[157] Singh and Shah[158] found that applying MJ to cadmium-stressed *O. sativa* resulted in significant improvements in CAT, SOD, and GR activity, as well as increased glutathione pools.

7.4.5 ORGANIC POLLUTANTS

Among the various anthropogenic sources of pollution, organic pollutants have the most damaging effect on our environment.[56] Jasmonates play a critical role in the mitigation of organic pollutants. JA may increase ethylene production, which is beneficial during floods. By altering the proteome profile of soybean plants, jasmonates significantly reduced the detrimental effects of flooding stress and encouraged soybean plant development.[159] In plants, ozone produces ROS, creating lesions and promoting programmed cell death. After ozone treatment, the JA concentration in wild-type Arabidopsis is considerably higher.[160] Furthermore, after exogenous MeJA treatment on tomato (*L. esculentum*) and (*Populus maximowiziix P. trichocarpa*) demonstrated decreased ozone susceptibility.[160] In tomato plants, an elevated ozone level triggered the jasmonic acid (JA) pathway, which considerably increased the rate of volatile compound emission to defend the plants from natural enemies.[161] Treatment with methyl jasmonate may be an effective natural method for safeguarding grapevine berries in the vineyard. In an experiment, Malabar spinach plants (*Basella alba. L*) were exposed to three medications (i.e., gemfibrozil, tolfenamic acid, and mefenamic acid) for fifteen days. The medications boosted JA production (3.1–9.4 times) while reducing MeJA levels by 18.3–38.1%. Because of the variation in jasmonates, the absorption of medications by plants acted as abiotic stress on plants, increasing the defense mechanism in plants.[162] Ma et al.[163] investigated a new epigenetic mechanism for the degradation of the herbicide ATZ in rice crops, mediated by the principal component *(OsCOI1a)* in the jasmonate-signaling pathway. ATZ treatment significantly promoted *OsCOI1a* protein localization in the nucleus. Overexpression of *OsCOI1a* (OE) gave considerable resistance to ATZ toxicity, resulting in better rice crop development and decreased ATZ accumulation (especially in grains). HPLC/Q-TOF-MS/MS analysis demonstrated a rise in ATZ-degraded compounds in the OE plants, suggesting that ATZ catabolism is active. According to bisulfite sequencing and chromatin immune-precipitation experiments, the ATZ treatment has significantly decreased DNA methylation at CpG context and histone *(H3K9me2)* marks in the upstream of *OsCOI1a.* Several biological, genetic, and chemical evidences have shown the causal links between DNA demethylation (hypomethylation), *OsCOI1a* expression, and subsequent ATZ detoxification as well as degradation in rice and the environment. Organic agriculture does not only depend upon synthetic chemical fungicides. The use of bioactive natural compounds as an alternative to chemical fungicides is a viable pest management method. Hordenine is a phenethylamine alkaloid found in barley. Its treatment reduced both grape downy mildew and strawberry anthracnose. It failed to inhibit the jasmonate-dependent defense system, resulting in more significant mycelial growth of phytopathogenic fungus in hordenine-treated plants. The development of new pesticides, such as bioactive natural compound-based insecticides, has been spurred by environmental concerns about pollution. Hordenine did not affect the formation of phytopathogenic fungus mycelium, but it enhanced plant defense by increasing the production of jasmonic acid, which is involved in jasmonate-dependent defense. The development of new pesticides, particularly bioactive natural compound-based insecticides, has been spurred by concerns about pollution. Hordenine has been hypothesized as a new elicitor of plant defense response in organic agriculture for use in controlling disease, namely downy mildew and anthracnose.[189] Signaling pathways of JA in Brassica and the Nicotiana species were launched in multiple studies by JA production caused by UVB treatment, which strengthened the plants' defensive mechanisms.[171–180] The red light/far-red light-mediated photomorphogenesis in A. thaliana and O. sativa was also regulated by JA signaling.[191–182]

7.5 PROSPECTS REGARDING UPREGULATION OF JASMONATES BIOSYNTHESIS AND MODIFICATION

Plant responses to biotic as well as abiotic stresses, and the development of plants, are regulated by jasmonates. The initially generated jasmonic acid converted into several metabolites, including the conjugate with isoleucine once synthesized from lipid constituents. A novel discovery has provided researchers with new information on how jasmonate signaling pathways influence stress responses and development. Breakthroughs identify JAZ proteins and their interactors (such as transcription factors and co-repressors), conjugation of enzyme jasmonate to amino acids, and the crystallization of the jasmonate receptor have occurred in the recent few years. The intricate nature of jasmonate signaling networks in development and stress responses, as well as hormone crosstalk, may now be addressed.[15] In an experiment, the upregulation of JA biosynthetic and the JA responsive genes was suggested in rice in response to *Xanthomonas oryzae* pv. *Oryzae*. It secretes enzymes that degrade cell walls such as ClsA and LipA. It shows that JA-Ile may play a role in innate immunity produced by cell wall disruption. Furthermore, a comparison of *ClsA-and LipA*-induced rice genes found critical rice functions that could be implicated in the development of DAMP-induced innate immunity.[166] In another work, MJ-induced alterations in triterpene saponins in *Panax ginseng C.A. Meyer* (ginseng) hairy roots were investigated and expression profiling of essential response genes. In another work, MJ-induced alterations in triterpene saponins in ginseng (Panax ginseng C.A. Meyer) hairy roots were investigated, and expression profiling of essential response genes. The contents of protopanaxadiol-type saponins increased by 5.5 to 9.7% compared to the control, whereas protopanaxatriol-type saponins increased by 1.85–3.82%. In ginseng hairy root cultures, the concentration of Rg1 ginsenoside after MJ elicitation was negatively influenced.[132] Exogenous methyl jasmonate (MeJA) has been shown to increase anthocyanin accumulation in strawberry fruit. The application of jarin-1 to MeJA-treated fruit is thought to upregulate PA biosynthesis-related genes, resulting in increased PA accumulation in strawberries.[167] Furthermore, in thin-layer explants of tobacco (*Nicotiana tabacum L. cv. Samsun*), the influence of MeJA on de novo shoot formation and polyamine metabolism was studied. According to the histological analysis, the production of meristemoids and shoot domes dropped as MJ concentrations increased although the incidence of cell hypertrophy increased. In the soluble portion of MJ-treated explants, the activities of the putrescine biosynthetic enzymes (ADC, EC4.1.1.19) and (ODC, EC4.1.1.17) increased up to threefold compared to control explants. Northern studies demonstrated that MJ also enhanced the mRNA levels of all three enzymes, which were generally dosage dependent. MJ primarily increased the activity of diamine oxidase (DAO, EC1.4.3.6) in the cell wall fraction. Polyamine metabolism was also found to be upregulated in connection to the morphogenic behavior of MJ-treated explants.[168] In wild-type tomato leaves and hairy roots, methyl jasmonate promotes JRE4 (jasmonate-responsive Ethylene 4) and Steroidal glycoalkaloid (SGA) biosynthetic genes, but not in (jai1), a loss-of-function mutant allele of the tomato COI1 gene. JRE4 overexpression in transgenic hairy roots of wild-type and jai1 mutant tomatoes boosted the expression of SGA biosynthesis genes, showing that JRE4 is the major transcription factor that operates downstream of the jasmonate signaling pathway.[169] Zhang et al.[162] proposed that in response to wounding, a dynamic JA wave (that cooperates with histone methylation) can upregulate a pulse of auxin synthesis and increase DNRR. Thejasmonate-mediated signaling pathway boosted the synthesis of nicotine and related pyridine alkaloids in the roots of tobacco plants. The ERF genes from the *NIC2-locus* are expressed in the root and induced by jasmonate in different ways depending on the member. As a result of gene duplication events, a cluster of highly homologous transcription factor genes with diverse transcriptional and functional functions emerged. Because the NIC2-locus ERFs are close homologs of *ORCA3*, a jasmonate-responsive transcriptional activator of indole alkaloid biosynthesis in *Catharanthus roseus* indicated that the NIC2/ORCA3 ERF subfamily was recruited independently to regulate jasmonate-inducible secondary metabolism in different plant lineages.[170] The role of jasmonates in Fragaria chiloensis fruit was also investigated using an in vitro ripening system that applied MeJA at two levels. On day 5, MeJA enhanced fruit ripening by temporarily increasing the solid-soluble content/titratable acidity ratio, anthocyanin accumulation, and softening. The findings reveal that JAs have a role in anthocyanin accumulation, cell wall alteration, and the manufacture of ethylene and JAs, all of which contribute to the ripening of non-climacteric fruits.[171]

7.6 CONCLUSION

Abiotic stressors significantly affect the molecular, biological, biochemical, and physiological processes, resulting in reduced efficiency and production. According to some estimates, abiotic stress reduces the yearly production of essential food crops by about 70%. Climate stress (both hot and cold), water stress (drought), salt, organic pollutants, ultraviolet radiation, and inorganic pollutants are among the most frequent abiotic stressors that harm the production and development of plants, among other things (heavy metal stress). Various methods have tried to improve the effects of abiotic stressors on plant development and production throughout the last few decades. The synthesis of jasmonates by plants in response to abiotic stress has lately been recognized as a viable alternative. Current research on jasmonates provides information on their evolution, whereas other oxylipins suggest that the production of jasmonates has remained constant. The goal of our study is to increase the function of jasmonates in reducing abiotic stresses.

KEYWORDS

- **jasmonic acid**
- **hormones**
- **endogenous**
- **exogenous**
- **redox oxidative potential**

REFERENCES

1. Dar, T. A.; Uddin, M.; Khan, M. M. A.; Hakeem, K. R.; Jaleel, H. Jasmonates Counter Plant Stress: A Review. *Environ. Exp. Bot.* **2015,** *115*, 49–57.
2. Mantri, N.; Patade, V.; Penna, S.; Ford, R.; Pang, E. Abiotic Stress Responses in Plants: Present and Future. In: *Abiotic Stress Responses in Plants: Metabolism, Productivity and Sustainability*, 2012; pp 1–19.
3. Basu, S.; Roychoudhury, A. Expression Profiling of Abiotic Stress-Inducible Genes in Response to Multiple Stresses in Rice (*Oryza sativa* L.) Varieties with Contrasting Level of Stress Tolerance. *BioMed. Res. Int.* **2014**.
4. Paul, S.; Roychoudhury, A. Transcript Analysis of Abscisic Acid-Inducible Genes in Response to Different Abiotic Disturbances in Two Indica Rice Varieties. *Theoret. Exp. Plant Physiol.* **2019,** *31* (1), 249–272.
5. Roychoudhury, A.; Paul, S.; Basu, S. Cross-Talk Between Abscisic Acid-Dependent and Abscisic Acid-Independent Pathways During Abiotic Stress. *Plant Cell Rep.* **2013,** *32* (7), 985–1006.
6. Masood, A.; Iqbal, N.; Khan, N. A. Role of Ethylene in Alleviation of Cadmium-Induced Photosynthetic Capacity Inhibition by Sulphur in Mustard. *Plant, Cell Environ.* **2012,** *35* (3), 524–533.
7. Khan, M. I. Application of Artificial Neural Network for Dimensional Stability of Repair Materials Under Local Hot Weather. *WIT Trans. Modell. Simulation* **2013,** *55*, 263–274.
8. Santisree, P.; Jalli, L. C. L.; Bhatnagar-Mathur, P.; Sharma, K. K. Emerging Roles of Salicylic Acid and Jasmonates in Plant Abiotic Stress Responses. In: *Protective Chemical Agents in the Amelioration of Plant Abiotic Stress*, 2020; pp 342–373.
9. Kachroo, A.; Kachroo, P. Salicylic Acid-, Jasmonic Acid- and Ethylene-Mediated Regulation of Plant Defense Signaling. *Genet. Eng.* **2007,** *28*, 55–83.
10. Wasternack, C. Action of Jasmonates in Plant Stress Responses and Development: Applied Aspects. *Biotechnol. Adv.* **2014,** *32* (1), 31–39.
11. Xue, R.; Zhang, B. Increased Endogenous Methyl Jasmonate Altered Leaf and Root Development in Transgenic Soybean Plants. *J. Genet. Genom.* **2007,** *34* (4), 339–346.

12. Hamberg, M.; Gardner, H. W. Oxylipin Pathway to Jasmonates: Biochemistry and Biological Significance. *Biochim. Biophys. Acta: Lipids Lipid Metabol.* **1992,** 1165 (1), 1–18.
13. Koda, Y. The Role of Jasmonic Acid and Related Compounds in the Regulation of Plant Development. *Int. Rev. Cytol.* **1992,** *135* (C), 155–199.
14. Lin, T.; Sharma, P.; Gonzalez, D. H.; Viola, I. L.; Hannapel, D. J. The Impact of the Long-Distance Transport of a BEL1-Like Messenger RNA on Development. *Plant Physiol.* **2013,** *161* (2), 760–772.
15. Wasternack, C.; Hause, B. Jasmonates: Biosynthesis, Perception, Signal Transduction and Action in Plant Stress Response, Growth and Development: An Update to the 2007 Review in Annals of Botany. *Ann. Bot.* **2013,** *111* (6), 1021–1058.
16. Munns, R. Genes and Salt Tolerance: Bringing Them Together. *New Phytol.* **2005,** *167* (3), 645–663.
17. Qadir, M. et al. Economics of Salt-Induced Land Degradation and Restoration. *Nat. Resources Forum* **2015,** *38*, 1–27.
18. Zhu, X.; Venosa, A. D.; Suidan, M. T.; Lee, K. Guidelines for the Bioremediation of Marine Shorelines and Freshwater Wetlands. *Environ. Protect.* **2001,** *163*.
19. Parihar, P.; Singh, S.; Singh, R.; Singh, V. P.; Prasad, S. M. Effect of Salinity Stress on Plants and Its Tolerance Strategies: A Review. *Environ. Sci. Pollut. Res.* **2015,** *22* (6), 4056–4075.
20. Isayenkov, S. V. Physiological and Molecular Aspects of Salt Stress in Plants. *Cytol. Genet.* **2012,** *46* (5), 302–318.
21. Geetanjali, M.; Neera, G. Salinity and Its Effects on the Functional Biology of Legumes. *Acta Physiol. Plantarum* **2008,** *30* (5), 595–618.
22. Roy, S. J.; Negrão, S.; Tester, M. Salt Resistant Crop Plants. *Curr. Opin. Biotechnol.* **2014,** *26*, 115–124.
23. Lisar, S. Y. S.; Motafakkerazad, R.; Rahm, I. M. M. Water Stress in Plants: Causes, Effects and Responses. *Water Stress* **2012**.
24. Ashkavand, P. et al. Application of SiO_2 Nanoparticles as Pretreatment Alleviates the Impact of Drought on the Physiological Performance of *Prunus mahaleb* (Rosa ceae). *Boletin de la Sociedad Argentina de Botanica* **2018,** *53* (2), 207–219.
25. Rao, K. V. M.; Raghavendra, A. S.; Reddy, K. J. Physiology and Molecular Biology of Stress Tolerance in Plants. *Physiol. Mol. Biol. Stress Tolerance Plants* **2006,** 1–345.
26. Yadav, S.; Modi, P.; Dave, A.; Vijapura, A.; Patel, D.; Patel, M. Effect of Abiotic Stress on Crops. *Sustain. Crop Product.* **2020**.
27. Salehi-Lisar, S. Y.; Bakhshayeshan-Agdam, H. Drought Stress in Plants: Causes, Consequences, and Tolerance. In: *Drought Stress Tolerance in Plants, Vol 1: Physiology and Biochemistry*, 2016; pp 1–16.
28. Arbona, V.; Manzi, M.; de Ollas, C.; Gómez-Cadenas, A. Metabolomics as a Tool to Investigate Abiotic Stress Tolerance in Plants. *Int. J. Mol. Sci.* **2013,** *14* (3), 4885–4911.
29. Hasanuzzaman, M.; Nahar, K.; Alam, M. M.; Roychowdhury, R.; Fujita, M. Physiological, Biochemical, and Molecular Mechanisms of Heat Stress Tolerance in Plants. *Int. J. Mol. Sci.* **2013,** *14* (5), 9643–9684.
30. Wahid, A.; Gelani, S.; Ashraf, M.; Foolad, M. R. Heat Tolerance in Plants: An Overview. *Environ. Exp. Bot.* **2007,** *61* (3), 199–223.

31. Bita, C. E.; Gerats, T. Plant Tolerance to High Temperature in a Changing Environment: Scientific Fundamentals and Production of Heat Stress-Tolerant Crops. *Front. Plant Sci.* **2013,** *4.*
32. The Core Writing Team IPCC. Climate Change 2014: Synthesis Report. Contribution of Working Groups I, II and III to the Fifth Assessment Report of the Intergovernmental Panel on Climate Change. *IPCC* **2015,** 151.
33. Zinn, K. E.; Tunc-Ozdemir, M.; Harper, J. F. Temperature Stress and Plant Sexual Reproduction: Uncovering the Weakest Links. *J. Exp. Bot.* **2010,** *61* (7), 1959–1968.
34. Krasensky, J.; Jonak, C. Drought, Salt, and Temperature Stress-Induced Metabolic Rearrangements and Regulatory Networks. *J. Exp. Bot.* **2012,** *63* (4), 1593–1608.
35. Király, L.; Hafez, Y. M.; Fodor, J.; Z. Király. Suppression of Tobacco Mosaic Virus-Induced Hypersensitive-Type Necrotization in Tobacco at High Temperature Is Associated with Downregulation of NADPH Oxidase and Superoxide and Stimulation of Dehydroascorbate Reductase. *J. Gen. Virol.* **2008,** *89* (3), 99–808.
36. Sharma, R. C.; Duveiller, E.; G. Ortiz-Ferrara. Progress and Challenge Towards Reducing Wheat Spot Blotch Threat in the Eastern Gangetic Plains of South Asia: Is Climate Change Already Taking Its Toll? *Field Crops Res.* **2007,** *103* (2), 109–118.
37. Ruelland, E.; Vaultier, M. N.; Zachowski, A.; Hurry, V.; Kader, J. C.; M. Delseny. Cold Signalling and Cold Acclimation in Plants. *Adv. Bot. Res.* **2009,** *49*, 35–150.
38. Kim, S. K.; Sohn, E. Y.; Joo, G. J.; Lee, I. J. Influence of Jasmonic Acid on Endogenous Gibberellin and Abscisic Acid in Salt-Stressed Chard Plant. *J. Environ. Biol.* **2009,** *30* (3), 333–338.
39. Korn, M.; T. Gärtner, Erban, A.; Kopka, J.; Selbig, J.; D. K. Hincha. Predicting Arabidopsis Freezing Tolerance and Heterosis in Freezing Tolerance from Metabolite Composition. *Mol. Plant* **2010,** *3* (1), 224–235.
40. Tiwari, S.; Patel, A.; Singh, M.; Prasad, S. M. Regulation of Temperature Stress in Plants. *Plant Life Under Chang. Environ.* **2020,** 25–45.
41. Ali, H.; Khan, E. Environmental Chemistry in the Twenty-First Century. *Environ. Chem. Lett.* **2017,** *15* (2), 329–346.
42. Saha, J. K.; Selladurai, R.; Coumar, M. V.; Dotaniya, M. L.; Kundu, S.; Patra, A. K. *Major Inorganic Pollutants Affecting Soil and Crop Quality*, 2017; pp 75–104.
43. Sytar, O.; Kumar, A.; Latowski, D.; Kuczynska, P.; Strzałka, K.; Prasad, M. N. V. Heavy Metal-Induced Oxidative Damage, Defense Reactions, and Detoxification Mechanisms in Plants. *Acta Physiol. Plantarum* **2013,** *35* (4), 985–999.
44. Mahmood, T.; Islam, K. R.; Muhammad, S. Toxic Effects of Heavy Metals on Early Growth and Tolerance of Cereal Crops. *Pak. J. Bot.* **2007,** *39* (2), 451–462.
45. F. A. Toxicity Comparison of Some Possible Toxic Metals (Cd, Cu, Pb, Se, Zn) on Young Seedlings of *Sinapis alba* L. *Plant Soil Environ.* **2004,** *50*, 33–38.
46. Saha, J. K.; Panwar, N.; Singh, M. V. Risk Assessment of Heavy Metals in Soil of a Susceptible Agro-Ecological System Amended with Municipal Solid Waste Compost. *J. Indian Soc. Soil Sci.* **2013,** *61* (1), 15–22.
47. Akito, M. et al. Reevaluation of Minamata Bay, 25 Years After the Dredging of Mercury-Polluted Sediments. *Marine Pollut. Bull.* **2014,** *89* (1–2), 112–120.
48. Roy, S.; Edwards, M. A. Preventing Another Lead (Pb) in Drinking Water Crisis: Lessons from the Washington D.C. and Flint MI Contamination Events. *Curr. Opin. Environ. Sci. Health* **2019,** *7*, 34–44.

49. Frank, J. J.; Poulakos, A. G.; Tornero-Velez, R.; Xue, J. Systematic Review and Meta-Analyses of Lead (Pb) Concentrations in Environmental Media (Soil, Dust, Water, Food, and Air) Reported in the United States from 1996 to 2016. *Sci. Total Environ.* **2019,** *694*.
50. Van Wezel, A. P.; Van Vlaardingen, P. Environmental Risk Limits for Antifouling Substances. *Aqua. Toxicol.* **2004,** *66* (4), 427–444.
51. Zhao, X.; Gao, B.; Xu, D.; Gao, L.; Yin, S. Heavy Metal Pollution in Sediments of the Largest Reservoir (Three Gorges Reservoir) in China: A Review. *Environ. Sci. Pollut. Res.* **2017,** *24* (26), 20844–20858.
52. Rezaie, B.; Anderson, A. Sustainable Resolutions for Environmental Threat of the Acid Mine Drainage. *Science of the Total Environment* **2020,** *717*.
53. Nick, S. T.; Bolandi, A.; Samuels, T. A.; Obare, S. O. Advances in Understanding the Transformation of Engineered Nanoparticles in the Environment. *Pure Appl. Chem.* **2014,** *86* (7), 1129–1140.
54. Cooke, C. A.; Martínez-Cortizas, A.; Bindler, R.; Sexauer Gustin, M. Environmental Archives of Atmospheric Hg Deposition—A Review. *Sci. Total Environ.* **2020,** *709*.
55. Adams, W. et al. Bioavailability Assessment of Metals in Freshwater Environments: A Historical Review. *Environ. Toxicol. Chem. 39* (1), 48–59.
56. Gaur, N.; Flora, G.; Yadav, M.; Tiwari, A. A Review with Recent Advancements on Bioremediation-Based Abolition of Heavy Metals. *Environ. Sci.: Process. Impacts* **2014,** *16* (2), 180–193.
57. Farrington, J. W.; Takada, H. Persistent Organic Pollutants (POPs), Polycyclic Aromatic Hydrocarbons (PAHs), and Plastics: Examples of the Status, Trend, and Cycling of Organic Chemicals of Environmental Concern in the Ocean. *Oceanography* **2014,** *27* (1), 196–213.
58. Venkatesan, A. K.; Halden, R. U. Wastewater Treatment Plants as Chemical Observatories to Forecast Ecological and Human Health Risks of Manmade Chemicals. *Sci. Rep.* **2014,** *4*.
59. El-Shahawi, M. S.; Hamza, A.; Bashammakh, A. S.; Al-Saggaf, W. T. An Overview on the Accumulation, Distribution, Transformations, Toxicity and Analytical Methods for the Monitoring of Persistent Organic Pollutants. *Talanta* **2010,** *80* (5), 1587–1597.
60. Chopra, A. K.; Sharma, M. K.; Chamoli, S. Bioaccumulation of Organochlorine Pesticides in Aquatic System—An Overview. *Environ. Monitor. Assessment* **2011,** *173* (1–4), 905–916.
61. Selendy, J. M. H. Water and Sanitation-Related Diseases and the Environment: Challenges, Interventions, and Preventive Measures. In: *Water and Sanitation-Related Diseases and the Environment: Challenges, Interventions, and Preventive Measures*, 2011.
62. Érseková, A.; Hilscherová, K.; Klánová, J.; Giesy, J. P.; Novák, J. Effect-Based Assessment of Passive Air Samples from four countries in Eastern Europe. *Environ. Monitor. Assessment* **2014,** *186* (6), 3905–3916.
63. Zhang, C.; Feng, Y.; Liu, Y. w.; Chang, H. q.; Li, Z. j.; Xue, J. m. Uptake and Translocation of Organic Pollutants in Plants: A Review. *J. Integr. Agric.* **2017,** *16* (8), 1659–1668.
64. Ramadass, K.; Megharaj, M.; Venkateswarlu, K.; Naidu, R. Toxicity and Oxidative Stress Induced by Used and Unused Motor Oil on Freshwater Microalga, *Pseudokirchneriella subcapitata*. *Environ. Sci. Pollut. Res.* **2015,** *22* (12), 8890–8901.

65. Nancy, B. A. et al. Induction of Microtubule Damage in Allium Cepa Meristematic Cells by Pharmaceutical Formulations of Thiabendazole and Griseofulvin. *Mutat. Res./Genet. Toxicol. Environ. Mutagen.* **2014,** *772*, 1–5.
66. Adamakis, I. D. S.; Panteris, E.; Cherianidou, A.; E. P. Eleftheriou. Effects of Bisphenol A on the Microtubule Arrays in Root Meristematic Cells of Pisum sativum L. *Mutat. Res./Genet. Toxicol. Environ. Mutagen.* **2013,** *750* (1–2), 111–120.
67. Zhang, B.; Zhang, H.; Jin, J.; Ni, Y.; Chen, J. PCDD/Fs-Induced Oxidative Damage and Antioxidant System Responses in Tobacco Cell Suspension Cultures. *Chemosphere* **2012,** *88* (7), 798–805.
68. Wang, J. et al. The Status and Prospect of Meiosis in Porphyra. *Periodical Ocean Univ. China* **2006,** *36*, 377–380.
69. Andrioli, N. B.; Mudry, M. D. Cytological and Cytogenetic Effects Induced by Thiabendazole on Allium Cepa Root Meristems. *BAG—J. Basic Appl. Genet.* **2011,** *22* (2), 17–23.
70. Wasternack, C.; Strnad, M. Jasmonates: News on Occurrence, Biosynthesis, Metabolism and Action of an Ancient Group of Signaling Compounds. *Int. J. Mol. Sci.* **2018,** *19* (9).
71. Schaller, F.; Schaller, A.; Stintzi, A. Biosynthesis and Metabolism of Jasmonates. *J. Plant Growth Regul.* **2004,** *23* (3), 179–199.
72. Gfeller, A.; Liechti, R.; Farmer, E. E. Arabidopsis Jasmonate Signaling Pathway. *Sci. Signal.* **2010,** *3* (109), cm4–cm4.
73. Footitt, S. et al. The COMATOSE ATP-Binding Cassette Transporter Is Required for Full Fertility in Arabidopsis. *Plant Physiol.* **2007,** *144* (3), pp. 1467–1480.
74. Lyons, R.; Manners, J. M.; Kazan, K. Jasmonate Biosynthesis and Signaling in Monocots: A Comparative Overview. *Plant Cell Rep.* **2013,** *32* (6), 815–827.
75. Li, C. et al. Role of β-Oxidation in Jasmonate Biosynthesis and Systemic Wound Signaling in Tomato. *Plant Cell* **2005,** *17* (3), 971–986.
76. León, J. Role of Plant Peroxisomes in the Production of Jasmonic Acid-Based Signals. *Subcell. Biochem.* **2013,** *69*, 299–313.
77. Huang, H.; Liu, B.; Liu, L.; Song, S. Jasmonate Action in Plant Growth and Development. *J. Exp. Bot.* **2017,** *68* (6), 1349–1359.
78. Gasperini, D. et al. Multilayered Organization of Jasmonate Signalling in the Regulation of Root Growth. *PLoS Genet.* **2015,** *11* (6).
79. Boter, M.; Golz, J. F.; Giménez-Ibañeza, S.; Fernandez-Barbero, G.; Franco-Zorrilla, J. M.; Solano, R. Filamentous Flower Is a Direct Target of JAZ3 and Modulates Responses to Jasmonate. *Plant Cell* **2015,** *27* (11), 3160–3174.
80. Poonam, S.; Kaur, H.; S. Geetika. Effect of Jasmonic Acid on Photosynthetic Pigments and Stress Markers in Cajanus cajan (L.) Millsp. Seedlings Under Copper Stress. *Am. J. Plant Sci.* **2013,** *04* (04), 817–823.
81. Attaran, E. et al. Temporal Dynamics of Growth and Photosynthesis Suppression in Response to Jasmonate Signaling. *Plant Physiol.* **2014,** *165* (3), 1302–1314.
82. Yin, Y.; Adachi, Y.; Nakamura, Y.; Munemasa, S.; Mori, I. C.; Y. Murata. Involvement of OST1 Protein Kinase and PYR/PYL/RCAR Receptors in Methyl Jasmonate-Induced Stomatal Closure in Arabidopsis Guard Cells. *Plant Cell Physiol.* **2016,** *57* (8), 1779–1790.

83. Guo, J.; Pang, Q.; Wang, L.; Yu, P.; Li, N.; Yan, X. Proteomic Identification of MYC2-Dependent Jasmonate-Regulated Proteins in *Arabidopsis thaliana*. *Proteome Sci.* **2012,** *10* (1).
84. Stenzel, I. et al. Allene Oxide Cyclase (AOC) Gene Family Members of *Arabidopsis thaliana*: Tissue-and Organ-Specific Promoter Activities and In Vivo Heteromerization. *J. Exp. Bot.* **2012,** *63* (17), 6125–6138.
85. Breeze, E. et al. High-Resolution Temporal Profiling of Transcripts During Arabidopsis Leaf Senescence Reveals a Distinct Chronology of Processes and Regulation. *Plant Cell* **2011,** *23* (3, 873–894.
86. Qi, T. et al. Regulation of Jasmonate-Induced Leaf Senescence by Antagonism Between bHLH Subgroup IIIe and IIId Factors in Arabidopsis. *Plant Cell* **2015,** *27* (6), 1634–1649.
87. Siddiqi, K. S.; Husen, A. Plant response to Jasmonates: Current Developments and Their Role in Changing Environment. *Bull. Natl. Res. Centre* **2019,** *43* (1).
88. Fonseca, S. et al. (+)-7-iso-Jasmonoyl-L-Isoleucine Is the Endogenous Bioactive Jasmonate. *Nat. Chem. Biol.* **2009,** *5* (5), 344–350.
89. Shumbe, L.; Chevalier, A.; Legeret, B.; Taconnat, L.; Monnet, F.; Havaux, M. Singlet Oxygen-Induced Cell Death in Arabidopsis Under High-Light Stress Is Controlled by OXI1 Kinase. *Plant Physiol.* **2016,** *170* (3), 1757–1771.
90. Munné-Bosch, S. et al. Adaptation of the Long-Lived Monocarpic Perennial Saxifraga longifolia to High Altitude. *Plant Physiol.* **2016,** *172* (2), 765–775.
91. Farhangi-Abriz, S.; Ghassemi-Golezani, K. Jasmonates: Mechanisms and Functions in Abiotic Stress Tolerance of Plants. *Biocatal. Agric. Biotechnol.* **2019,** *20.*
92. Per, T. S.; Khan, M. I. R.; Anjum, N. A.; Masood, A.; Hussain, S. J.; Khan, N. A. Jasmonates in Plants Under Abiotic Stresses: Crosstalk with Other Phytohormones Matters. *Environ. Exp. Bot.* **2018,** *145*, 104–120.
93. Wasternack, C.; Strnad, M. Jasmonate Signaling in Plant Stress Responses and Development—Active and Inactive Compounds. *New Biotechnol.* **2016,** *33* (5), 604–613.
94. Ahmad, P. et al. Jasmonates: Multifunctional Roles in Stress Tolerance. *Front. Plant Sci.* **2016,** *7.*
95. Kazan, K. Diverse Roles of Jasmonates and Ethylene in Abiotic Stress Tolerance. *Trends Plant Sci.* **2015,** *20* (4), 219–229.
96. Sharma, M.; Laxmi, A. Jasmonates: Emerging Players in Controlling Temperature Stress Tolerance. *Front. Plant Sci.* **2016,** *6.*
97. Pedranzani, H. et al. Salt Tolerant Tomato Plants Show Increased Levels of Jasmonic Acid. *Plant Growth Regul.* **2003,** *41* (2), 149–158.
98. Ellouzi, H.; K. Ben Hamed, Cela, J.; M. Müller, Abdelly, C.; Munné-Bosch, S. Increased Sensitivity to Salt Stress in Tocopherol-Deficient Arabidopsis Mutants Growing in a Hydroponic System. *Plant Signal. Behav.* **2013,** *8* (2).
99. De Domenico, S. et al. Oxylipin Dynamics in *Medicago truncatula* in Response to Salt and Wounding Stresses. *Physiol. Plantar.* **2019,** *165* (2), 198–208.
100. Zhang, H. et al. Transcript Profile Analysis Reveals Important Roles of Jasmonic Acid Signalling Pathway in the Response of Sweet Potato to Salt Stress. *Sci. Rep.* **2017,** *7.*

101. Yoon, J. Y.; Hamayun, M.; S.-Lee, K.; Lee, I.-J. Methyl Jasmonate Alleviated Salinity Stress in Soybean. *J. Crop Sci. Biotechnol.*, *12* (2), 63–68.
102. Ghassemi-Golezani, K.; Hosseinzadeh-Mahootchi, A. Improving Physiological Performance of Safflower Under Salt Stress by Application of Salicylic Acid and Jasmonic Acid. *WALIA J.* **2015,** *31* (S1), 104–109.
103. Qiu, Z. B.; Guo, J. L.; Zhu, A. J.; Zhang, L.; Zhang, M. M. Exogenous Jasmonic Acid Can Enhance Tolerance of Wheat Seedlings to Salt Stress. *Ecotoxicol. Environ. Saf.* **2014,** *104* (1), 202–208.
104. Faghih, S.; Ghobadi, C.; Zarei, A. Response of Strawberry Plant cv. 'Camarosa' to Salicylic Acid and Methyl Jasmonate Application Under Salt Stress Condition. *J. Plant Growth Regul.* **2017,** *36* (3), 651–659, Feb.
105. Farhangi-Abriz, S.; Ghassemi-Golezani, K. How Can Salicylic Acid and Jasmonic Acid Mitigate Salt Toxicity in Soybean Plants? *Ecotoxicol. Environ. Saf.* **2018,** *147*, 1010–1016.
106. Jiang, M.; Xu, F.; Peng, M.; Huang, F.; Meng, F. Methyl Jasmonate Regulated Diploid and Tetraploid Black Locust (*Robinia pseudoacacia* L.) Tolerance to Salt Stress. *Acta Physiol. Plantarum* **2016,** *38* (4).
107. Ali, H.; Khan, E. Environmental Chemistry in the Twenty-First Century. *Environ. Chem. Lett.* **2017,** *15* (2), 329–346, Jun.
108. Farhangi-Abriz, S.; Ghassemi-Golezani, K. Improving Amino Acid Composition of Soybean Under Salt Stress by Salicylic Acid and Jasmonic Acid. *J. Appl. Bot. Food Qual.* **2016,** *89*, 243–248.
109. Vered, T.; Gad, G. New Insights into the Shikimate and Aromatic Amino Acids Biosynthesis Pathways in Plants. *Mol. Plant* **2010,** *3* (6), 956–972.
110. Ghassemi-Golezani, K.; Farhangi-Abriz, S. Changes in Oil Accumulation and Fatty Acid Composition of Soybean Seeds under Salt Stress in Response to Salicylic Acid and Jasmonic Acid. *Russian J. Plant Physiol.* **2018,** *65* (2), 229–236.
111. Walia, H.; Wilson, C.; Condamine, P.; Liu, X.; Ismail, A. M.; Close, T. J. Large-Scale Expression Profiling and Physiological Characterization of Jasmonic Acid-Mediated Adaptation of Barley to Salinity Stress. *Plant Cell Environ.* **2007,** *30* (4), 410–421.
112. Zhao, Y. et al. A Wheat Allene Oxide Cyclase Gene Enhances Salinity Tolerance via Jasmonate Signaling. *Plant Physiol.* **2014,** *164* (2), 1068–1076.
113. Garcia-Abellan, J. O. et al. The Tomato Res Mutant Which Accumulates JA in Roots in Non-Stressed Conditions Restores Cell Structure Alterations Under Salinity. *Physiol. Plant.* **2015,** *155* (3), 296–314.
114. Ghassemi-Golezani, K.; Farhangi-Abriz, S. Foliar Sprays of Salicylic Acid and Jasmonic Acid Stimulate H+-ATPase Activity of Tonoplast, Nutrient Uptake and Salt Tolerance of Soybean. *Ecotoxicol. Environ. Saf.* **2018,** *166*, 18–25.
115. Sheteawi, S. Improving Growth and Yield of Salt-Stressed Soybean by Exogenous Application of Jasmonic Acid and Ascobin. *Int. J. Agric. Biol.* **2007,** *9* (3), 473–478.
116. Azooz, M. M.; Metwally, A.; Abou-Elhamd, M. F. Jasmonate-Induced Tolerance of *Hassawi okra* Seedlings to Salinity in Brackish Water. *Acta Physiol. Plantarum* **2015,** *37* (4), 1–13.
117. Harpreet, K.; Poonam, S. et al. Sugar Accumulation and Its Regulation by Jasmonic Acid in *Brassica napus* L. Under Salt Stress. *J. Stress Physiol. Biochem.* **2013,** *9* (4), 53–64.

118. Nafie, E.; Hathout, T.; Al Mokadem, A. S. Jasmonic Acid Elicits Oxidative Defense and Detoxification Systems in *Cucumis melo* L. Cells. *Braz. J. Plant Physiol.* **2011,** *23* (2), 161–174.
119. Ahmad, P.; Ahanger, M. A.; Alyemeni, M. N.; Wijaya, L.; Alam, P.; Ashraf, M. Mitigation of Sodium Chloride Toxicity in Solanum Lycopersicum l. By Supplementation of Jasmonic Acid and Nitric Oxide. *J. Plant Interact.* **2018,** *13* (1), 64–72.
120. Mir, M. A.; John, R.; Alyemeni, M. N.; Alam, P.; Ahmad, P. Jasmonic Acid Ameliorates Alkaline Stress by Improving Growth Performance, Ascorbate Glutathione Cycle and Glyoxylase System in Maize Seedlings. *Sci. Rep.* **2018,** *8* (1).
121. Kiribuchi, K. et al. Involvement of the Basic Helix-Loop-Helix Transcription Factor RERJ1 in Wounding and Drought Stress Responses in Rice Plants. *Biosci. Biotechnol. Biochem.* **2005,** *69* (5), 1042–1044.
122. Buchner, O. et al. Drought Affects the Heat-Hardening Capacity of Alpine Plants as Indicated by Changes in Xanthophyll Cycle Pigments, Singlet Oxygen Scavenging, α-tocopherol and Plant Hormones. *Environ. Exp. Bot.* **2017,** *133*, 159–175.
123. Liu, C. H.; Shi, X.; Zhang, W. L.; Lou, S. Q. Nitric Oxide Participates in the Signal Transduction of Stomatal Closure Induced by Jasmonic Acid in *Vicia faba* L. *Chin. Sci. Bull.* **2005,** *50*, 453–458.
124. Shan, C.; Zhou, Y.; Liu, M. Nitric Oxide Participates in the Regulation of the Ascorbate-Glutathione Cycle by Exogenous Jasmonic Acid in the Leaves of Wheat Seedlings Under Drought Stress. *Protoplasma* **2015,** *252* (5), 1397–1405.
125. Abdelgawad, Z. A.; Khalafaallah, A. A.; Abdallah, M. M. Impact of Methyl Jasmonate on Antioxidant Activity and Some Biochemical Aspects of Maize Plant Grown Under Water Stress Condition. *Agric. Sci.* **2014,** *05* (12), 1077–1088.
126. Shan, C.; Liang, Z. Jasmonic Acid Regulates Ascorbate and Glutathione Metabolism in Agropyron Cristatum Leaves Under Water Stress. *Plant Sci.* **2010,** *178* (2), 130–139.
127. Daszkowska-Golec, A.; Szarejko, I. Open or Close the Gate—Stomata Action Under the Control of phytohormones in Drought Stress Conditions. *Front. Plant Sci.* **2013,** *4*.
128. NA. A. AHarrison, M.; NA, Khan, Nazar, R.; Iqbal, N. Cross-Talk Between Phytohormone Signaling Pathways Under Both Optimal and Stressful Environmental Conditions. In: *Phytohormones and Abiotic Stress Tolerance in Plants*, 2012; pp 49–76.
129. Savchenko, T. et al. Functional Convergence of Oxylipin and Abscisic Acid Pathways Controls Stomatal Closure in Response to Drought. *Plant Physiol.* **2014,** *164* (3), 1151–1160.
130. Bideshki, A.; Arvin, M. J. Interactive Effects of Methyl Jasmonate (MJ) and Indole-3-Butyric Acid (IBA) on Growth and Biochemical Parameters, Bulb and Allicin Yield of Garlix (Allium sativum L.) Under Drought Stress in Iran. *Int. J. Agric.: Res. Rev.* **2013,** *3* (2), 349–360.
131. Ali, M. S.; Baek, K. H. Jasmonic Acid Signaling Pathway in Response to Abiotic Stresses in Plants. *Int. J. Mol. Sci.* **2020,** *21* (2).
132. Kim, D. H.; Doyle, M. R.; Sung, S.; Amasino, R. M. Vernalization: Winter and the Timing of Flowering in Plants. *Annu. Rev. Cell Dev. Biol.* **2009,** *25*, 277–299.
133. Balbi, V.; Devoto, A. Jasmonate Signalling Network in *Arabidopsis thaliana*: Crucial Regulatory Nodes and New Physiological Scenarios. *New Phytol.* **2008,** *177* (2), 301–318.

134. De Ollas, C.; Hernando, B.; Arbona, V.; Gómez-Cadenas, A. Jasmonic Acid Transient Accumulation Is Needed for Abscisic Acid Increase in Citrus Roots Under Drought Stress Conditions. *Physiol. Plantarum* **2013,** *147* (3), 296–306.
135. Alam, M. M.; Nahar, K.; Hasanuzzaman, M.; Fujita, M. Exogenous Jasmonic Acid Modulates the Physiology, Antioxidant Defense and Glyoxalase Systems in Imparting Drought Stress Tolerance in Different Brassica Species. *Plant Biotechnol. Rep.* **2014,** *8* (3), 279–293.
136. Wu, H.; Wu, X.; Li, Z.; Duan, L.; Zhang, M. Physiological Evaluation of Drought Stress Tolerance and Recovery in Cauliflower (*Brassica oleracea* L.) Seedlings Treated with Methyl Jasmonate and Coronatine. *J. Plant Growth Regul.* **2012,** *31* (1), 113–123.
137. Todaka, D.; Shinozaki, K.; Yamaguchi-Shinozaki, K. Recent Advances in the Dissection of Drought-Stress Regulatory Networks and Strategies for Development of Drought-Tolerant Transgenic Rice Plants. *Front. Plant Sci.* **2015,** *6.*
138. Mohamed, H. I.; Latif, H. H. Improvement of Drought Tolerance of Soybean Plants by Using Methyl Jasmonate. *Physiol. Mol. Biol. Plants* **2017,** *23* (3), 545–556.
139. Fu, J.; Wu, H.; Ma, S.; Xiang, D.; Liu, R.; Xiong, L. OSJAZ1 Attenuates Drought Resistance by Regulating JA and ABA Signaling in Rice. *Front. Plant Sci.* **2017,** *8.*
140. Sánchez-Romera, B. et al. Enhancement of Root Hydraulic Conductivity by Methyl Jasmonate and the Role of Calcium and Abscisic Acid in This Process. *Plant, Cell Environ.* **2014,** *37* (4), 995–1008.
141. Hu, Y.; Jiang, L.; Wang, F.; Yu, D. Jasmonate Regulates the Inducer of CBF Expression-C-Repeat Binding Factor/DRE Binding Factor1 Cascade and Freezing Tolerance in Arabidopsis. *Plant Cell* **2013,** *25* (8).
142. Bokszczanin, K. L.; Fragkostefanakis, S. Perspectives on Deciphering Mechanisms Underlying Plant Heat Stress Response and Thermotolerance. *Front. Plant Sci.* **2013,** *4.*
143. Clarke, S. M.; Cristescu, S. M.; Miersch, O.; Harren, F. J. M.; Wasternack, C.; Mur, L. A. J. Jasmonates Act with Salicylic Acid to Confer Basal Thermotolerance in *Arabidopsis thaliana. New Phytol.* **2009,** *182* (1), 175–187.
144. Acharya, S. M.; Assmann, B. R. Hormone Interactions in Stomatal Function. *Plant Mol. Biol.* **2008,** *69* (4), 451–462.
145. Bandurska, H.; Stroiński, A.; Kubiś, J. The Effect of Jasmonic Acid on the Accumulation of ABA, Proline and Spermidine and Its Influence on Membrane Injury Under Water Deficit in Two Barley Genotypes. *Acta Physiol. Plantarum* **2003,** *25* (3), 279–285.
146. Hou, S.; Lin, L.; Lv, Y.; Xu, N.; Sun, X. Responses of Lipoxygenase, Jasmonic Acid, and Salicylic Acid to Temperature and Exogenous Phytohormone Treatments in *Gracilariopsis lemaneiformis* (Rhodophyta). *J. Appl. Phycol.* **2018,** *30* (6), 3387–3394.
147. Farooq, M. A. et al. Methyl Jasmonate Regulates Antioxidant Defense and Suppresses Arsenic Uptake in *brassica napus* L. *Front. Plant Sci.* **2016,** *7.*
148. Per, T. S.; Khan, N. A.; Masood, A.; Fatma, M. Methyl Jasmonate Alleviates Cadmium-Induced Photosynthetic Damages Through Increased S-Assimilation and Glutathione Production in Mustard. *Front. Plant Sci.* **2016,** *7.*
149. Maksymiec, W.; Wianowska, D.; Dawidowicz, A. L.; Radkiewicz, S.; Mardarowicz, M.; Krupa, Z. The Level of Jasmonic Acid in *Arabidopsis thaliana* and *Phaseolus coccineus* Plants Under Heavy Metal Stress. *J. Plant Physiol.* **2005,** *162* (12), 1338–1346.

150. Maksymiec, W.; Krupa, Z. Effects of Methyl Jasmonate and Excess Copper on Root and Leaf Growth. *Biol. Plantarum* **2007,** *51* (2), 322–326, Jun.
151. Aftab, T.; Khan, M. M. A.; Idrees, M.; Naeem, M.; Moinuddin; Hashmi, N. Methyl Jasmonate Counteracts Boron Toxicity by Preventing Oxidative Stress and Regulating Antioxidant Enzyme Activities and Artemisinin Biosynthesis in *Artemisia annua. Protoplasma* **2011,** L.; *248* (3), 601–612.
152. Yan, Z.; Chen, J.; Li, X. Methyl Jasmonate as Modulator of Cd Toxicity in *Capsicum frutescens* var. Fasciculatum Seedlings. *Ecotoxicol. Environ. Saf.* **2013,** *98*, 203–209.
153. Yao, J. X.; Tao, L.; Zhi, M. Y. Aluminum-Induced Cell Wall Peroxidase Activity and Lignin Synthesis Are Differentially Regulated by Jasmonate and Nitric Oxide. *J. Agric. Food Chem.* **2008,** *56* (20), 9676–9684.
154. Yan, Z.; Zhang, W.; Chen, J.; Li, X. Methyl Jasmonate Alleviates Cadmium Toxicity in *Solanum nigrum* by Regulating Metal Uptake and Antioxidative Capacity. *Biol. Plantarum* **2015,** *59* (2), 373–381.
155. Chen, J.; Yan, Z.; Li, X. Effect of Methyl Jasmonate on Cadmium Uptake and Antioxidative Capacity in *Kandelia obovata* Seedlings Under Cadmium Stress. *Ecotoxicol. Environ. Saf.* **2014,** *104* (1), 349–356.
156. Gallego, S. M. et al. Unravelling Cadmium Toxicity and Tolerance in Plants: Insight into Regulatory Mechanisms. *Environ. Exp. Bot.* **2012,** *83*, 33–46.
157. Farooq, M. A. et al. Physiological and iTRAQ-Based Quantitative Proteomics Analysis of Methyl Jasmonate–Induced Tolerance in Brassica napus Under Arsenic Stress. *Proteomics* **2018,** *18* (10).
158. Singh, I.; Shah, K. Exogenous Application of Methyl Jasmonate Lowers the Effect of Cadmium-Induced Oxidative Injury in Rice Seedlings. *Phytochemistry* **2014,** *108*, 57–66.
159. Kamal, A. H. M.; Komatsu, S. Jasmonic Acid Induced Protein Response to Biophoton Emissions and Flooding Stress in Soybean. *J. Proteomics* **2016,** *133*, 33–47.
160. Rao, M. V.; Lee, H. I.; Creelman, R. A.; Mullet, J. E.; Davis, K. R. Jasmonic Acid Signaling Modulates Ozone-Induced Hypersensitive Cell Death. *Plant Cell* **2000,** *12* (9), 1633–1646.
161. Cui, H.; Wei, J.; Su, J.; Li, C.; Ge, F. Elevated O_3 Increases Volatile Organic Compounds via Jasmonic Acid Pathway That Promote the Preference of Parasitoid *Encarsia formosa* for Tomato Plants. *Plant Sci.* **2016,** *253*, 243–250.
162. Zhang, T.; Qiu, J.; Chen, G.; Xu, J.; Zhu, F.; Ouyang, G. Uptake of Pharmaceuticals Acts as an Abiotic Stress and Triggers Variation of Jasmonates in Malabar Spinach (Basella alba. L). *Chemosphere* **2019,** *236*.
163. Ma, L. Y. et al. Identification of a Novel Function of a Component in the Jasmonate Signaling Pathway for Intensive Pesticide Degradation in Rice and Environment Through an Epigenetic Mechanism. *Environ. Pollut.* **2021,** *268*.
164. Svyatyna, K.; Riemann, M. Light-Dependent Regulation of the Jasmonate Pathway. *Protoplasma* **2012,** *249* (SUPPL.2), 137–145.
165. Mewis, I. et al. UV-B Irradiation Changes Specifically the Secondary Metabolite Profile in Broccoli Sprouts: Induced Signaling Overlaps with Defense Response to Biotic Stressors. *Plant Cell Physiol.* **2012,** *53* (9), 1546–1560.
166. Ranjan, A. et al. Upregulation of Jasmonate Biosynthesis and Jasmonate-Responsive Genes in Rice Leaves in Response to a Bacterial Pathogen Mimic. *Funct. Integrative Genom.* **2015,** *15* (3), 363–373.

167. Delgado, L. D et al. Application of a JA-ile Biosynthesis Inhibitor to Methyl Jasmonate-Treated Strawberry Fruit Induces Upregulation of Specific MBW Complex-Related Genes and Accumulation of Proanthocyanidins. *Molecules* **2018,** *23* (6).
168. Biondi, S.; Scaramagli, S.; Capitani, F.; Maddalena Altamura, M.; Torrigiani, P. Methyl Jasmonate Upregulates Biosynthetic Gene Expression, Oxidation and Conjugation of Polyamines, and Inhibits Shoot Formation in Tobacco Thin Layers. *J. Exp. Bot.* **2001,** *52* (355), 231–242.
169. Abdelkareem, A.; Thagun, C.; Nakayasu, M.; Mizutani, M.; Hashimoto, T.; Shoji, T. Jasmonate-Induced Biosynthesis of Steroidal Glycoalkaloids Depends on COI1 Proteins in Tomato. *Biochem. Biophys. Res. Commun.* **2017,** *489* (2), 206–210.
170. Shoji, T.; Kajikawa, M.; Hashimoto, T. Clustered Transcription Factor Genes Regulate Nicotine Biosynthesis in Tobacco. *Plant Cell* **2010,** *22* (10), 3390–3409.
171. Concha, C. M.; Figueroa, N. E.; Poblete, L. A.; Oñate, F. A.; Schwab, W.; Figueroa, C. R. Methyl Jasmonate Treatment Induces Changes in Fruit Ripening by Modifying the Expression of Several Ripening Genes in *Fragaria chiloensis* Fruit. *Plant Physiol. Biochem.* **2013,** *70*, 433–444.

CHAPTER 8

JA-Mediated Defenses to Promote Thermotolerance in Plants: What Is the State of Our Knowledge?

HANAN AHMED HASHEM

Department of Botany, Faculty of Science, Ain Shams University, Cairo, Egypt

ABSTRACT

Global climate change and rising human population put pressure on the world's food production systems. Since new arable land is scarce, improved productivity must come from existing land, which entails lowering losses caused by abiotic and biotic stresses, as well as effective harvesting and storage procedures. High temperature (HT) stress has emerged as one of the most damaging stressors among the ever-changing environmental factors. During the last century, global surface temperatures rose by roughly 0.6°C. The climate models predict that global mean temperatures will rise (over 1990 levels) by 0.4 to 1.1°C by 2025, 0.8 to 2.6°C by 2050, and 1.4 to 5.8°C by 2100. Adaptation strategies are required to counteract the detrimental impacts of climate change and variability. One possible option is to enhance the resistance of current genotypes to HT stress. Jasmonates (JAs), which include jasmonic acid (JA) and methyl jasmonate (MeJA), are new phytohormones that play key roles in plant defense and responses to biotic and abiotic stressors. Jasmonates have received a lot of interest recently since they have been shown in studies to reduce the damage caused by HT stress to plants. The major physiological mechanisms by

Jasmonates and Plant Defense: An Eco-Friendly and Sustainable Approach for Stress Tolerance.
Uzma Azeem, & Khalid Rehman Hakeem (Eds.)

which JAs attenuate HT stress to plants are thought to include activating the antioxidant defense system, improving osmotic control, safeguarding photosynthesis, and interacting with other phytohormones, particularly ethylene and abscisic acid (ABA). Elevating JA levels in plants might be a valuable strategy for improving crop thermotolerance through the development of new varieties. Understanding the functions of JAs in HT stress responses would be extremely beneficial in elucidating the mechanism by which plants respond to HT stress and exploring novel methods to relieving crop stress. This chapter examined all the current findings on the roles and mechanisms of JAs in HT stress mitigation.

8.1 INTRODUCTION

A primary goal of plant biology is to acquire a comprehensive knowledge of how plants survive and reproduce in their environment.[1] Jasmonic acid (3-oxo-2-[2′-*cis*-pentyl]-cyclopentane-1-acetate; JA) and its methyl ester (JA-Me) are cyclopentanone compounds that are widely used and are thought to be plant growth regulators.[2] Following the description of their physiological effects, JA and its methyl ester drew interest. They, like other phytohormones, have a variety of biological actions through inhibition, promotion, and induction. There have been reports of new functions for JA in the physiology of nonstressed plants. JA is thought to be a tuber-inducing agent and may be implicated in potato tuberization.[3] Furthermore, it may control nitrogen partitioning in soybeans by causing the buildup of vegetative storage proteins (VSP) in the leaves.[4] Exogenous JA treatment induced premature senescence in *Arabidopsis*-connected and -detached leaves, showing that JA might be involved in the senescence pathway.[5] Besides that, JA has been shown to promote sugarcane cutting elongation, differentiation in plant tissue culture, adventitious root formation, seed dormancy breaking, pollen germination, flower and fruit development, tendril coiling, stomatal closure, ethylene biosynthesis, and protein biosynthesis.[6,7] There are three cardinal points for the types of temperatures required by plants: (i) The lowest temperature at which plant growth and development ceases; (ii) The optimal temperature for optimum plant growth and development; (iii) The highest temperature at which plant growth and development ceases. Temperature requirements of various plants vary for the cardinal points due to exposure time, plant age, prior history, and other externalities. In general, the air temperature or ambient temperature has a variety of effects on the crop. One of the primary environmental variables affecting agricultural yield is high temperature (HT) stress. The abrupt increase in ambient maximum temperature by 5–7°C in a few days, with a concomitant increase in lowest temperature, generates "heat stress"(HS) on plants. The plant's natural physiology is disrupted, and plant maturity is hastened. To cope with the unusual stress, plants may lose leaf/flower/fruit, dry up nonproductive tillers, or even become sterile. In practice, such HS results in massive crop losses. Heat stress is expected to become more prevalent in various parts of the world because of global climate change.[8] It is widely believed that phytohormones are signaling molecules that play critical roles not only in regulating plant growth and development but also in modulating physiological processes in response to a variety of biotic and abiotic stimuli, including HT stress.[9,10] Jasmonates (JAs), particularly JA and methyl jasmonate (MeJA), have gotten a lot of interest since they have been shown in recent research to reduce the damage caused by HT stress to plants.[11,12] It has been revealed that JAs can provide thermotolerance in *Arabidopsis* by boosting the expression of JA-related genes. JA production and signaling pathways.[13] Increases in JAs content in rice lodicules caused by JA or MeJA application have been shown to dramatically enhance the opened-spikelet rate when subjected to HT stress during anthesis.[12] The physiological mechanism through which JAs improve resistance to abiotic stressors is mostly discussed in plants such as *Arabidopsis*, rice, wheat, and tomato.[11,12,14,15] This chapter reviewed all the available data up today about the roles and mechanisms of JAs in mitigating HT stress.

8.2 THE MAJOR IMPACT OF HEAT STRESS ON PLANTS

High-temperature stress has been shown to have an impact on plant growth and development, as well as many physiological and yield processes. Excess accumulation of reactive oxygen species (ROS) induces oxidative stress, irreversible denaturation of proteins end up causing misfolding and aggregation, and alterations to the lipid membrane resulting in impaired membrane permeability are all biochemical and physiological consequences of HS.[16] Furthermore, the photosynthetic mechanism is extremely susceptible to HS.[17,18] High temperatures can affect photosynthesis by regulating stomatal closure and reducing CO_2 passage into mesophyll tissue, or by directly affecting metabolic processes. High-temperature stress impairs photosynthetic electron transport, ribulose-1,5-bisphosphate (RuBP) regeneration ability, and ribulose 1,5-bisphosphate carboxylase/oxygenase (rubisco) protein concentration and activity.[19,20] In general, reduced conductance via stomata is the major source of photosynthetic reduction during the early phases of a stress period. Because of the reduced rubisco activase enzyme, HT stress inhibits stomatal conductance and net photosynthesis.[21] High temperature greatly affects the subcellular level as well as anatomical structures at the tissue and cellular levels. The primary alterations at the subcellular level include chloroplast morphology (Figure 8.1), swelling of the stromal lamellae, and a tiny mass of vacuoles that modify the structural organization of thylakoids and create antenna without photosystem II, reducing photosynthetic and respiratory activities.[22] These cumulative impacts may reduce plant height, branching, leaf area, root growth, dry matter yield, fruit size, and harvest index.[8]

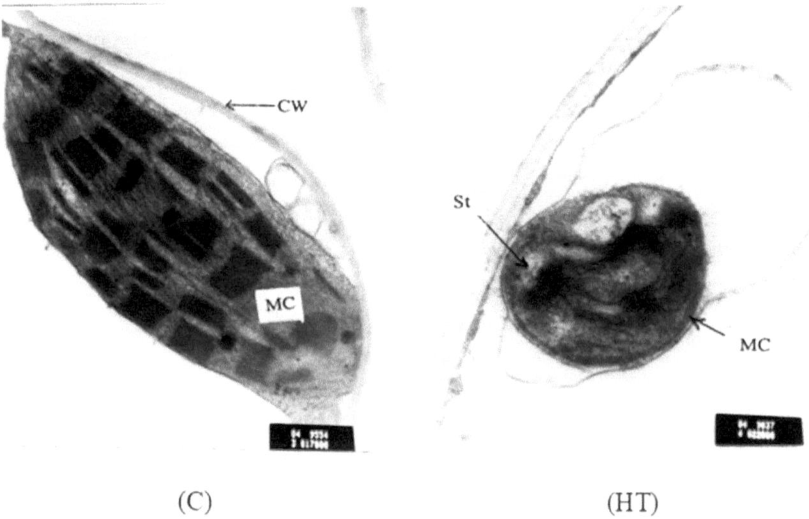

FIGURE 8.1 Transmission electron microscopy (TEM) analysis of leaf tissues from *Zea mays* L. plants under control condition (C) or HS (50°C) (HT). CW, cell wall; MC, mesophyll cell; St, starch (Magnification 17,000× and 22,000×, respectively).[23]

With increasing temperature, the rates of photosynthesis and respiration rise until a maximum photosynthesis level is reached. It is usually about 22–24°C and stays that way up to 30–32°C. When the HT peak is reached, different enzymes become inactivated, reducing photosynthetic effectiveness. When the temperature surpasses 46°C, the plant goes into irreversible wilting due to excessive respiration and evaporation. The extent to which HS causes crop loss may devastate the income of impoverished farmers living in heat-prone locations. Although there may be enough soil moisture in many situations, the negative water balance caused by HS resulted in plants withering in a few days. Plants have been exposed to more than one stress at a time in their natural habitat; as a result, they have evolved defenses against a mixture of stresses by responding to them in diverse ways. High temperatures, high transpiration rates, and high light (HL) intensity are frequently observed combined, affecting plant production.[24] Figure 8.2 summarizes the effects of HS on plants.

To deal with HS, plants use antioxidants, proteins, ion transporters, or osmoprotectants. Plants engage their signaling cascades and transcriptional machinery in response to HS to prevent physiological and biochemical changes.[14] The ROS that are produced during stress are scavenged by antioxidants or osmoprotectants.[25] Another heat tolerance approach is the use of heat shock proteins (HSPs).[26] Heat stress (HS) is an additional environmental factor that may be associated with HL intensities. High-temperature stress and HL intensities were shown to increase the accumulation of JA and its derivatives, an amino acid conjugate of JA (jasmonoyl isoleucine: JA-Ile), as well as the expression of over 2200 distinct transcripts (including many that are JA-associated). A mutant lacking in JA biosynthesis (allene oxide synthase) was more sensitive to combined HL + HS. According to Balfagón et al.,[27] JA has an essential role in plant adaptation to HT and HL intensity stress. However, little is known about the role and mechanism of action of JAs in increasing HT tolerance in plants.

8.3 BIOSYNTHESIS OF JASMONATES

Jasmonates are produced in plants via the octadecanoid pathway.[2] In addition to JA/JA-Isoleucine, the JA precursor *cis*-(+)-12-oxo-phytodienoic acid (OPDA) was found, demonstrating a strong positive correlation to JA production.[28] A combination of conventional biochemical and genetic techniques has now successfully identified many of the enzymes involved in JA production.[29] The first committed step in JA production is the release of linolenic acid from phospholipids, which is an abundant octadecanoid in higher plant membranes, particularly chloroplast membranes.[30] Lipoxygenase

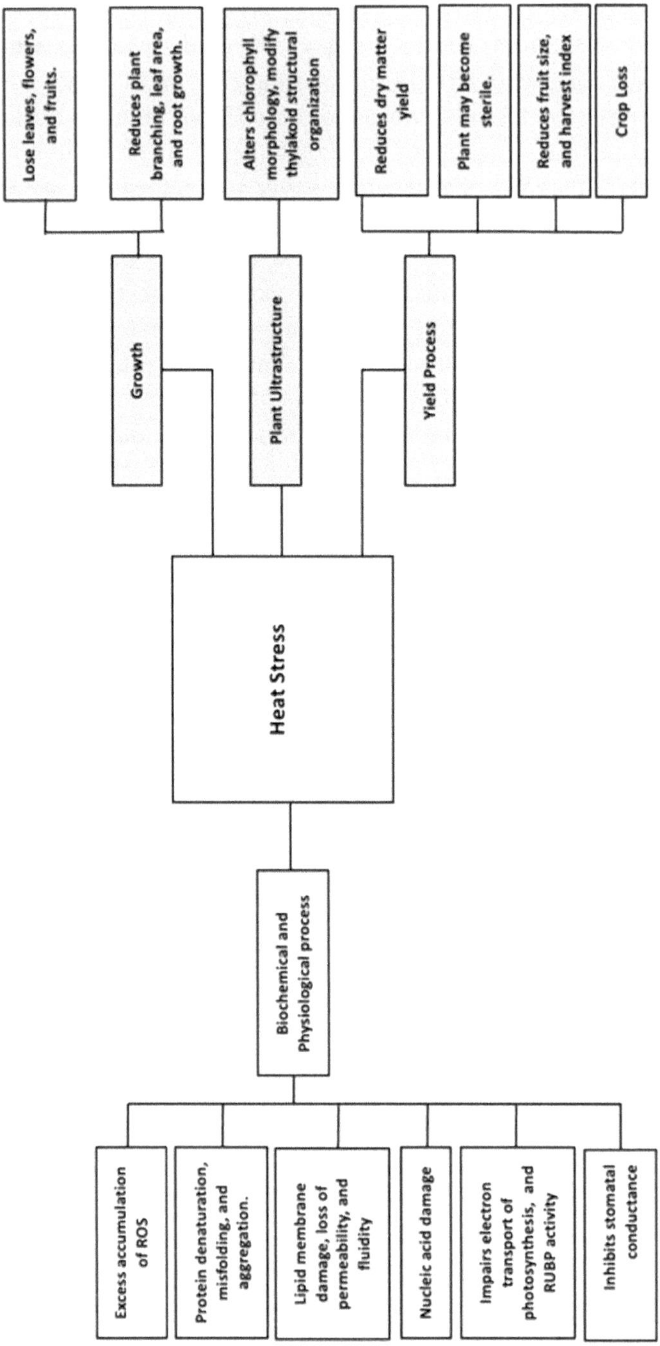

FIGURE 8.2 Major detrimental impacts of HS on plants.

(LOX) then oxygenates linolenic acid.[31] The resultant 13-hydroperoxide is transformed to 12-oxo-phytodienoic acid (12-oxo-PDA) in the chloroplast by the sequential action of allene oxide synthase (AOS) and allene oxide cyclase (AOC).[32] The reduction of the cyclopentenone ring of OPDA-by-OPDA reductase was the following stage in the process (OPR). In peroxisomes, OPR most likely catalyzes this process.[33] Where oxidation enzymes finish JA production.[34] Thus, JA biosynthesis begins in the chloroplast and ends in the peroxisome.[32] Jasmonic acid is subsequently catabolized further to produce its volatile analog, MeJA, as well as a variety of conjugates.[35] Lipoxygenases are nonheme iron-containing enzymes classified into three types (LOX1, LOX2, and LOX3). They are encoded in a gene family with organ-specific expression, with LOX1 mostly expressed in tubers and roots, LOX2 in leaves, and LOX3 in leaves and roots.[36] The chloroplastic LOX2 is involved in JA production in *Arabidopsis.*[37]

8.4 JASMONIC ACID PERCEPTION AND SIGNALING

The perception of the stress and the transmission of the resultant signal are the initial steps in adaptation to environmental stress. Organisms and/or individual cells appear to be equipped with sensors and signal transducers that detect and transmit signals caused by environmental changes. Furthermore, the sensors and signal transducers appear to be specialized to only certain types of environmental stress.[38] The signal transduction mechanism that leads to the stimulation of hormone production at HTs is yet unknown. To explain how the biological activities of these compounds are transmitted to the promoters of various genes that they activate, Reinbothe et al.[39] hypothesized that JAs interact with receptors in the cell, activating a signaling pathway that results in changes in transcription, translation, and other responses mediated by JA. According to Rojo et al.,[40] a staurosporine-sensitive protein kinase is implicated in a signaling cascade that leads to JA-induced gene expression. Devoto et al.[41] discovered that the *Arabidopsis* gene coronatine insensitive 1 (COI1) is required for the growth inhibition caused by JA, insect herbivory resistance, and disease resistance. They further demonstrated that COI1 is required for the production of about 84% of the 212 genes activated by JA and 53% of the 104 genes repressed by JA utilizing microarray gene transcription profiling of wild-type and COI1 mutant plants. Furthermore, many sources of evidence point to that JA may regulate plant heat response via a subset of JA-inducible

WRKY superfamily transcription factors (WRKY transcription factors are unique to plants and are a new type of transcriptional regulatory factors with N-terminal ends containing conserved WRKYGQR amino acid sequences).[42] In the JA signal perception and transduction network, many transcription factors (TFs), repressors, certain genes are upregulated and down-regulated, and members of ubiquitin–proteasome compounds are included. GC-MS and HPLC tests in *Arabidopsis thaliana* indicated that (+)-7-*iso*-JA-L-Ile is the sole natural and direct JA-signaling ligand in plants.[43] Skp1 (S-phase kinase-associated protein 1)/Cullin/F-box (SCF) complex is a proteinous ubiquitin–proteasome complex. The COI1 locus encodes an F-box protein that interacts with its other equivalents, SKP1, Cullin, and Rbx proteins, to create an E3 ubiquitin ligase.[44] As a result, COI1 has been identified as the JA receptor in the signaling cascade. In 2007, researchers in independent work[45,46] discovered a new family of proteins in *Arabidopsis* known as JASMONATE ZIM (JAZ) domain proteins, which were discovered to act as the primary suppressor of JA signals and to act as a substrate for the SCFCOI1 E3 ubiquitin ligase complex. *A. thaliana* contains approximately 12 JAZ proteins, each with an N-terminal domain, a highly conserved C-terminal JAZ domain that mediates interaction with COI1 and several transcription factors, and a conserved protein–protein interaction domain, the ZIM (or TIFY) domain that aids in JAZ dimerization and interaction with novel JAZ interactors (NINJA).Furthermore, COI1 binding to the Jas domain of JAZ proteins in the presence of JA-Ile results in the formation of the coreceptor complex[47] SCFCOI1, E3 ubiquitin ligase complex, JAZ degrons (JAZ1–JAZ12), and IP5 constitute the coreceptor complex and have been identified as genuine JAs receptors.[48] The JA-associated VQ-Motif GENE1 (JAV1, also known as VQ22) genes are crucial in JA signal and plant defense. JAV1 belongs to the plant-specific protein family with the fixed short amino acid sequence motif FxxhVQx-hTG, where "x" can be any amino acid and "h" can be any hydrophobic group.[49] This family is involved in a variety of plant defensive strategies, tolerance to stress, and growth and development.[49] Several corepressors also perform roles in the signal transduction pathway at the same time. In transcriptional regulators, corepressors suppress transcription start. TOPLESS (TPL) and TPL-related proteins (TPRs) are members of the Groucho/Tup1 corepressor family, which induces histone deacetylation and demethylation, resulting in chromatin modification and, eventually, gene suppression.[50] TPL interacts with JAZ proteins via an

ERF-associated amphiphilic repression (EAR) motif. JAZ proteins that lack a repression motif interact with TPL via a new variety of specific JAZ (NINJA) adaptor proteins.[51] JAR1 (JASMONOYL ISOLEUCINE CONJUGATE SYNTHASE1) is an enzyme that is essential in the formation of the most reactive JA molecule active as a receptor ligand, and was crystallized in 2012.[52] MYC2, the master regulator of numerous biological processes, is a bHLH (basic helix–loop–helix proteins) transcription factor that regulates JA transcription. MYC2 can stimulate transcription of early JA-responsive genes such as JAZ and MYC2. Other transcription factors that regulate JA-mediated signaling include MYC3, MYC4, MYB, GL3, EGL3 AP, GL1, and others. Under normal conditions, the JA concentration is minimal, and the JAV1 and JAZ proteins bind to different TF, limiting their function. Under stress, JA content rises, and JAZ proteins are degraded, resulting in active TF that upregulates genes implicated in stress responses.[53] Previous research clearly shows that numerous JA/MeJA components/genes, including receptors, have been discovered, but much more needs to be explored. Although JAs have been identified to play a variety of roles in HS tolerance, the whole genetic cascade of JAs involved has only been partially established, which might be examined by contrasting plants that are enriched or deficient in JAs following stress exposure.

8.5 JASMONATES AND REACTIVE OXYGEN SPECIES

Reactive oxygen species formation is a unifying feature of a wide range of abiotic stresses. For example, the atmospheric pollutant ozone interacts with cellular components in the leaf apoplast to create ROS, which in turn induces active ROS generation by the plant cell itself, a phenomenon known as an "oxidative burst." Plants are continuously subjected to changes in external conditions, which affect the availability of nutrients and water, potentially leading to the production of ROS in different cellular compartments. A combination of strong light and cold, for example, may disrupt photosynthetic activities and promote ROS production in chloroplasts.[54] ROS include superoxide ($^{\cdot}O_2^-$), hydrogen peroxide (H_2O_2), hydroxyl radical (OH), singlet oxygen ($^1O_2$), and the air pollutant ozone (O_3). One kind of ROS can be changed into another kind, for example, O_3 can break and be transformed into H_2O_2, $^{\cdot}O_2^-$, and $^1O_2$. $^{\cdot}O_2^-$ is rapidly dismutated to H_2O_2 via SOD activity, which may then combine with Fe^{2+}

to generate highly reactive $^{\cdot}OH$. The coaccumulation of H_2O_2 and $^{\cdot}O_2^-$ increases the formation of $^{\cdot}OH$.[55] All ROS are not made equal, and each one leaves a distinct "oxidative imprint" on the cell. Individual ROS's chemical composition, site of synthesis, regulation of ROS accumulation via scavenging and detoxification, and signaling components that interact with ROS are all inputs that are combined to generate a specific response. $^{\cdot}OH$, for example, rapidly interacts with all sorts of biological components, $^{\cdot}O_2$ predominantly with protein Fe-S centers, and $^1O_2$ is especially reactive with polyunsaturated fatty acid double bonds.[56] The mechanism by which JAs protect plants from the devastating impacts of HT stress is not entirely known. In general, HT stress may increase ROS while decreasing antioxidants in plant organs, and excessive ROS production and buildup may damage cell membranes and trigger programmed cell death.[57] It has been proposed that H_2O_2 is the most stable and least reactive ROS among the ROS, and that it may readily pass the membrane to cause cell harm.[58,59] Catalase (CAT) is widely regarded as the first antioxidant enzyme to scavenge H_2O_2 produced by various pathways under normal and stressful circumstances, and it plays an important role in maintaining redox equilibrium in plant cells.[60] Antioxidant metabolites such as reduced ascorbic acid (AsA) and reduced glutathione (GSH) are critical for controlling ROS levels during abiotic stress via direct or indirect scavenging.[61] AsA is thought to be the most versatile antioxidant molecule since it can scavenge all kinds of ROS and regenerate other antioxidants.[62] It has been reported that JAs can activate the plant defense system, principally by enhancing the activity of antioxidative enzymes such as CAT and GSH peroxidase and decreasing the levels of other defensive compounds like AsA and heat shock proteins (HSPs), thereby suppressing ROS generation and reducing programmed cell death under stress.[11,12,15] According to Chen et al.,[63] photothermosensitive genic male sterile (PTSGMS) rice (*Oryza sativa* L.) lines with higher levels of JA and MeJA in the stigmas under HT stress had lower H_2O_2 levels, higher AsA levels, and CAT activity. The contents of JA and MeJA were inversely associated with the contents of H_2O_2 but favorably connected with the contents of AsA and CAT activity. To combat the toxicity of active oxygen species, plant cells have a very effective antioxidative defense that includes both nonenzymatic and enzymatic components.[64] The impact of JA on antioxidant machinery in plants is primarily responsible for its function in reducing the ROS accumulated during abiotic stress, particularly HT stress. Hashem[65] discovered that JA treatment resulted in a highly significant increase in glutathione content as well as CAT, peroxidase, and ascorbate peroxidase activities, as well as a highly significant decrease in membrane–lipid peroxidation [as measured by malondialdehyde (MDA) content] in soybean plants in comparison to the control samples. Moreover, Xiang and Oliver[66] discovered that JA administration enhanced mRNA levels and glutathione production capability in *Arabidopsis.* Wang[67] investigated the impact of MeJA on changes in oxygen-scavenging enzyme activities and membrane–lipid peroxidation and discovered that MeJA treatment lowered peroxidase activity while maintaining greater CAT, superoxide dismutase, and AsA levels. Furthermore, MeJA treatment decreased transpiration and membrane–lipid peroxidation as measured by MDA concentration. On the other hand, Hurkman and Tanaka[68] found that after 30 min of treatment with NaCl, salicylate, methyl salicylate, or MeJA, peroxidase mRNA levels dropped to less than control levels. However, plants have a variety of peroxidase isozymes, and it is possible that a distinct, yet to be identified, transcript of this gene family accumulates in barley roots in response to these treatments.

8.6 JASMONIC ACID AND CELL OSMOREGULATION

Accumulation of compatible osmolytes is one of the most significant defensive mechanisms that occurs in response to excessive light and HS, and it is most likely mediated by phytohormones, which transduce stress signals throughout tissues.[69] Proline, glycine–betaine, polyamines, and sugars are the most prevalent osmolytes that play important roles in osmoregulation. These metabolites maintain the osmotic differences between the cell's surroundings and the cytoplasm. Furthermore, they protect plant cells from oxidative stress by preventing the generation of damaging ROS such as hydroxyl ions, superoxide ions, hydrogen peroxide, and other free radicals.[70] Torres et al.[69] discovered that sun-damaged apple tissue had lower relative water content (<9–10%) and greater negative tissue water potentials (<28%) than undamaged tissue, as well as osmolytes production and accumulation (i.e., carbs), indicating dehydration stress and osmoregulation. Interestingly, there is evidence of synchronized cooperation between ABA and JA in responding to water limitation (dehydration) to defend plants from its detrimental consequences; particularly, JA would play a role in regulating ABA production and signaling network.[71] Changes in tissue water potential and relative water content, both of which indicate dehydration stress, have previously been seen in sun-damaged fruit.[69] The buildup of suitable solutes or osmolytes, which have a protective role in plants, is one of the methods used by plants when subjected to temperature stress. The JA amendment had a favorable effect on plant water relations, owing to improvements in osmoregulation. Proline is usually regarded as an osmotic solute.[72] Under HS, proline accumulates and aids in osmotic correction as well as playing a protective function.[73] Proline was responsible for increased pressure potential in heat-stressed sugarcane leaves.[74] Proline aids in osmotic adjustment raise cell protoplasm concentration to maintain proper membrane function under heat stress and improves plant adaptation. Its capacity to serve as an osmoprotectant aids in the stabilization of the antioxidant system and therefore the detrimental effect of ROS.[75] High JA levels resulted in an eight-fold increase in proline content in barley cv. Alfa seedlings compared to controls. Similar findings were found in young apple trees[76] and tobacco seedlings.[77] Sugar reduction in heat-stressed plants may be due to higher assimilate use rather than production.[78] Sugar levels increased in some cases, which was attributed to the upregulation of starch hydrolyzing enzymes such as α-amylase during seed filling.[79] Thus, HS reduces glucose buildup by altering metabolic pathways; the alterations are crop-specific and dependent on HS exposure. Chen et al.[80] demonstrated that when rice was treated with MeJA, both the photosynthetic rate and absorption of P32 and N15 were increased, whereas chlorophyll breakdown was delayed. When MeJA was applied during the mid and late phases of flowering, it enhanced amylase activity and promoted assimilate transfer from flag leaf to panicle. When compared to untreated plants, accumulated assimilates in panicles treated with MeJA rose by eight-fold. Comprehensive research including numerous genes (such as SUS, sucrose synthase, and SUTs, sucrose transporters) linked to starch-sugar inter-conversions in heat-stressed plants of diverse crops is needed in the future to understand the influence of JAs on carbohydrate accumulation patterns.

8.7 JASMONIC ACID AND SECONDARY METABOLITES

Secondary metabolites in plants are compounds that play no basic function in the maintenance of life processes in plants but are necessary for the plant to interact with its environment for adaptation and defense. However, we are only now getting started to grasp the critical function they play in plant growth and development. Secondary metabolites (e.g., terpenes, flavonoids, and alkaloids) are produced from primary metabolites in higher plants (e.g., carbohydrates, lipids, and amino acids). They serve a vital role in plant defense against herbivores and pathogens, and they frequently protect plants from environmental stresses.[81] Many plant species have been shown to produce a variety of compounds (alkaloids, terpenoids, phenolic phytoalexins, coumarins, and taxanes) in response to MeJA and JA.[82] Jasmonates have been linked to secondary metabolite buildup, which is also part of the defensive response. Several plant species are distinguished by inherent emissions of volatile organic molecules such as isoprene, mono- and sesquiterpenes.[83–85] Under non-stressed conditions, these emissions usually consume 1–2% of photosynthetic carbon, but under stress conditions, such as HTs stress, these emissions can consume up to 10% of photosynthetic carbon.[86] In fact, these constitutive emissions have been linked to improved HS tolerance, either directly by improving membrane heat resistance[87,88] or indirectly by acting as membrane-dissolved antioxidants that can quench free radicals formed during HS.[89] Xu et al.[90] applied HS to *Aquilaria sinensis* cell suspension cultures and investigated how JA influenced sesquiterpene accumulation. Sesquiterpene molecules accumulate as gene expression in the JA biosynthesis pathway is dramatically upregulated. In addition, as compared to H_2O_2 and salicylic acid (SA), MeJA treatment to *A. sinensis* had the strongest impact on sesquiterpene production. These findings imply that JA is an essential signal transducer in a cascade of intracellular signals caused by HS, and that JA eventually plays a key part in the accumulation of sesquiterpene molecules.[90] Grape berries treated with HL intensity (2500 mol/m^2/s^1), HT (40°C), JA (200 mM), menadione (120 mM), and ABA (3.026 mM) displayed significant metabolic fluctuations, indicating preveraison barriers. Flavonoid accumulation increased the berry's resistance to cue-induced changes during the veraison stage.[91]

8.8 JASMONIC ACID AND PROTEINS

The responses of many plants to JAs, whether applied externally or produced internally, appear to be linked with changes in gene expression. At least three major JA effects on gene expression have been reported: (1) the induction of novel abundant polypeptides known as JA-induced proteins (JIPs),[92] (2) the selective repression of synthesis of several polypeptides presents prior to JA treatment,[93] and (3) the overall downregulation of protein production that occurs in long-term MeJA-treated leaf tissues.[94] According to Herrmann et al.,[95] most JIPs appear to be expressed in a genus- and species-specific manner. Furthermore, there are qualitative and quantitative variations in the amount to which MeJA inhibits the synthesis of preexisting proteins.[94] Reinbothe et al.[96] discovered that most JIPs exhibit activity and/or expression patterns that imply a function in plant defense and stress reactions in barley plants. Proteins involved in photosynthetic carbon assimilation are those whose production is inhibited or even shut down by MeJA. Genes that appear to have no effect on expression appear to encode proteins with housekeeping activities.

Proteinase inhibitors,[97] thionine,[98] ribosome–inactivating proteins, and phenylpropanoid metabolism enzymes are among the JA-induced plant defense proteins that have been discovered.[99] All these proteins are produced in response to exogenous JA application or endogenous JA increase caused by environmental stress. In this respect, Sadder et al.[100] discovered that heat causes the activation of an unusual set of plant genes in response to stress. Proteinase inhibitors and HSPs were identified in upregulated genes in the heat-tolerant tomato line.

8.8.1 HEAT SHOCK PROTEINS

When plants are moved from typical growing temperatures to temperatures five or more degrees higher, gene expression changes dramatically. Most normal protein synthesis is suppressed, while new transcription and translation of a low number of "heat shock proteins" (HSPs) is begun.[101] Plants produce HSPs, which inhibit protein denaturation and aid in the refolding of damaged proteins.[102] HSPs are widely acknowledged to give thermotolerance or thermoprotection against otherwise deadly temperatures. Each organism has an "ideal" temperature for eliciting a complete HS response.[103] Every organism has a certain degree of basic thermotolerance. The amount of acquired thermal tolerance was observed to be connected to the cellular level of HSP.[104] Marmiroli et al.[105] discovered that the absence of HSPs is associated with the loss of thermotolerance. The major HSPs synthesized in eukaryotes are divided into five conserved classes: HSP 100: members of this family have molecular weights ranging from 100 to 104 kDa.[106] HSP 90 has a molecular weight of 85–90 kDa,[102] HSP 70 has a molecular weight of 67–78 kDa,[107] HSP 60 has a molecular weight of 58–65 kDa,[108] and small HSPs (sm HSPs) of 17–30 kDa.[109] Most HSP families' important roles are now relatively well known. Many of the known functions are related to aspects of protein metabolism, such as stabilizing native proteins,[110] preventing aggregation of denatured proteins, refolding stress-denatured proteins,[111] participating in the folding and assembly of nascent polypeptides into functional forms,[112] facilitating translocation across membranes, and degradation of non-native and incompetent polypeptides.[112] All these functions fall under the molecular chaperone concept.[107] Heat shock response represents an example of inducible gene expression. This reaction involves transcriptional activation mediated by the heat shock transcriptional factor (HSF). Its activation for DNA binding is completely reliant on HS and demands a protein transition from monomeric (inactive) to homotrimeric (active) state.[113] Previous research suggested that JA treatment might stimulate the production of HSP in plants. Hamilton and Coleman[114] discovered that applying 50 μg of MeJA to a single leaf enhanced the amount of HSP70 as well as a single sHSP (21 kD) in treated leaves. Furthermore, they discovered that the application of MeJA can enhance HSP accumulation in tissues that were not directly exposed to the stress. Exogenous MeJA therapy can enhance antioxidant production, decrease LOX activity, and promote HSP family transcription.[115] The effects of JAs on the heat shock response in C6 glioma cells were studied by Oh et al.[116] They show that MeJA, but not JA, promotes heat shock protein 72 (HSP72) activation via heat shock factor I (HSF1) activation in C6 glioma cells without compromising cell survival. Tian et al.[117] recently published research that deepens understanding of the molecular connection between HS/HSFs and the JA signaling pathway. HsfA1b may transform into a functional homo-trimer in response to HS and stimulate *A. thaliana* oxophytodienoate-reductase 3 (AtOPR3) expression. As a result, JA biosynthesis and accumulation are enhanced. Following then, the JA-mediated signaling pathway initiates a cascade that increases dehydration-responsive element-binding protein 2A (DREB2A) expression and heat tolerance in plants. Furthermore, decreased transcript levels of JA marker genes in HSP90 RNAi lines indicated the importance of HSP90 and HSP70 in JA-COR (coronatine) reactions. Furthermore, pretreatment of WT *Arabidopsis* with an HSP inhibitor reduced COR—induced gene expression. Given that HSP proteins have a wide range of substrates, including transcription factors, E3 ligases, and kinases, the findings showed that COI1 is a client protein of SGT1b–HSP70–HSP90 chaperone complexes. This increases SGT1b–HSP70–HSP90 chaperone complexes' functional capability in regulating JA responses.[118]

8.9 THE CROSS-TALK BETWEEN JASMONIC ACID AND OTHER PHYTOHORMONES IN MODULATING HT STRESS

All phytohormones interact with one another, affecting biochemical, physiological, and morphological alterations in plant tissue exposed to HS, which include, initially, antioxidant metabolites and enzymes, as well as secondary metabolites, were unregulated, specifically flavanols, osmolytes (carbohydrates, amino acids, and so on), and osmolytes (carbohydrates, amino acids, and so on).[119] To moderate biotic and abiotic stressors, it has been hypothesized that JAs can work either synergistically or antagonistically with other important phytohormones such as brassinosteroids (BRs), ethylene (ET), and abscisic acid (ABA).[120] High temperatures were found to cause JA and ET accumulation in the model plant *A. thaliana*.[121, 122] The effects of JA and ET on plant heat sensitivity are inverse. The constitutive expresser of pathogen-related genes, PR1 (cpr5-1) mutant, in which the signaling pathways of SA, JA, and ET are constitutively active, exhibits increased HS tolerance.[123] When cpr5-1 was crossed with a mutant defective in the JA biosynthesis pathway (i.e., jar1-1) or the JA signaling pathway (i.e., coi1-1), thermotolerance was impaired, indicating that at least JA is necessary for heat tolerance.[123] JAs have been shown to stimulate ethylene production by increasing the activity of ethylene-producing enzymes in tomatoes.[124] It has been hypothesized that the JA-ethylene interaction is mostly mediated by the ethylene-activated transcription factor EIN3 (ethylene insensitive 3) and its similar homolog EIN3-Like1 (EIL1).[125] Under HT stress, a coordinated influence between JA and ABA has been proposed, which is substantiated by the fact that foliar spray of JA enhanced ABA concentration in plant cells, promoting stomata closure and conserving water content.[126] However, in rice response to nematode invasion and saline stress, antagonism between JA and ABA was reported.[127] Numerous studies have found that JAs and BRs have an antagonistic interaction when it comes to controlling plant development.[128] MeJA, for example, has been shown to inhibit lamina joint inclination in rice leaves by repressing BR biosynthesis and signaling,[129] and MeJA application significantly inhibits the expression of BR biosynthesis-related genes, decreases endogenous BR levels, and activates the JA-dependent rice immune response against root-knot nematodes in rice roots.[130] It has been noted, however, that the application of vitamin E (Ve) + vitamin C (Vc) + MeJA + BR gave the best effect among the four treatments examined in alleviating the negative impact of HT stress during anthesis on rice pollen fertility and germination (the other treatments: BR + triazoles + MeJA; Ve + Vc; and MeJA).[131] Under HT stress, a PTSGMS rice line with greater BRs contents also has higher JA or MeJA contents, and they are strongly linked.[132] When MeJA or 24-EBL (24-epibrassinolide) were applied to heat-stressed PTSGMS rice lines during anthesis, they significantly increased the expression of genes related to JAs and BRs biosynthesis or signaling in the pistils, whereas the results were reversed when neomycin (an inhibitor of JA-mediated reactions) or brassinazole were applied. Furthermore, adding MeJA and 24-EBL increased the pistil vigor of PTSGMS rice lines, as evidenced by greater AsA content and lower H_2O_2 concentration in the pistil than either MeJA or 24-EBL alone.[132] As a result, we believe that there could be a synergistic interaction between JAs and BRs in modulating rice responses to HT stress. More convincing evidence, however, is needed to back up such a claim. Salicylic acid (SA), another type of new phytohormone, has been proposed to improve innate thermotolerance and produce membrane thermoprotection.[133] However, an antagonistic relationship between JA and SA has been found in signaling pathways in response to HT stress. Many studies have revealed that endogenous SA buildup inhibits JA-dependent defenses, promoting SA-dependent defense over JA-dependent resistance.[134]

8.10 CONCLUSION

The physiological mechanism through which JAs improve resistance to abiotic stressors, particularly HT stress, may be divided into two parts (Figure 8.3). To begin, JAs have the potential to activate plants' defense systems, primarily by increasing the activity of antioxidative enzymes such as CAT, superoxide dismutase (SOD), guaiacol peroxidase, glutathione peroxidase, and increasing other defensive compounds such as AsA and HSPs, thereby suppressing ROS generation and reducing programmed cell death under stress. Second, JAs may improve osmotic control by boosting the production of certain osmoregulators, such as proline, phenolic compounds, and soluble carbohydrates, and therefore minimize stress damage to plant organs and increase plant cell growth. JAs may also interact with other phytohormones such as ABA, ethylene, and BRs to control the effect of HT stress on plant growth and development. There have been few studies on the impact of JAs in the response to HT. More study is required to fully comprehend the molecular mechanisms involved in the effect of JA on plant thermotolerance. Functional genomics and bioinformatics techniques can also be used to investigate JA-mediated signaling networks and cellular responses.

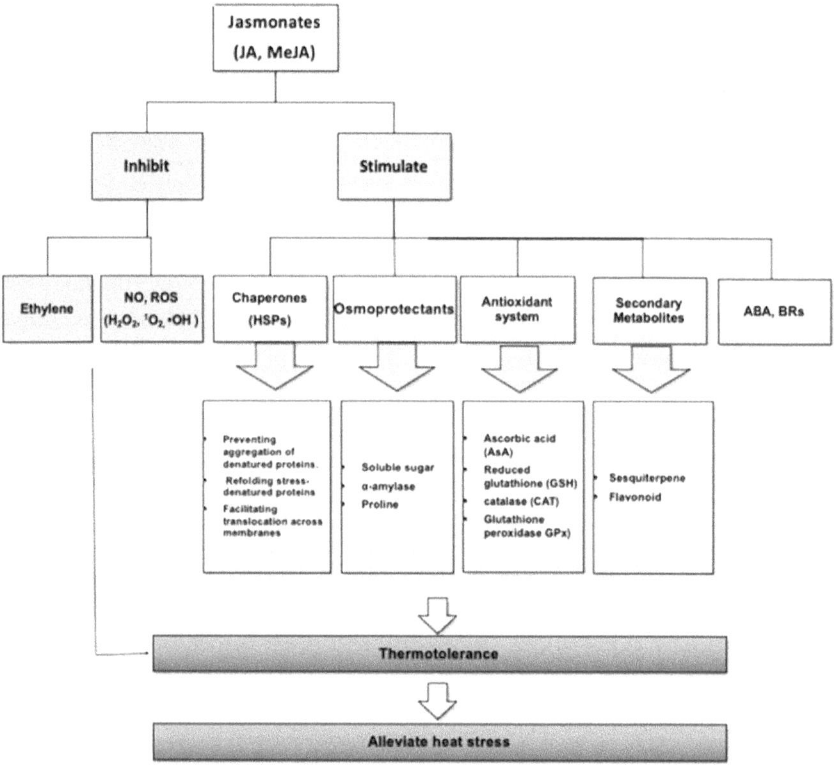

FIGURE 8.3 Schematic diagram of JAs' potential mode of action in alleviating HS in plants.

KEYWORDS

- **heat stress**
- **jasmonates**
- **heat shock proteins**
- **osmoprotectants**
- **reactive oxygen species**
- **signal transduction**

REFERENCES

1. Stintzi, A.; Weber, H.; Reymond, P.; Browse, J.; Farmer, E.; Plant Defence in the Absence of Jasmonic Acid; The Role of Cyclopentanones. *Proc. Natl. Acad. Sci. USA* **2001,** *98*, 12837–12842.
2. Creelman, R. A.; Mullet, J. E.; Biosynthesis and Action of Jasmonates in Plants. *Ann. Rev. Plant Physiol. Plant Mol. Biol.* **1997,** *48*, 355–381.
3. Koda, Y.; Kikuta, Y.; Tazaki, H.; Tsujino, Y.; Sakamura, S.; Yoshihara, T. Potato Tuber Inducing Activities of Jasmonic Acid and Related Compounds. *Phytochemistry* **1991,** *30*, 1435–1438.
4. Mason, H. S.; Mullet, J. E. Expression of Two Soybean Vegetative Storage Protein Genes During Development and in Response to Water Deficit, Wounding and Jasmonic Acid. *Plant Cell* **1990,** *2*, 569– 579.
5. He, Y.; Fukushige, H.; Hildebrand, D. F.; Gan S. Evidence Supporting a Role of Jasmonic Acid in *Arabidopsis* Leaf Senescence. *Plant Physiol.* **2002,** *128*, 876–884.
6. Sasaki, Y.; Asamizu, E.; Shibata, D.; Nakamura, Y.; Kaneko, T.; Awai, K.; Masuda, T.; Shimada, H.; Takamiya, K.; Tabata, S.; Ohta, H. Genome-Wide Expression-Monitoring of Jasmonate-Responsive Genes of *Arabidopsis* Using cDNA Arrays. *Biochem. Soci. Trans.* **2000,** *28*, 863–864.
7. Horiuchi, J.; Arimura, G.; Ozawa, R.; Shimoda, T.; Takabayashi, J.; Nishioka, T. Exogenous ACC Enhances Volatiles Production Mediated by Jasmonic Acid in Lima Bean Leaves. *FEBS Lett.* **2001,** *509*, 332–336. http://www.faculty.missouristate.edu/c/cma545f/jasmonicacid.htm
8. Djanaguiraman, M.; Prasad, P. V.; High Temperature Stress. In: *Plant Genetic Resources and Climate Change*; Jackson, M., Ford-Lloyd, B. V., Perry, M. L., Eds.; CABI, 2014; pp 201–220.
9. Wani, S. H.; Kumar, V.; Shriram, V.; Sah, S. K. Phytohormones and Their Metabolic Engineering for Abiotic Stress Tolerance in Crop Plants. *Crop J.* **2016,** *4*, 162–176. DOI: 10.1016/j.cj.2016.01.010
10. Zhou, R.; Hu, Q.; Pu, Q.; Chen, M.; Cao, Y. Spermidine Enhanced Free Polyamine Levels and Expression of Polyamine Biosynthesis Enzyme Gene in Rice Spikelet

Under Heat Tolerance Before Heading. *Sci. Rep.* **2020,** *10*, 8976. DOI: 10.1038/s41598-020-64978-2

11. Chen, J.; Xu, Y.; Fei, K.; Wang, R.; He, J.; Fu, L., Shao, S.; Li, K.; Zhu, K.; Zhang, W.; Wang, Z.; Yang, J. Physiological Mechanism Underlying the Effect of High Temperature During Anthesis on Spikelet-Opening of Photo-Thermo-Sensitive Genic Male Sterile Rice Lines. *Sci. Rep.* **2020,** *10*, 2210. doi: 10.1038/s41598-020-59183-0
12. Yang, J. C.; Fei, K. Q.; Chen, J.; Wang, Z. Q.; Zhang, J. H. Jasmonates Alleviate Spikelet, Penning Impairment Caused by High Temperature Stress During Anthesis of Photo-Thermo-Sensitive Genic Male Sterile Rice Lines. *Food Energy Secur.* **2020,** *9*, e233. DOI: 10.1002/fes3.233
13. Kazan, K. Diverse Roles of Jasmonates and Ethylene in Abiotic Stress Tolerance. *Trends Plant Sci.* **2015,** *20*, 219–229. DOI: 10.1016/j.tplants.2015.02.001
14. Hasanuzzaman, M.; Nahar, K.; Alam, M. M.; Roychowdhury, R.; Fujita, M. Physiological, Biochemical, and Molecular Mechanisms of Heat Stress Tolerance in Plants. *Int. J. Mol. Sci.* **2013,** *14*, 9643–9684. DOI: 10.3390/ijms14059643
15. Farhangi-Abriz, S.; Ghassemi-Golezani, K. Jasmonates: Mechanisms and Functions in Abiotic Stress Tolerance of Plants. *Biocat. Agric. Biotech.* **2019,** *20*, 101210. DOI: 10.1016/j.bcab.2019.101210
16. Lippmann, R.; Babben, S.; Menger, A.; Delker, C.; Quint, M. Development of Wild and Cultivated Plants Under Global Warming Conditions. *Curr. Biol.* **2019,** *29*, R1326–R1338. DOI: 10.1016/j.cub.2019.10.016
17. Wang, X.; Xu, C.; Cai, X.; Wang, Q.; Dai, S. Heat-Responsive Photosynthetic and Signaling Pathways in Plants: Insight from Proteomics. *Int. J. Mol. Sci.* **2017,** *18*, 2191. DOI: 10.3390/ijms18102191
18. Hu, S.; Ding, Y.; Zhu, C. Sensitivity and Responses of Chloroplasts to Heat Stress in Plants. *Front. Plant Sci.* **2020,** *11*, 375. DOI: 10.3389/fpls.2020.00375
19. Wise, R. R.; Olson, A. J.; Schrader, S. M.; Sharkey, T. D. Electron Transport Is the Functional Limitation of Photosynthesis in Field-Grown Pima Cotton Plants at High Temperature. *Plant Cell Environ* **2004,** *27*, 717–724.
20. Sharkey, T. D. Effects of Moderate Heat Stress on Photosynthesis: Importance of Thylakoid Reactions, Rubisco Deactivation, Reactive Oxygen Species, and Thermotolerance Provided by Isoprene. *Plant, Cell Environ* **2005,** *28*, 269–277.
21. Morales, D.; Rodriguez, P.; Dellamico, J.; Nicolas, E.; Torrecillas, A.; Sanchez-Blanco, M. J. High-Temperature Preconditioning and Thermal Shock Imposition Affects Water Relations, Gas Exchange and Root Hydraulic Conductivity in Tomato. *Biol. Plant* **2003,** *47*, 203–208.
22. Lipiec, J.; Doussan, C.; Nosalewicz, A.; Kondracka, K. Effect of Drought and Heat Stresses on Plant Growth and Yield: A Review. *Int. Agrophys* **2013,** *27* (4), 463–477. DOI: https://doi.org/10.2478/intag-2013-0017
23. Hashem, H. A. *Molecular Physiological Studies on Heat Shock Protein Expression on a Stressed Plant*; M.Sc. Thesis, Faculty of Science, Ain Shams University, Cairo, Egypt, 2000.
24. Wardlaw, I. F.; Wrigley, C. W. Heat Tolerance in Temperate Cereals: An Overview. *Aust. J. Plant Physiol.* **1994,** *21*, 695–703.

25. Khan, M. I. R.; Khan, N. *Reactive Oxygen Species and Antioxidant System in Plants: Role and regulation Under Abiotic Stress*; Springer Nature: New York, 2017; pp 978–981.
26. Wang, K.; Zhang, X.; Goatley, M.; Ervin, E. Heat Shock Proteins in Relation to Heat Stress Tolerance of Creeping Bent Grass at Different N Levels. *PLoS One* **2014,** *9*, e102914.
27. Balfagón, D.; Sengupta, S.; Gómez-Cadenas,A.; Fritschi, F. B.; Azad, R. K.; Mittler, R., Zandalinas, S. I. Jasmonic Acid Is Required for Plant Acclimation to a Combination of High Light and Heat Stress. *Plant Physiol.* **2019,** *181* (4), 1668–1682. https://doi.org/10.1104/pp.19.00956
28. Gleason, C.; Leelarasamee, N.; Meldau, D.; Feussner, I. OPDA Has Key Role in Regulating Plant Susceptibility to the Root-Knot Nematode *Meloidogyne hapla* in *Arabidopsis*. *Front. Plant Sci.* **2016,** *7*, 1565. DOI: 10.3389/fpls.2016.01565
29. Creelman, R. A.; Mulpuri, R. The Oxylipin Pathway in *Arabidopsis*. In: Somerville, C. R., Meyerozitz E. M., Eds.; *The Arabidopsis Book*; American Society of Plant Biologists: Rockville, 2002; pp 1–24.
30. Ziegler, J.; Keinanen, M.; Baldwin, I. Herbivore-Induced Allene Oxide Synthase Transcripts and Jasmonic Acid in *Nicotiana attenuate*. *Phytochemistry* **2001,** *58*, 729–738.
31. Perez-Gilabert, M.; Lopez-Nicolas, J. M.; Garcia, C. F. Purification of a Novel Lipoxygenase from Eggplant (*Solanummelongena*) Fruit Chloroplast. *Physiol. Plant* **2001,** *111*, 276–282.
32. Theodoulou,F. L.; Job, K.; Slocombe, S. P.; Footitt, S.; Holdsworth, M.; Baker, A.; Larson, T.; Graham, I. Jasmonic Acid Levels Are Reduced in Comatose ATP-Binding Cassette Transporter Mutants: Implications for Transport of Jasmonate Precursors into Peroxisomes. *Plant Physiol.* **2005,** *137*, 835–840.
33. Schneider, K.; Kienow, L.; Schmelzer, E.; Colby, T.; Bartsch, M.; Miersch, O.; Wasternack, C.; Kombrink, E.; Stuible, H-P. A New Type of Peroxisomal Acyl-Coenzyme A Synthase from *Arabidopsis thaliana* has the Catalytic Capacity to Activate Biosynthesis Precursors of Jasmonic Acid. *J. Biol. Chem.* **2005,** *280* (14), 13962–13972.
34. Cruz Castillo, M.; Martinez, C.; Buchala, A.; Metraux, J.; Leon J. Gene-Specific Involvement of β-Oxidation in Wound–Activated Responses in *Arabidopsis*. *Plant Physiol.* **2004,** *135*, 85–94.
35. Feussner, I.; Wasternack, C. The Lipoxygenase Pathway. *Ann. Rev. Plant Biol.* **2002,** *53*, 275–297.
36. Royo, J.; Vancanneyt,G.; Pérez, A.; Sanz, C.; Störmann, K.; Rosahl, S.; Sanchez-Serrano J. Characterization of Three Potato Lipoxygenases with Distinct Enzymatic Activities and Different Organ-Specific and Wound-Regulated Expression Patterns. *J. Plant Cell.* **1996,** *271*, 21012–21019.
37. Bell, E.; Creelman, R. A.; Mullet, J. E. A Chloroplast Lipoxygenase Is Required for Wound-Induced Jasmonic Acid Accumulation in *Arabidopsis*. *Proc. Natl. Acad. Sci. USA* **1995,** *92*, 8675–8679.
38. Murata, N.; Los, D. A.; Membrane Fluidity and Temperature Perception. *Plant Physiol.* **1997,** *115*, 875–879.

39. Reinbothe, S.; Reinbothe, C.; Neumann, D.; Apel, K. A Plastid Enzyme Arrested in the Step of Precursor Translocation In Vivo. *PNAS* **1996,** *93*, 12026–12030.
40. Rojo, E.; Titarenko, E.; León, J.; Berger, S.; Vancanneyt, G.; Sánchez-Serrano, J. Reversible Protein Phosphorylation Regulates Jasmonic Acid-Dependent and –Independent Wound Signal Transduction Pathways in *Arabidopsis thaliana. Plant J.* **1998,** *13*, 153–165.
41. Devoto, A.; Ellis, C.; Magusin, A.; Chang, H.; Chilcott, C.; Zhu, T.; Turner, J. Expression Profiling Reveals Coi1 to be a Key Regulator of Genes Involved in Wound-and Methyl Jasmonate-Induced Secondary Metabolism, Defence, and Hormone Interactions. *Plant Mol. Biol.* **2005,** *58*, 497–513.
42. Dang, F. F.; Wang, Y. N.; Yu, L.; Eulgem, T.; Lai, Y.; Liu, Z. Q.; Wang, X.; Qiu, A. L.; Zhang, T. X.; Lin, J.; Chen, Y. S.; Guan, D. Y.; Cai, H. Y.; Mou, S. L.; He, S. L. CaWRKY40, a WRKY Protein of Pepper, Plays an Important Role in the Regulation of Tolerance to Heat Stress and Resistance to *Ralstonia solanacearum* Infection. *Plant Cell Environ.* **2013,** *36*, 757–774. DOI: 10.1111/pce.12011
43. Fonseca, S.; Chini, A.; Hamberg, M.; Adie, B.; Porzel, A.; Kramell, R. (+)-7-Iso-Jasmonoyl-L-Isoleucine Is the Endogenous Bioactive Jasmonate. *Nat. Chem. Biol.* **2009,** *5*, 344–350.
44. Xie, D. X.; Feys, B. F.; James, S.; Nieto-Rostro, M.; Turner, J. G. COI1: An *Arabidopsis* Gene Required for Jasmonate-Regulated Defense and Fertility. *Science* **1998,** *280*, 1091–1094. DOI: 10.1126/science.280.53 66.1091
45. Chini, A.; Fonseca, S.; Fernández, G.; Adie, B.; Chico, J. M.; Lorenzo, O. The JAZ Family of Repressors Is the Missing Link in Jasmonate Signaling. *Nature* **2007,** *448*, 666–671. DOI: 10.1038/nature06006
46. Yan, Y.; Stolz, S.; Chételat, A.; Reymond, P.; Pagni, M.; Dubugnon, L. A Downstream Mediator in the Growth Repression IMB of the Jasmonate Pathway. *Plant Cell* **2007,** *19*, 2470–2483. DOI: 10.1105/tpc.107. 050708
47. Yan, J.; Zhang, C.; Gu, M.; Bai, Z.; Zhang, W. The Arabidopsis CORONATINE INSENSITIVE1 Protein Is a Jasmonate Receptor. *Plant Cell* **2009,** *21*, 2220–2236. DOI: 10.1105/tpc.109.065730
48. Sheard, L. B.; Tan, X.; Mao, H.; Withers, J.; Ben-Nissan, G.; Hinds, T. R. Jasmonate Perception by Inositol-Phosphate-Potentiated COI1-JAZ Co-receptor. *Nature* **2010,** *468*, 400–405. DOI: 10.1038/nature09430
49. Jing, Y.; Lin, R. The VQ Motif-Containing Protein Family of Plant-Specific Transcriptional Regulators. *Plant Physiol.* **2015,** *169*, 371–378.
50. Macrae, R.; Long, J. A. Interactions Between TOPLESS and Histone Modifying Enzymes. In: *22nd International Conference on Arabidopsis Research*; The North American *Arabidopsis* Steering Committee: Madison, 2011.
51. Pauwels, L.; Barbero, G. F.; Geerinck, J.; Tilleman, S.; Grunewald, W.; Pérez, A. C.; NINJA Connects the Co-repressor TOPLESS to Jasmonate Signaling. *Nature* **2010,** *464*, 788–791. DOI: 10.1038/nature08854
52. Westfall, C. S.; Zubieta, C.; Herrmann, J.; Kapp, U.; Nanao, M.H; Jez, J. M. Structural Basis for Pre-Receptor Modulation of Plant Hormones by GH3 Proteins. *Science* **2012,** *336*, 1708–1711.

53. Raza, A.; Charagh, S.; Zahid, Z. et al. Jasmonic Acid: A Key Frontier in Conferring Abiotic Stress Tolerance in Plants. *Plant Cell Rep.* **2020**. https://doi.org/10.1007/s00299-020-02614-z
54. Brosché, M.; Merilo, E.; Mayer, F.; Pechter, P.; Puzõrjova, I.; Brader, G.; Kangasjärvi, J.; Kollist, H. Natural Variation in Ozone Sensitivity Among *Arabidopsis thaliana* Accessions and Its Relation to Stomatal Conductance. *Plant Cell Environ.* **2010,** *33*, 914–25. DOI: 10.1111/j.1365-3040.2010.02116.x.
55. Mittler, R.; Vanderauwera, S.; Gollery, M.; Van Breusegem, F. Reactive Oxygen Gene Network of Plants. *Trends Plant Sci.* **2004,** *9*, 490–498.
56. Møller, I. M.; Jensen, P. E.; Hansson, A. Oxidative Modifications to Cellular Components in Plants. *Annu. Rev. Plant Biol.* **2007,** *58*, 459–481.
57. Hoang, T. V.; Vo, K. T. X.; Mizanor, M. M.; Choi, S. H.; Jeon, J. S. Heat Stress Transcription Factor *OsSPL7* Plays a Critical Role in Reactive Oxygen Species Balance and Stress Responses in Rice. *Plant Sci.* **2019,** *289*, 110273. DOI: 10.1016/j.plantsci.2019.110273
58. Yang, T.; Poovaiah, B. W. Hydrogen Peroxide Homeostasis: Activation of Plant Catalase by Calcium/Calmodulin. *Proc. Natl. Acad. Sci. USA* **2002,** *99*, 4097–4102. DOI: 10.1073/pnas.052564899
59. Quan, L. J.; Zhang, B.; Shi, W. W.; Li, H. Y. Hydrogen Peroxide in Plants: A Versatile Molecule of the Reactive Oxygen Species Network. *J. Integr. Plant Biol.* **2008,** *50*, 2–18. DOI: 10.1111/j.1744-7909.2007. 00599.x
60. Foyer, C. H.; Noctor, G. Oxidant and Antioxidant Signaling in Plants: A Re-Evaluation of the Concept of Oxidative Stress in a Physiological Context. *Plant Cell Environ.* **2005,** *28*, 1056–1071. DOI: 10.1111/j.1365-3040.2005.01327.x
61. Foyer, C. H. Reactive Oxygen Species, Oxidative Signaling and the Regulation of Photosynthesis. *Environ. Exp. Bot.* **2018,** *154*, 134–142.
62. Ai, L.; Li, Z. H.; Xie, Z. X.; Tian, X. L.; Eneji, A. E.; Duan, L. S.; Coronatine Alleviates Polyethylene Glycol-Induced Water Stress in Two Rice (*Oryza sativa* L.) Cultivars. *J. Agron. Crop. Sci.* **2008,** *194*, 360–368. DOI: 10.1111/j.1439-037X.2008.00325.x
63. Chen, J.; Miao, W.; Fei, K.; Shen, H.; Zhou, Y.; Shen, Y.; Li, C.; He, J.; Zhu, K.; Wang, Z.; Yang, J. Jasmonates Alleviate the Harm of High-Temperature Stress During Anthesis to Stigma Vitality of Photothermosensitive Genetic Male Sterile Rice Lines. *Front. Plant Sci.* **2021,** *12*, 634–659. DOI: 10.3389/fpls.2021.634959
64. Loggini, B.; Scartazza, A.; Brugnoli, E.; Navari-Izzo, F. Antioxidative Defence System, Pigment Composition, and Photosynthetic Efficiency in Two Wheat Cultivars Subjected to Drought. *Plant Physiol.* **1999,** *119*, 1091–1100.
65. Hashem, H. A. *Physiological and Molecular Actions of Jasmonic Acid on Soybean Plant*; Ph.D. Thesis, Faculty of Scince, Ain Shams University, Cairo, Egypt, 2006.
66. Xiang, C. B.; Olver, D. J. Glutathione Metabolic Genes Coordmately Respond to Heavy Metals and Jasmonic Acid in *Arabidopsis*. *Plant Cell.* **1998,** *10*, 1539–1550.
67. Wang, S. Y. Methyl Jasmonate Reduces Water Stress in Strawberry. *J. Plant Growth Regul.* **1999,** *18*, 127–134.
68. Hurkman, W. J.; Tanaka, C. K. Effect of Salt Stress on Germin Gene Expression in Barley Roots. *Plant Physiol.* **1996,** *110*, 971–977.

69. Torres, C. A.; Sepulveda, A.; González-Talice, J.; Yuri, J. A.; Razmilic, I. Fruit Water Relations and Osmoregulation on Apples (*Malus domestica* Borkh.) with Different Sun Exposures and Sun-Injury Levels on the Tree. *Sci. Hort.* **2013,** *161*, 143–152. DOI: 10.1016/j.scienta.2013.06.035
70. Sharma, A.; Shahzad, B.; Kumar, V.; Kohli, S. K.; Sidhu, G.; Bali, A. S.; Handa, N.; Kapoor, D.; Bhardwaj, R.; Zheng, B. Phytohormones Regulate Accumulation of Osmolytes Under Abiotic Stress. *Biomolecules* **2019,** *9* (7), 285. https://doi.org/10.3390/biom9070285
71. De Ollas, C.; Dood, I. C. Physiological Impacts of ABA-JA Interactions Under Water-Limitation. *Plant Mol. Biol.* **2016,** *91*, 641–650. DOI: 10.1007/s11103-016-0503-6
72. Lutt, S.; Klinet, J. M.; Bouharmot, J. Effects of Various Salts and of Mannitol on Ion and Proline Accumulation in Relation to Osmotic Adjustment in Rice (*Oryza sativa* L.) Varieties Differing in Salinity Resistance. *J. Exp. Bot.* **1996,** *46*, 1843–1852.
73. Per, T. S.; Khan, N. A.; Reddy, P. S. B.; Masood, A.; Hasanuzzaman, M.; Khan, M. I. R.; Anjum, N. A. Approaches in Modulating Proline Metabolism in Plants for Salt and Drought Stress Tolerance: Phytohormones, Mineral Nutrients and Transgenics. *Plant Physiol. Biochem.* **2017,** *115*, 126–140.
74. Wahid, A.; Gelani, S.; Ashraf, M.; Foolad, M. R. Heat Tolerance in Plants: An Overview. *Environ. Exp. Bot.* **2007,** *61*, 199–223.
75. Reddy, P. S.; Jogeswar, G.; Rasineni, G. K.; Maheswari, M.; Reddy, A. R.; Varshney, R. K.; Kavi Kishor, P. B. Proline Over-Accumulation Alleviates Salt Stress and Protects Photosynthetic and Antioxidant Enzyme Activities in Transgenic Sorghum [*Sorghum bicolor* (L.) Moench]. *Plant Physiol. Biochem.* **2015,** *94*, 104–113.
76. Yun-Cong, Y.; Dapeng, Z.; Younian, W.; Yao, Y. C.; Zhang, D. P.; Wang, Y. N. Preliminary Study of MeJA Effect in the Increasing Drought-Tolerance of Young Apple Tree. *Acta Hort. Sin.* **1999,** *26*, 333–334.
77. Han, J. F.; Lu, H. S.; Chen, P. H.; Yan, S. Q. Effect of Jasmonic Acid on the Germination of Seed and Primary Seedling Growth of Tobacco. *Plant Physiol. Commun.* **1999,** *35*, 183–195.
78. Asthir, B.; Koundal, A.; Bains, N. S. Putrescine Modulates Antioxidant Defense Response in Wheat Under High Temperature Stress. *Biol. Plant.* **2012,** *56*, 757–761. DOI: 10.1007/s10535-012-0209-1
79. Tanamachi, K.; Miyazaki, M.; Matsuo, K.; Suriyasak, C.; Tamada, A.; Matsuyama, K.; Iwaya-Inoue, M.; Ishibashi, Y. Differential Responses to High Temperature During Maturation in Heat-Stress Tolerant Cultivars of Japonica Rice. *Plant Prod. Sci.* **2016,** *19*, 300–308. DOI: 10.1080/1343943X.2016.1140007
80. Chen, R. M.; Li, H. N.; Pan, R. C.; Liu, Z. S. The Regulatory Effect of Methyl Jasmonate on the Photosynthetic Rate and the Transportation of Assimilate in Rice. *ActaBotSin* **1993,** *35*, 600–605.
81. Ramakrishna, A.; Ravishankar, G. A. Influence of Abiotic Stress Signals on Secondary Metabolites in Plants. *Plant Signal Behav.* **2011,** *6* (11), 1720–1731. https://doi.org/10.4161/psb.6.11.17613
82. Van der Fits, L.; Memelink, J. ORCA3, a Jasmonate Responsive Transcriptional Regulator of Plant Primary and Secondary Metabolism. *Science* **2000,** *289*, 295–297. DOI: 10.1126/science.289.5477.295.

83. Fuentes, J. D.; Wang, D. On the Seasonality of Isoprene Emissions from a Mixed Temperate Forest. *Ecol. Appl.* **1999,** *9* (4), 1118–1131. DOI: 10.2307/2641382.
84. Harrison, S. P.; Morfopoulos, C.; Dani, K. G. S.; Prentice, I. C.; Arneth, A.; Atwell, B. J.; Barkley, M. P.; Leishman, M. R.; Loreto, F.; Medlyn, B. E.; Niinemets, Ü.; Possell, M.; JosepPeñuelas, J.; Wright, I. J.; Volatile Isoprenoid Emissions from Plastid to Planet. *New Phytol.* **2013,** *197*, 49–57. DOI: 10.1111/nph.12021.
85. Fineschi, S.; Loreto, F.; Staudt, M.; Peñuelas, J. Diversification of Volatile Isoprenoid Emissions from Trees: Evolutionary and Ecological Perspectives. In: *Biology, Controls and Models of Tree Volatile Organic Compound Emissions: Tree Physiology*; Niinemets, Ü., Monson, R. K., Eds., Vol. 5; Springer: Berlin, 2013, pp 1–20.
86. Fatouros, N. E.; Lucas-Barbosa, D.; Weldegergis, B. T.; Pashalidou,F. G.; Van Loon, J. J. A.; Dicke, M.; Harvey, J. A.; Gols, R.; Huigens, M. E. Plant Volatiles Induced by Herbivore Egg Deposition Affect Insects of Different Trophic Levels. *PLOS One* **2012,** *7* (8). DOI: 10.1371/journal.pone.0043607
87. Sharkey, T. D.; Chen, X. Y.; Yeh, S.; Isoprene Increases Thermotolerance of Fosmidomycin-Fed Leaves. *Plant Physiol.* **2001,** *125* (4), 2001–2006. DOI: 10.1104/pp.125.4.2001
88. Llusia, J.; Penuelas, J.; Asensio, D.; Munne-Bosch, S. Airborne Limonene Confers Limited Thermotolerance to *Quercus ilex*. *Physiol Plant.* **2005,** *123* (1), 40–48. DOI: 10.1111/j.1399-3054.2004.00426.x.
89. Vickers, C. E.; Gershenzon, J.; Lerdau, M. T.; Loreto, F. A Unified Mechanism of Action for Volatile Isoprenoids in Plant Abiotic Stress. *Nat. Chem. Biol.* **2009,** *5*, 283–291.
90. Xu, Y-H.; Liao, Y-C.; Zhang, Z.; Liu, J.; Sun, P-W.; Gao, Z-H.; Sui, C.; Wei, J.-H. Jasmonic Acid Is a Crucial Signal Transducer in Heat Shock Induced Sesquiterpene Formation in *Aquilaria sinensis*. *Sci. Rep.* **2016,** *6*, 21843.
91. Degu, A.; Ayenew, B.; Cramer, G. R.; Fait A. Polyphenolic Responses of Grapevine Berries to Light, Temperature, Oxidative Stress, Abscisic Acid and Jasmonic Acid Show Specific Developmental-Dependent Degrees of Metabolic Resilience to Perturbation. *Food Chem.* **2016,** *212*, 828–836.
92. Parthier, B.; Lehmann, J.; Lerbs, S.; Lerbs, W.; Weidhase, R.; Wollgiehn, R. Hormone and Light Actions in the Differentiation Program of Chloroplasts. In: *Biology of Plant Growth Control*; Fox, E., Jacobs, M., Eds.; Alan R. Liss: New York, 1987; pp 391–400.
93. Reinbothe, S.; Reinbothe, C.; Partheir, B. Methyl Jasmonate Responses Translation Initiation of a Specific Set of mRNAs in Barley. *Plant J.* **1993**, *4*, 459–467.
94. Reinbothe, S.; Mollenhauer, B.; Reinbothe, C. JIPs and RIPs: The Regulation of Plant Gene Expression by Jasmonate in Response to Environmental Cues and Pathogen. *Plant Cell* **1994,** *6*, 1197–1209.
95. Herrmann, G.; Lehmann, J.; Peterson, A.; Sembdner, G.; Weidhase, R.; Parthier B. Species-and Tissue-Specificity of Jasmonate-Induced Abundant Proteins. *J. Plant Physiol.* **1989,** *134*, 703–709.
96. Reinbothe, S.; Reinbothe, C.; Heintzen, C.; Seidenbecher C.; Partheir, B. A Methyl Jasmonate-Induced Shift in the Length of the 5' Untranslated Region Impairs Translation of the Plastid rbcL Transcript in Barley. *EMBO J.* **1993**, *12*, 1505–1512.

97. Ussuf, K.; Laxmi, N.; Mitra, R. Proteinase Inhibitors: Plant-Derived Genes of Insecticidal Protein for Developing Insect-Resistant Transgenic Plants. *Curr. Sci.* **2001,** *8*, 847–853.
98. Vignutelli, A.; Wasternack, C.; Apel, K.; Bohlmann, H.; Systemic and local induction of an *Arabidopsis* thionin gene by wounding and pathogen. *Plant J.* **1998,** *14*, 285–295.
99. Devoto, A.; Turner, J. Regulation of Jasmonate-Mediated Plant Responses in *Arabidopsis*. *Annu. Bot.* **2003,** *92*, 329–337.
100. Sadder, M.; Alsadon, A.; Wahb-Allah, M. Transcriptomic Analysis of Tomato Lines Reveals Putative Stress-Specific Biomarkers. *Turk. J. Agric. For.* **2014,** *38*, 700–771.
101. Vierling, E. Heat Shock Protein Function and Expression in Plants. In: *Plant Biology. Stress Responses in Plants: Adaptation and Acclimation Mechanisms*; Alsher and Cumming, Ed.; Wiley-Liss (A John Wiley & Sons INC., Publication), 1990; pp 357–375.
102. Boston, R. S.; Viitanen, P. V.; Vierling, E. Molecular Chaperones and Protein Folding in Plants. *Plant Mol. Biol.* **1996,** *32*, 191–222. DOI: 10.1007/BF00039383
103. Key, J. L.; Lee, Y. R. J.; Goekjian, V.; Nagao, R. T. The Low Molecular Weight Heat Shock Proteins of Soybean Seedlings. In: *Cell Biology: Biochemical and Cellular Mechanisms of Stress Tolerance in Plants: High Temperature Stress*; Cherry J. H., Ed.; NATO ASI Series, Springer-Verlag, 1994; pp 115–141.
104. Mc Alister, L.; Finkelstein, D. B. Heat Shock and Thermal Resistance in Yeast. *Biochem. Biophys. Res. Comm.* **1980,** *93*, 819–824.
105. Marmiroli, N.; Lorenzoni, C.; Cattevelli, L.; Stanca, A. M.; Terzi, V. Induction of Heat Shock Proteins and Acquisition of Thermotolerance in Barley (*Hordeum vulgare* L.) Variation Association with Growth Habit and Plant Development. *J. Plant Physiol.***1989**, *135*, 267–273.
106. Pareek, A.; Singla, S. L.; Grover, A. Immunological Evidence for Accumulation of Two High-Molecular-Weight (104 and 90 kDa) HSPs in Response to Different Stresses in Rice and in Response to High Temperature Stress in Diverse Plant Genera. *Plant Mol. Biol.* **1995,** *29*, 293–301.
107. Li, Q. B.; Haskell, D. W.; Guy, C. L. Coordinate and Non-Coordinate Expression of the Stress 70 Family and Other Molecular Chaperones at High and Low Temperature in Spinach and Tomato. *Plant Mol. Biol.* **1999,** *39*, 21–34.
108. Corrales, F. J.; Fersht A. R. Toward a Mechanism for GroEL-GroES Chaperone Activity: An ATPase-Gated and Pulsed Folding and Annealing Cage. *Proc. Natl. Acad. Sci. USA* **1996,** *93*, 4509–4512.
109. Lee, B. H.; Tanaka, Y.; Iwasaki T.; Yamamoto, N.; Kayano, T.; Miyao, M. Evolutionary Origin of Two Genes for Chloroplast Small Heat Shock Protein of Tobacco. *Plant Mol. Biol.* **1998,** *37*, 1035–1043.
110. Anderson, J. V.; Guy, C. L. Spinach Leaf 70-Kilodalton Heat-Shock Cognate Stabilizes Bovinc Adrenal Glucose-6-Phosphate Dehydrogenase In Vitro Without Apparent Stable Binding. *Planta* **1995,** *196*, 303–310.
111. Lee; G. J.; Pokala N.; Vierling E. Structure and In Vitro Molecular Chaperone Activity of Cytosolic Small Heat Shock Proteins from Pea. *J. Biol. Chem.* **1995,** *270*, 10432–10438.
112. Chiang, H. L.; Terlecky, S. R.; Plant, C. P.; Dice, J. F. A Role for a 70-Kilodalton Heat Shock Protein in Lysosomal Degradation of Intracellular Proteins. *Science* **1989,** *246*, 382–385.

113. Hübel, A.; Schöffl, F. *Arabidopsis* Heat Shock Factor: Isolation and Characterization of the Gene and the Recombinant Protein. *Plant Mol. Biol.* **1994,** *26*, 353–362.
114. Hamilton, E. W.; Coleman, J.S. Heat-Shock Proteins Are Induced in Unstressed Leaves of *nicotiana attenuata* (Solanaceae) When Distant Leaves Are Stressed. *Am. J. Bot.* **2001,** *88* (5), 950–955.
115. Ruan, J.; Zhou, Y.; Zhou, M.; Yan, J.; Khurshid, M.; Weng, W.; Cheng, J.; Zhang, K. Jasmonic Acid Signaling Pathway in Plants. *Int. J. Mol. Sci.* **2019,** *20* (10), 2479. https://doi.org/10.3390/ijms20102479
116. Oh, S. Y.; Kim, J. H.; Park, M. J.; Kim, S. M.; Yoon, C. S.; Joo, Y. M.; Kang, H. S. Induction of Heat Shock Protein 72 in C6 Glioma Cells by Methyl Jasmonate Through ROS-Dependent Heat Shock Factor 1 Activation. *Int. J. Mol. Med.* **2005,** *16*, 833–839. https://doi.org/10.3892/ijmm.16.5.833
117. Tian, X.; Wang, F.; Zhao, Y.; Lan, T.; Yu, K.; Zhang, L.; Qin, Z.; Hu, Z.; Yao, Y.; Ni, Z.; Sun, Q.; Rossi, V.; Peng, H.; Xin, M. Heat Shock Transcription Factor A1b Regulates Heat Tolerance in Wheat and *Arabidopsis* Through OPR3 and Jasmonate Signaling Pathway. *Plant Biotech. J.* **2020,** *18* (5), 1109–1111. https://doi.org/10.1111/pbi.13268
118. Sharma, M.; Laxmi, A. Jasmonates: Emerging Players in Controlling Temperature Stress Tolerance. *Front. Plant Sci.* **2016,** *6*, 1129. DOI: 10.3389/fpls.2015.01129
119. Torres, C. A.; Sepúlveda, G.; Kahlaoui, B. Phytohormone Interaction Modulating Fruit Responses to Photooxidative and Heat Stress on Apple (*Malus domestica* Borkh.). *Front. Plant Sci.* **2017,** *8*, 2129. DOI: 10.3389/fpls.2017.02129
120. Per, T. S.; Khan, M. I. R.; Anjum, N. A.; Masood, A.; Hussain, S. J.; Khan, N. A. Jasmonates in Plants Under Abiotic Stresses: Crosstalk with Other Phytohormones Matters. *Environ. Exp. Bot.* **2018,** *145*, 104–120.
121. Larkindale, J.; Knight, M. R. Protection Against Heat Stress-Induced Oxidative Damage in *Arabidopsis* Involves Calcium, Abscisic Acid, Ethylene, and Salicylic Acid. *Plant Physiol.* **2002,** *128*, 682–695. DOI: 10.1104/pp.010320
122. Larkindale, J.; Hall, J. D.; Knight, M. R.; Vierling, E. Heat Stress Phenotypes of *Arabidopsis* Mutants Implicate Multiple Signaling Pathways in the Acquisition of Thermotolerance. *Plant Physiol.* **2005,** *138*, 882–897. DOI: 10.1104/ pp.105.062257
123. Clarke, S. M.; Cristescu, S. M.; Miersch, O.; Harren, F. J. M.; Wasternack, C.; Mur, L. A. J. Jasmonates Act with Salicylic Acid to Confer Basal Thermotolerance in *Arabidopsis thaliana*. *New Phytol.* **2009,** *182*, 175–187. DOI: 10. 1111/j.1469-8137.2008.02735.x
124. Czapski, J.; Saniewski, M. Stimulation of Ethylene Production and Ethylene-Forming Enzyme Activity in Fruits of the Non-Ripening Nor and Rin Tomato Mutants by Methyl Jasmonate. *J. Plant Physiol.* **1992,** *139*, 265–268.
125. Zhang, X.; Zhu, Z.; An, F.; Hao, D.; Li, P.; Song, J.; Yi, C.; Guo, H. Jasmonate-Activated MYC2 Represses ETHYLENE INSENSITIVE3 Activity to Antagonize Ethylene-Promoted Apical Hook Formation in *Arabidopsis*. *Plant Cell* **2014,** *26*, 1105–1117.
126. Acharya, B. R.; Assmann, S. M. Hormone Interactions in Stomatal Function. *Plant Mol. Biol.* **2009,** *69*, 451–462.
127. Nahar, K.; Kyndt, T.; Nzogela, Y. B.; Gheysen, G. Abscisic Acid Interacts Antagonistically with Classical Defense Pathways in Rice-Migratory Nematode Interaction. *New Phytol.* **2012,** *196*, 901–913.

128. Kim, B.; Fujioka, S.; Kwon, M.; Jeon, J.; Choe, S. *Arabidopsis* Brassinosteroid-Overproducing Gulliver3-D/Dwarf4-D Mutants Exhibit Altered Responses to Jasmonic Acid and Pathogen. *Plant Cell Rep.* **2013,** *32*, 1139–1149.
129. Gan, L.; Wu, H.; Wu, D.; Zhang, Z.; Guo, Z.; Yang, N.; Xia, K.; Zhou, X.; Oh, K.; Matsuoka, M.; Ng, D.; Zhu, C. Methyl Jasmonate Inhibits Lamina Joint Inclination by Repressing Brassinosteroid Biosynthesis and Signaling in Rice. *Plant Sci.* **2015,** *241*, 238–245.
130. Nahar, K.; Kyndt, T.; Hause, B.; Höfte, M.; Gheysen, G. Brassinosteroids Suppress Rice Defense Against Root-Knot Nematodes Through Antagonism with the Jasmonate Pathway. *Mol. Plant Microbe Interact.* **2013,** *26*, 106–115.
131. Fahad, S.; Hussain, S.; Saud, S.; Hassan, S.; Ihsan, Z.; Shah, A. N.; Wu, C.; Yousaf, M.; Nasim, W.; Alharby, H.; Alghabari, F.; Huang, J. L. Exogenously Applied Plant Growth Regulators Enhance the Morpho-Physiological Growth and Yield of Rice Under High Temperature. *Front. Plant Sci.* **2016,** *7*, 1250.
132. Chen, J.; Fei, K.; Zhang, W.; Wang, Z.; Zhang, J.; Yang, J. Brassinosteroids Mediate the Effect of High Temperature During Anthesis on the Pistil Activity of Photo-Thermosensitive Genetic Male-Sterile Rice Lines. *Crop J.* **2021,** *9*, 109–119.
133. Kotak, S.; Larkindale, J.; Lee, U.; Von Koskull-Döring, P.; Vierling, E.; Scharf, K. D. Complexity of the Heat Stress Response in Plants. *Curr. Opin. Plant Biol.* **2007,** *10*, 310–316.
134. Van der Does, D.; Leon-Reyes, A.; Koornneef, A.; Van Verk, M. C.; Rodenburg, N.; Pauwels, L.; Goossens, A.; Körbes, A. P.; Memelink, J.; Ritsema, T.; Van Wees, S. C. M.; Pieterse, C. M. J. Salicylic Acid Suppresses Jasmonic Acid Signaling Downstream of SCFCOI1-JAZ by Targeting GCC Promoter Motifs via Transcription Factor ORA59. *Plant Cell* **2013,** *25*, 744–761.

CHAPTER 9

Signaling and Manipulation of Jasmonate-Dependent Defenses by Microbial Pathogens (Fungi, Bacteria, and Viruses) and Insects

ASHHAD PARVEZ[1], SYED KAMRAN AHMED[2], ANWAR SHAHZAD[1], P. Q. RIZVI[2], and UZMA AZEEM[3]

[1]*Department of Botany, Aligarh Muslim University, Aligarh, Uttar Pradesh, India*

[2]*Department of Plant Protection, Aligarh Muslim University, Aligarh, Uttar Pradesh, India*

[3]*Department of Botany, Government College Malerkotla, Malerkotla, Punjab, India*

ABSTRACT

Plants are capable of defending themselves against various biotic and abiotic stressors through means of biochemical, physical, and mechanical strategies. However, the selection of defense relies basically on the type of stress and stressors. In the case of biotic stressors, namely, insects and microbial pathogens, biochemical mechanisms, along with other defenses contribute a significant role in defense mechanisms. Jasmonates (JAs) are lipid-derived cyclopentanones that are the primary defensive biochemicals that include jasmonic acids (JA) and their derivatives. As one of the major plant immunity hormones, JAs are known to positively regulate the plant

Jasmonates and Plant Defense: An Eco-Friendly and Sustainable Approach for Stress Tolerance. Uzma Azeem, & Khalid Rehman Hakeem (Eds.)

defenses against necrotrophic pathogens, chewing insects and mechanical wounding responses, nevertheless, they negatively regulate plant defenses to biotrophic pathogens. Moreover, JA is a potent inducer of defensive proteinase inhibitors (PI) and secondary metabolites in plants that participates in various defense activities against insects and pathogen attacks. In addition to their role in defense against herbivory and pathogens, JAs also promote the micro-tuber formation, bulb let formation, somatic embryogenesis, callus formation, shoots, and root initiation in plants. This chapter discusses all possible JA JA-related chemical happenings in the plants concerned with defense in insect herbivory and pathogen attacks.

9.1 INTRODUCTION

In general, plants harbor a prolific source of nutrients for various organisms comprising insects, viruses, fungi, bacteria, protists, and vertebrates.[1] Even though deficient in the immune system as good as of animals, plants have established a magnificent assortment of defenses based on structural, chemical, and protein levels, which are meant to spot invasive organisms and inhibit them before they start to cause massive damage.[2] Humans depend almost exclusively on plant products, and they disseminate many imperative nonfood products including dyes, inks, wood, medicines, plastics, textiles, soaps, cosmetics, rubber, and industrial chemicals.[3] Exploring how plants protect themselves from a range of herbivores and pathogens is a prerequisite to keeping our food supply and acquiring plant species with a high disease-resistance variety.

9.2 HERBIVORY

Herbivory is amongst the most significant aspects affecting plant robustness and, thus, it is considered a key selection pressure affecting the plant trait's evolution on a large scale.[46] It is an interrelation among two distinct organisms, its proper comprehension requires the study of both the plants and the herbivores.[7] By that elucidation, many fungi, protists, bacteria, animals, and some parasitic plants can be apprised as herbivores. An herbivore is adapted to eat only plants and sometimes it is also said to be vegetarian, and it can be terrestrial or aquatic. Herbivores vary in size, selectivity, and feeding habits. Some, for instance, many Lepidopteran larvae, are confined to consuming particular species or genera of plants and so they are categorized according to the plant part they eat. Frugivores, the fruit-eating herbivores; folivores, the leaves-eating herbivores, and nectarivores, specifically nectar-feeding herbivores.[8] Among herbivorous insects include granivores, seed-eating herbivores; palynivores, pollen-eating herbivores; mucivores, the plant fluid-feeding herbivores, and xylophages, the herbivores that feed on wood or rhizophages, the root-eating herbivores.[9]

9.3 PLANT PATHOGENS

Generally, viruses, fungi, mycoplasmas, and bacteria are said to be plant pathogens, that are capable of bringing ailments to a healthy plant.[10] This ailment commonly referred to as disease is an abnormal physiological process that disturbs the plant's normal metabolic rate which eventually disrupts its growth, structure, function, and several other activities.[11] It can be caused either due to living agents or biotic agents such as bacteria and fungi or due to environmental agents or abiotic factors such as drought, nutrient deficiency, lack of oxygen, excessive water and temperature, pollution, or ultraviolet (UV) radiation.[1] These pathogens contrast with each other concerning the type of plant, age of the organ or tissue, and site of its infection. The realm of plant pathogens and their specificity is very wide; some pathogens contain a single target of a plant species, while other pathogens can a highjack plant genus, and eventually, there are some other pathogens with an extensive range of hosts that affect many families of higher plants.[12] Some pathogens infect, multiply, and complete their life cycles within a living host, whereas others kill the host in the course of the development of the infection.[13]

9.4 PLANT DEFENCE AGAINST PATHOGENS

Plants are naturally shielded with mechanisms of withstanding diseases caused by pathogens and abiotic stresses, namely, drought, salinity, extreme temperatures, and toxins.[13] This shield includes a wide variety of constitutive and induced defenses.[1] In constitutive defense, the plant body synthesizes the defense compound constitutively. Whenever there is an attack by the insect, the defense system is activated. So, it provides an extra degree of advantage in the form of protection, but growth is compromised because the plant invests more biomass and energy in the defense. Likewise, in induced defense, the defense compound is produced including the production of pathogen degrading enzymes, toxic chemicals, and deliberate cell suicide, only after the attack of the insect, so it uses less energy and biomass for the defense strategies but it is found to be responsible for high risk of attack.[1,14]

9.4.1 SNEAK ATTACK VERSUS BRUTE FORCE

According to Liu et al.,[15] several pathogens act as silent burglars who want to enter another's premises by using specific instruments to incapacitate another's premises security system without endorsement and being detected to steal money. Likewise, there are many pathogens associated with their host to suppress the defense system and nutritional flow without causing immense damage. Pathogens with such a long-term feeding relationship with the host without killing it are termed biotropism, that is, powdery mildew fungi (*Blumeria graminis*) and Mycorrhiza.[16] On the contrary, some pathogens regularly use or produce the cell tissue degrading enzymes or toxins by which they obtain the nutrients from the dead and decayed cells and tissues. Such types of pathogens have been termed necrotrophs,[17,18] that is, *Botrytis cinerea*, *Alternaria solani*, *Plectosphaerella cucumerina*, *Monilinia* spp., *Alternaria brassicicola*, etc. However, there are some pathogens that convert into necrotrophs from biotrophs called hemibiotrophs,[17] that is, including the bacterium *Cladosporium fulvum*, *Septoria tritici*, *Pyrenopeziza brassicae*, *Pseudomonas syringae*, and the oomycete *Phytophthora infestans* (potato blight).[19] Most biotrophic and hemibiotrophic pathogens possess a narrow range of host plants because of the slightly different sets of specific genes and molecular mechanisms.[20] The host range of any pathogen denotes a plant species on which a pathogen is capable of colonization. For example, the brome mosaic virus (BMV) infects grasses such as barley but not legumes.[21] While a plant when colonized by a pathogen without showing peculiar disease symptoms is said to be a nonhost plant species for that pathogen and such pathogens are termed as nonpathogens. An example is the saprophytic bacterial species *Pseudomonas putida*.[22] Therefore, such interactions between pathogens and their host plants can be classified as compatible and incompatible responses. A compatible response causes disease, while an incompatible response leads to petite or does not cause disease. Although a plant species is susceptible to a specific pathogen, some individuals are genetically resistant and, are capable of recognizing the pathogen and defending themselves. For instance, some tomato cultivars are susceptible to *P. syringae* (a compatible response), while others such as cultivar the Rio Grande exhibit disease resistance (an incompatible response).[1] Disease resistance ranges from immunity (no symptoms) to greatly resistant (mild disease) to greatly susceptible (severe disease).[23]

9.4.2 PLANT DEFENSE AGAINST HERBIVORES

Plants evolve ample ways of defense against herbivores to ameliorate their anticipation of existence and reproduction. These defenses may include mechanical shielding on the plant surface, reducing the plant digestibility to animals by producing a complex of polymers, and producing the toxins or poisonous substances that kill or repel the herbivores. Defenses can either be constitutive or induced. Constitutive defense is always present and activated in the plant, while Induced defense is produced or activated when the plant sense any damage or stress.[24] These mechanisms are also been referred to as host plant resistance by the plant breeders. Plants also have the property that escalates the probability to attract antagonists of herbivores, such as *Acacias* or *Mimosa* providing ants with nutriment from sap and special fruits and habitation in hollow thorns, while the ants give protection from other harmful insects and browsing mammals. Plants may void semiochemicals and odors that magnetize natural enemies and may provide food and dwellings to uphold their presence.[25] In some instances, herbivory is stimulated by plants to aid in reproduction; for example, production of nectar to allure bees, which is vital for pollination.[26]

9.4.3 ELICITORS

The identification of insects attacks plants' ability to recognize chemical cues produced by the insects.[27] These chemical cues released by the insects are said to be elicitors. These are insect drives compounded by which a plant can recognize the insect attack. The first identified herbivore-associated elicitor (HAE) compound was fatty acid (FA) amide.[28,29] These HAEs are diverse in structure, ranging from enzymes to modified forms of lipids and from fragments of cell walls to peptides discharged from digested plant proteins (Table 9.1). Refs. [30, 33] has resulted in the stimulation of distinct plant responses to defend against or tolerate attack from insect herbivores; these responses can range from changes in metabolism which include the variations in volatile emissions[34,35] and, expressions of the gene for changes in the pattern of growth and development.[36,37] Similar to microbial effectors, some HAEs can also counteract the defense response of plants.[38,39]

TABLE 9.1 Elicitors Against the Herbivory.

Class I	Class II	Class III
Lytic enzymes	**FA–amino acid conjugates (FACs)**	**Bruchins (mono- and diesterified with 3-hydroxypropanoic acid)**
β-glucosidase isolated from *Pieris brassicae*	Larval Sphingidae, Noctuidae, and Geometridae	From oviposition fluid of pea and cowpea weevils (Bruchidae)
Glucose oxidases in *Helicoverpa zea* saliva	Volicitin [*N*-(17-hydroxylinolenoyl)-L-glutamine], from *Spodoptera exigua*	The oviposition of elm leaf beetles, *Xanthogalerucaluteola*, elicits the emission of specific VOCs from the field elm host trees
Alkaline phosphatase in whitefly (*Bemisia tabaci*) saliva	–	–
Aide array of watery digestive enzymes from aphid saliva	–	–

9.4.3.1 MECHANISM OF ELICITATION

On the cellular level, the compound that is being released by the herbivores and their interaction with the plants is said to be a potential elicitor such as the two-insect fluid/secretions that perpetually come in association with the site of injury is the oral exudation of lepidopteran species and the oviposition fluid of weevil beetles from which several herbivore specific elicitors have been isolated. Nevertheless, some shreds of evidence also suggest that the production of elicitor which is found in the oral secretions is due to the microbes that exist in the digestive organs of herbivores.[40] The initial step in the herbivore elicitation mechanism remains to be illuminated. However, the host resistance gene (R-gene) products have been profoundly found for herbivore elicitors in pathogen recognition. At the site of the attack, Glucose oxidase increases the H_2O_2 production and may potentiate elicitation by forming reactive oxygen species (ROS).[41] R-gene receptor recognizes the pathogen's antivirulence gene (avr-gene) product and shows avr specificity. Binding of elicitor with its receptor, NADPH oxidase and nitric oxide synthase (NOS) both are activated and have various defense strategies of activation. Particularly, Nicotinamide adenine dinucleotide phosphate (NADPH) oxidase out-turn in the formation of super-oxy anion ($O_2{}^{*-}$) which further participates in the production of other relative oxygen species (Figure 9.1). There are several other mechanisms for the elicitation and according to Engelberth et al.[42] channel forming peptide alamethicin, formed by the *Trichoderma viridae* (parasitic fungi), elicited the release of volatile in Lima beans that are on a par with those elicited when the JA cascade is antagonized by salicylic acid (SA) cascade. Unfortunately, there has been no evidence of finding the channel-forming peptides in the saliva of insect herbivores. *Nicotiana attenuata Manduca* system is said to draw a connection from the elicitor to succeeding signaling steps. Fatty acid–amino acid conjugates (FACs) in the oral secretion of two Manduca species extract jasmonic acid (JA) and ethylene outbursts, which are entangled in the modification of *N. attenuata* wound responses[37,43,45] NOS and ROS activate a range of different defense responses such as the following.

Phytoalexins

These are the low-molecular-weight antimicrobial compounds that are produced upon pathogen attack and are also well known for their environmental activities, namely, pisatin in pea plants.[46]

Hypersensitive Response

This type of response is found in all higher plants, when any plant part gets infected with the pathogen, the said infected spot triggers a hypersensitive response that results

Systemic Acquired Resistance

When a plant survives the attack of a pathogen, at the site of infection it develops increased resistance against many pathogenic species in the whole plant body. The phenomenon is known as systemic acquired resistance (SAR) for which SA and H_2O_2 are responsible. H_2O_2 mediates local response whereas SA (a benzoic acid derivative) is responsible for the response to the neighboring or adjacent plants as it is the phloem mobile molecule and acts as a systemic signal. It can have volatile properties when it converts into methyl salicylate (conjugate form).[48]

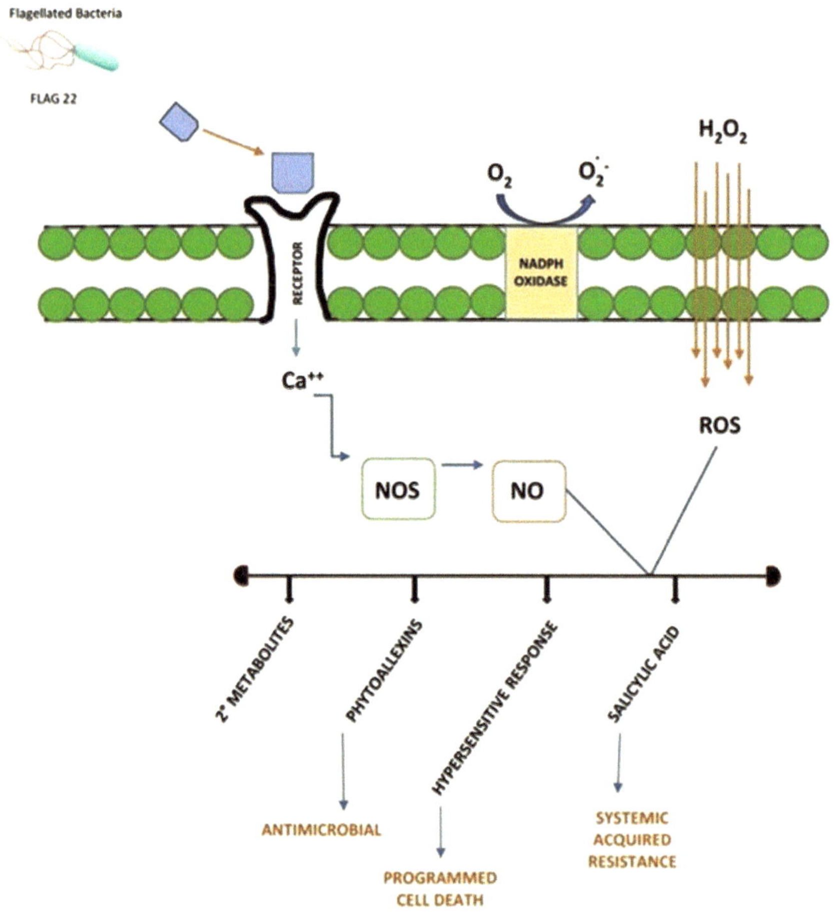

FIGURE 9.1 Mechanism of elicitation induced systemic response.

Sometimes many plants induce their defense responses even after being exposed to nonpathogenic microbe, the phenomenon is called as induced systemic resistance (ISR). So, the plant is getting a certain degree of advantage and has increased preparedness against the real pathogens, but the growth of the plant will be compromised.[49] Example of ISR is rhizobium–mycorrhizal fungi relationship.

The signaling molecule that participates in ISR are as follows:

1. JA
2. Ethylene

9.5 JASMONIC ACID

9.5.1 UNCOVERING OF JASMONATES (JA)

JAs, also called as jasmonates, that are oxylipins, cognates of oxygenated FA or lipid-derived cyclopentanones. In addition to already known plant hormones viz., auxin, cytokinins, gibberellin (GA), abscisic acid, and ethylene (ET),[50] the JAs were newcomers in the family of plant growth hormones. In the year 1962, the methyl-ester or methyl-jasmonate of JA (MeJA) was isolated and studied as a paramount fragrance from the ethereal oil (a volatile substance) of Spanish Jasmine flowers (*J. grandiflorum*), and the free acid was isolated eventually from the filtrates of fungus culture, that is, *Botryodiplodia theobromae.* Later on, JA was further isolated from *Cucurbita pepó* (Pumpkin) and *Vicia faba* (Faba bean). In 1980, these naturally occurring JA and similar compounds were synthesized in the laboratory and bioassay of rice seedlings.[51]

9.5.2 EFFECT OF JASMONIC ACIDS UNDER IN VITRO CONDITIONS

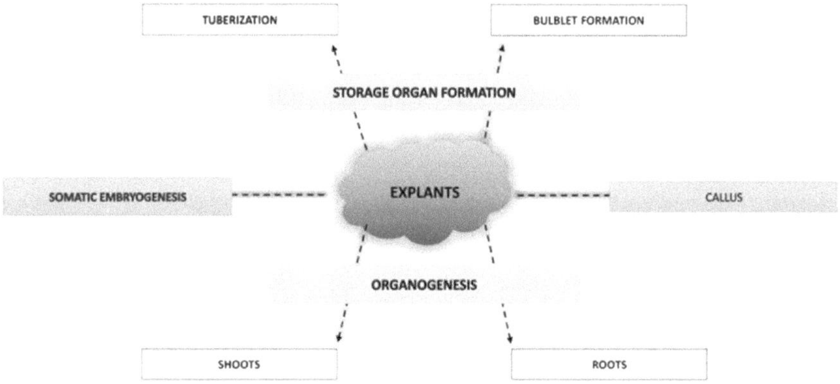

FIGURE 9.2 JA induced modifications in plants under in vitro conditions.

Tuberization: JA promotes micro-tuber formation.[52]

Bulblet Formation: JA affects and stimulates the formation of bulbils as well as the total fresh and dry weight and has a promoting and inhibitory effect on the size of the bulblet.[53]

Somatic Embryogenesis: JA promotes plant regeneration, and it has an inhibitory effect on the number of somatic embryos, germination of embryos as well as on differentiation if it is provided above and below its optimum requirement.[54]

Callus Formation: JA has a promoting as well as inhibitory effect on the proliferation of callus along with its cell cycle progression and it inhibits the growth and viability of callus if it is provided above and below its optimum requirement.[55]

Shoots: JA promotes the number of leaves and the fresh and dry weight but has both the promoting and inhibitory effect on the proliferation of shoots and their elongation.[56]

Roots: JA has the promoting and inhibitory effect on root initiation, and its formation along with its elongation but it inhibits the biomass of roots.[56]

9.5.3 PHYSIOLOGICAL FUNCTION OF JAs

With the advancement of plant sciences, diverse physiological functions of JAS have been reported. Initially, the promotion of senescence and inhibition of seedling growth was described as primary functions of JAs.[57,58] Later on, some volatile MeJAs that functions as an effective inducer of defensive proteinase inhibitors (PI) were found in tomato leaves.[59] This was followed by the discovery of the effect of MeJAs in hampering the root growth of *Arabidopsis thaliana*, the first JA-response mutant named JA resistant 1 (jar1).[60] In plant cell cultures, JAs are considered effective initiators of secondary metabolites.[61,62] These interesting findings inspired a plethora of studies on the biological functions of JAs across the globe. The responses of mutant variants of JA in tomato and *A. thaliana* express the most significant outcome to furnish the biological synthesis and signaling line of communication of JAs which has been addressed in detail separately. Analysis of mutants in association with advanced levels of Biochemical, genomic, molecular, and biochemical approximation has also eloquently advanced our comprehension of the distinctive functions of JAs. Considering the vital plant immunity hormones, JAs have positive regulation in plant defense in response to mechanical wounding, chewing insects, and necrotrophic pathogens, whereas they have negative feedback on plant defense against biotrophic pathogens. Grafting experiments on tomato mutants altered by JA synthesis devour a scenario where JAs may perform as long-distance mobile signals in governing systemic plant immunity.[63] As a PGR, JAs generally repress vegetative growth and stimulates reproductive development. In many cases, JAs in collaboration with growth hormones stimulates plant growth. For example, primary root growth inhibition through JA is obtained via interaction with auxin to moderate the activity of root meristem.[51] Recent examples have shown that the all functions of JA along with its transcription factors (TFs) interface with JAZ repressor subsets to harmonize the range of JA-responsive gene expressions.[64] In spite of the fact that JAs lack in the animal kingdom, exhaustive studies have discovered that JAs show selective and direct repressive effects on numerous tumor cells of human beings without upsetting the normal cells. This has paved the way for the possibility of bringing out anticancer drugs through JAs.[65]

9.5.4 BIOSYNTHESIS OF JA

9.5.4.1 PIONEERING JA BIOSYNTHESIS

The first pathway of JA was proposed by Vick and Zimmerman[66] followed by the addition of some important details about regulation, its subcellular locations, and enzymes involved in the pathway reactions by Browse,[67,68] Wasternack,[69] and Wasternack and Hause.[65] The pathway starts in the chloroplast by LOX (lipoxygenase) catalyzed oxygenation of -α-linolenic acid (Figure 9.3). As a result, 13-HOPT, that is,13(S)-hydro-peroxy-octadecatrienoic acid serves as a substrate for allene oxide synthase (AOS), which creates an unstable polyunsaturated FA, 12,13-EOT allene oxide, [12,13(S)-epoxy-octadecatrienoic acid] which is exclusively cyclized to the 9 and 13S isomers of 12-OPDA (12-oxo-phytodienoic acid) in presence of AOC (allene oxide cyclase), and then exported to the peroxisome. In peroxisomes, OPDA converts to OPDA reductase3 that yields 3-oxo-2-(2(Z)-pentenyl)-cyclopentane-1-octanoic acid. Three cycles of -oxidation omit 6 carbon atoms from the carboxyl side chain resulting in the biosynthesis of JA.[70] In a parallel hexadecanoid pathway, the chloroplast having the same enzymes act on 6,9,12-HTA (hexadecatrienoic acid) in the ratio of 16:3 for the formation of dn-OPDA (dinor-oxophytodienoic acid).

Presuming that dn-OPDA serves as a substrate for 12-oxophytodienoate reductase 3, which could be further metabolized to JA by -oxidation enzymes in the peroxisome.[71]

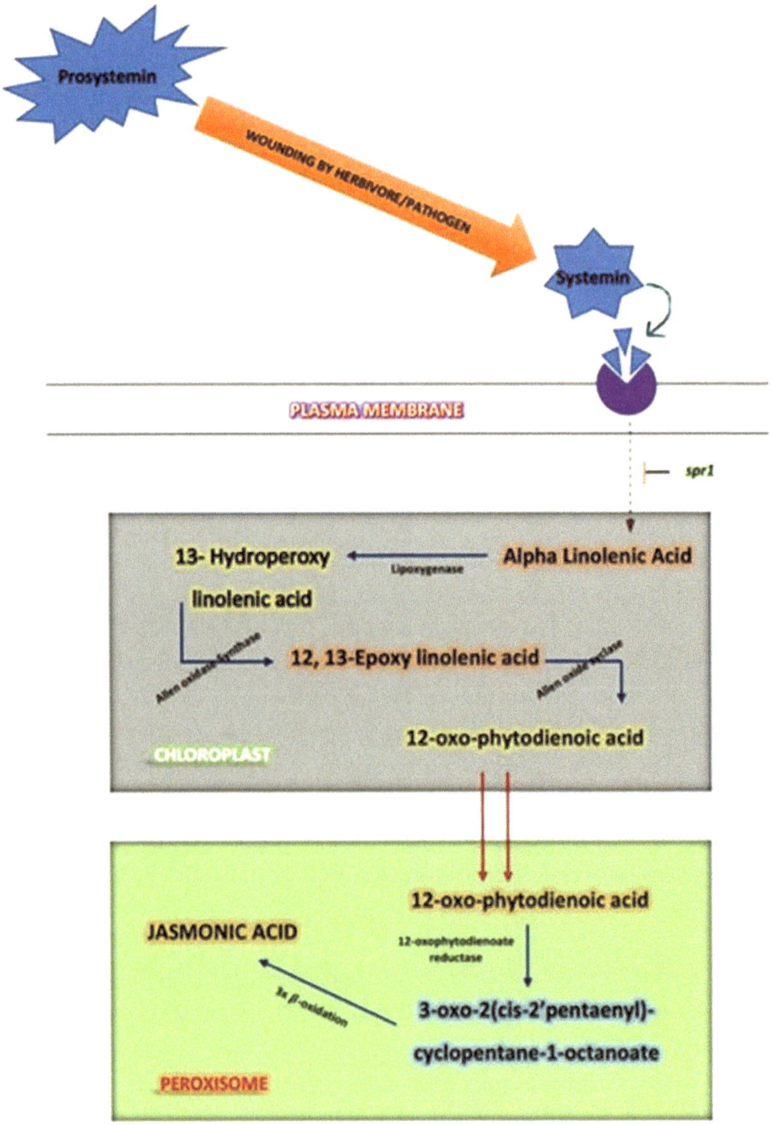

FIGURE 9.3 Octadecanoid pathway of JA biosynthesis.

9.5.4.2 A-LINOLENIC ACID SYNTHESIS FROM LINOLEIC ACID

The precursor of JA is α-α-linolenic acid, a product of linoleic acid formulated through a desaturation reaction which is catalyzed by an enzyme fatty acid desaturase (FAD). They consist of three omega-3 FA desaturation enzymes, that is, FAD-3, FAD-7, and FAD-8 found in the genome of *Arabidopsis*, that form trienoic FAs (16:3 and 18:3) from dienoic FAs (16:2 and 18:2) and these enzymes found in-operative in JA synthesis because it does not end up to a detectable trienoic FA.[72] These mutants typically exhibit characteristics of the male phenotype, readily stored by exogenous (18:3) and JA,[72,73] producing foremost evidence that JA plays a vital role in the development of male gametophytes in *Arabidopsis*. Further elaborating the role of FADs in the biosynthesis of JA has come from the delineation of the suppressor of pro-systemin mediated responses-2 (spr2) mutant of tomato.[74] The mutation of spr2 impedes wound-induced JA biosynthesis and defense gene expression. Spr2 introduces a chloroplast omega-3 FAD which is the homolog of the FAD7 gene product of *Arabidopsis* that manifests normal growth, development, and reproduction despite having lower trienoic FAs which were demonstrated during map-based cloning studies. Further characterization of the tomato spr2 mutants and triple mutants of *Arabidopsis* should provide additional understanding of the JA defense response and functions in different plant species.

9.5.4.3 DISSEMINATION OF A-LINOLENIC ACID VIA GALACTO-LIPIDS

It is widely accepted that the process of releasing the 18:3 precursor from membrane lipids is essential for starting the production of JA. Characterization of the mutant *Arabidopsis* delayed in anther dehiscence1 (dad1), which displays a male sterile phenotype, was a significant advancement in this stage. A chloroplastic phospholipase A1 (PLA1) that releases 18:3 from the sn1 site of chloroplast membrane glycerolipid is encoded by the DAD1 gene.[75] However, levels of JA build up upon wounding are generally normal in the dad1 mutant, indicating that DAD1-dependent JA biosynthesis is primarily involved in reproductive development. The dad1 mutant is also deficient in anther dehiscence and pollen formation. As demonstrated in Ref. [75], DONGLE (DGL), a homolog of DAD1 that is produced at high levels in leaves but not in flowers, siliques, or roots, differs from DAD1 in this regard.[76] A DGL-overexpressing mutant showed higher baseline JA accumulation and increased expression of JA-responsive genes, showing that DGL-dependent JA production is primarily engaged in early wound response. Expression of DGL was transiently stimulated by wounding.[76] These investigations demonstrate the idea that the regulation of JA biosynthesis, which is involved in wound response or development, is variable.

9.5.4.4 OXYGENATION OF A-A-LINOLENIC ACID BY 13-LOX

After release from galactolipids, free18:3 is converted to 13-HPOT in reactions catalyzed by 13-LOX enzymes. The 13-LOXs (LOX2, LOX3, LOX4, and LOX6) encoded in the genome of *Arabidopsis* are capable of oxygenating 18:3 in vitro.[77] Out of these 13-LOXs, LOX2 principally involved wound-induced JA production, and LOX3 and LOX4 in the development of fertile flowers.[78,79] On the contrary, LOX6 is mainly responsible for wound-induced and drought-induced JA production.[80] In tomato, pro-systemin mediated responses suppressor, spr8 mutant encodes tomato lipoxygenase D (TomLox-D), which 13-LOX catalyzes the hydro-peroxidation of α-linolenic acid in JA biosynthesis. Interestingly, the spr8 plants have growth and flower development largely normal, pointing to the involvement of TomLox-Dis specifically in the wound response.[81] The structural protein in TomLoxD exhibits high sequence similarity with maize TASSELSEED-1 (TS1) protein, which is vital in sex determination.[82,81] All the abovementioned investigations advocate distinct physiological functions of different lipoxygenase isoforms. Moreover, hyper action of Tom Lox D leads to enhanced expression of wound-responsive genes and increased resistance against insects and necrotrophic pathogens along with a minor impact on general plant growth, nullifying the association fitness cost with overexpression of TomLoxD. However, continuous mounting of defenses usually results in a high cost in the form of severe impacts on plant growth.[83] Thus, the finding that overexpression of TomLoxD does not lead to a significant fitness cost suggests that the TomLoxD gene has potential utility in crop protection.[81]

9.5.4.5 DEHYDRATION OF 13-HPOT VIA ALLENE OXIDE SYNTHASE (AOS)

From 13-HPOT to 12,13-EOT (unstable) conversion is catalyzed by the plastid localized 13-AOS, which is an atypical cytochrome P450 enzyme (CYP74A).[84] AOS in the *Arabidopsis* genome is usually a single gene, whereas it is two in the tomato genome.[85] These AOS genes from *Arabidopsis* and tomato are involved in wound-induced JA biosynthesis as they get induced by wounding, directly evident from the characterization of the *Arabidopsis* AOS mutant.[86] Additionally, this AOS mutant exhibits a severe male sterile phenotype that did not recover during development but gets rescued by exogenous MeJA spraying.[86] These findings are strong shreds of evidence that, for wound response and reproductive development of *Arabidopsis*, AOS-dependent JA biosynthesis is an important event.

9.5.4.6 OPDA SYNTHESIS BY AOC

The allylic epoxides are the products of AOS that lead to the formation of -ketols and -ketols as a result of hydrolysis of allylic epoxides. This AOC takes part in the cyclization of allylic epoxides to produce OPDA (a racemic mixture of positive and negative *cis* enantiomers).[87,88] AOC also plays a vital role in the establishment of the enantiomeric structure of the JA precursor as it catalyzes the stereospecific cyclization of allene oxide into the *cis* (+) enantiomer OPDA.[70] Encoding of AOC through genes takes place by a single gene in tomato and by four genes in *Arabidopsis*.[89,90] These four AOC genes in *Arabidopsis* are highly and redundantly expressed in the above soil tissues, namely, leaves and flowers. Further, AOC3 and AOC4 were found highly expressed in the meristematic region of the root. It was proved by using the corresponding promoter, glucuronidase (GUS) fusion lines.[87,91] These findings indicated the possibility that these four AOC genes act redundantly in JA biosynthesis in Arial as well as in root tissues.[87,91]

9.5.4.7 BIOSYNTHESIS OF JA IN PEROXISOMES

The OPDA is the end product of the plastid-localized JA biosynthesis pathway. To synthesize OPC-6:0 and OPC-8:0, the reduction process of the dnOPDA and OPDA cyclopentenone ring is catalyzed by OPR (peroxisomal OPDA reductase) respectively.[87,70] It is only OPR3 (mutant) out of six *Arabidopsis* OPRs, which plays an indispensable role in JA biosynthesis.[92] According to Koo et al.[93] OPCL1 (carboxyl-CoA ligase) excites the OPC-8:0. As a response to physical damage injury caused by insects, a null mutant opcl1 found in *Arabidopsis* exhibit increased OPC-8:0 accumulation with coupled with reduced quantity of JAs indicating that there is a clear-cut involvement of OPCL1 in the biosynthesis of JA. Nevertheless, a reduction in the level of OPCL1 results in the lessening of wound-induced JA accumulation by around 50%. It shows that there is an involvement of CoA-ligase isozymes to activate the OPC-8:0 during the biosynthesis of JA.[93,94] Octanoic acid side chains of OPC-8:0 get shortened due to the involvement of the three rounds of -oxidation to yield (+)-7-iso-JA.[95] The multifunctional protein (MFP), 3-ketoacyl-CoA thiolase (KAT), and acyl-CoA oxidase (ACX) are responsible for catalyzing the final steps of JA synthesis. These are core enzymes of the oxidation rounds. Experimental proof for the role of these enzymes (MPF, KAT, and ACX) in the biosynthesis of JA emanated from the characterization of *Arabidopsis* and tomato mutants.[95,97] *Arabidopsis e*nzyme ACX1-A has biochemically shown that it can catalyze the first step in the oxidation of OPC-8:0-CoA.[96] As has been observed ACX1 tomato mutant is consistent with this observation, the acx1 tomato mutant does not take part in wound-induced JA biosynthesis and is compromised in defense response to wounding and insect attack.[96] Similarly, depletion of altered inflorescence meristem 1 (AMI1), which is an important peroxisomal MPF protein found in Arabidopsis also responsible for the impaired wound-induced accumulation of JAs and defense gene expression.[95,98]

9.6 DERIVATIVES AND METABOLITES OF JA

9.6.1 JA DERIVATIVES AND THEIR TYPES

Jasmonic acid goes for several alterations to form concatenation of metabolites like conjugation with isoleucine to form JA-Ile complex, hydroxylation to 12-OH-JA, methylation to form MeJA complex, hydroxylation to 12-OH-JA and JA glucose ester formation, during the cytosol process.[99,103] Such derivatives always vary in their biological functions (Figure 9.4).

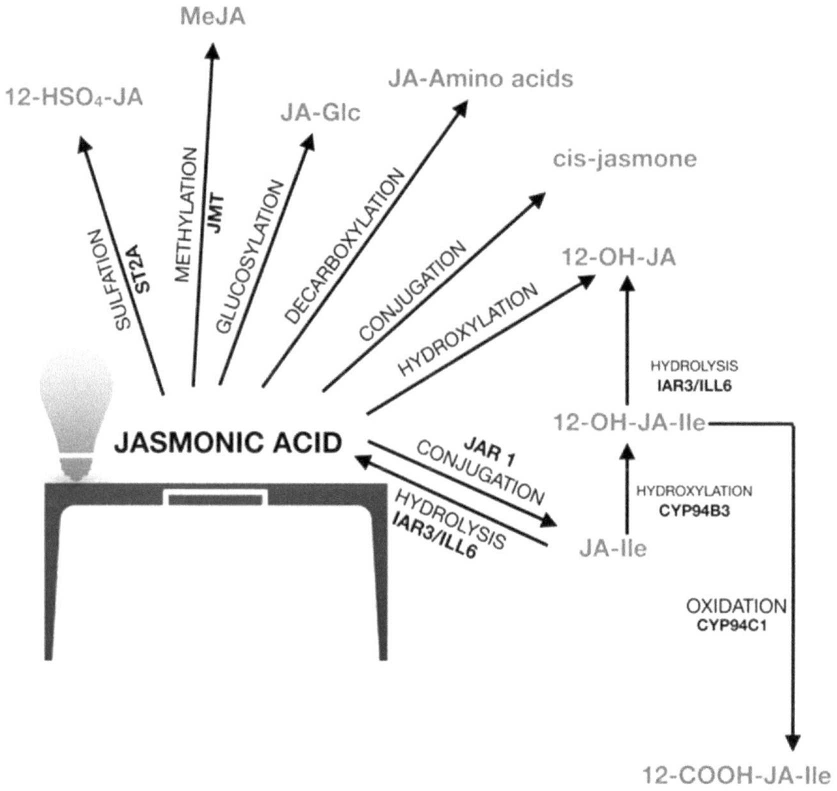

FIGURE 9.4 Jasmonic acid (JA) metabolites and their derivatives.

Methylation to form MeJA complex, decarboxylation to from *cis*-jasmone, glucosylation to form JA-Glc complex, conjugation with amino acids, hydroxylation to form 12-OH-JA complex, and formation of isoleucine to give JA-Ile complex, are manifested. The hormone's active form is JA-Ile. Conversion of JA-Ile into its inactive form is regulated by the two biochemical pathways which include hydrolysis and oxidation process. The enzymes involved are CYP94B3 codes for jasmonyl-L-isoleucine-12-hydroxylase (JA-Ile-12-hydroxylase); 12-hydroxyjasmonoyl-L-amino acid 12-hydroxylase/FA hydroxylase (CYP94C1), 12-hydroxy-jasmonoyl-isoleucine (12-OH-JA-Ile) carboxylase; jasmonoyl-L-amino acid hydrolase (IAR3/ILL6) codes for amidohydrolases; jasmonate resistance 1 (JAR1) codes for JA-Ile synthase (jasmonoyl-isoleucine synthetase); jasmonic acid carboxyl methyltransferase (JMT) codes for methylation of JA to MeJA; and sulfotransferase 2 A (ST2A) encodes 12-hydroxyjasmonic acid sulfotransferase (12-OH-JA sulfotransferase).

9.7 JASMONATE SIGNALING

9.7.1 A CORE JA SIGNALING UNIT

Decades of research in the model system of Arabidopsis have revealed a central JA signaling module made up of a group of JAZ proteins that act as transcriptional repressors, the F-box protein COI1, which forms a functional Skp-Cullin-F-box (SCF) E3 ubiquitin ligase complex, and a battery of TFs that differentially regulate various JA responses. Rapid synthesis of JA-Ile in response to environmental stimuli or developmental signals is detected by the F-box protein COI1, which then recruits JAZ repressors for degradation by the 26S proteasome. JAZ protein degradation frees TFs to regulate target gene expression, leading to the regulation of JA-regulated defense and growth[51] (Figure 9.5).

9.7.2 ROLE OF COI1-BASED SCF COMPLEX IN JA SIGNALING

The discovery of the COI1 (coronatine insensitive) mutant in *Arabidopsis* which is insensitive to JA) helped further in decoding JA signaling in plants.[104,105] This mutant exhibited defects in JA response to wound responses, pest and pathogen resistance, fertility, root inhibition, and secondary metabolite biosynthesis.[106,107] According to Xie et al.,[108] COI1 resembling transport inhibitor response 1 (TIR1) is involved in auxin signaling, suggests the possibility of COI1 acting as part of an SCF E3 ubiquitin ligase to mediate JA signaling. Indeed, COI1 is established to have an association with components of the SCF complex (ASK1, ASK2, Cullin1, and Rbx1).[109,110] COI1 offers specificity to the SCFCOI1 complex through binding target proteins and modulating the transcriptional response to JA.[104] Yan et al.,[81] opined that the dynamic balance between the SCFCOI1- mediated stabilization and degradation of COI1 (26S proteasome-mediated) maintains this protein at a level suitable for exerting its biological function. Exact homologs of COI1 were further reported from *Oryza sativa*, *Solanum lycopersicum*, *N. attenuata*, etc.[111,112]

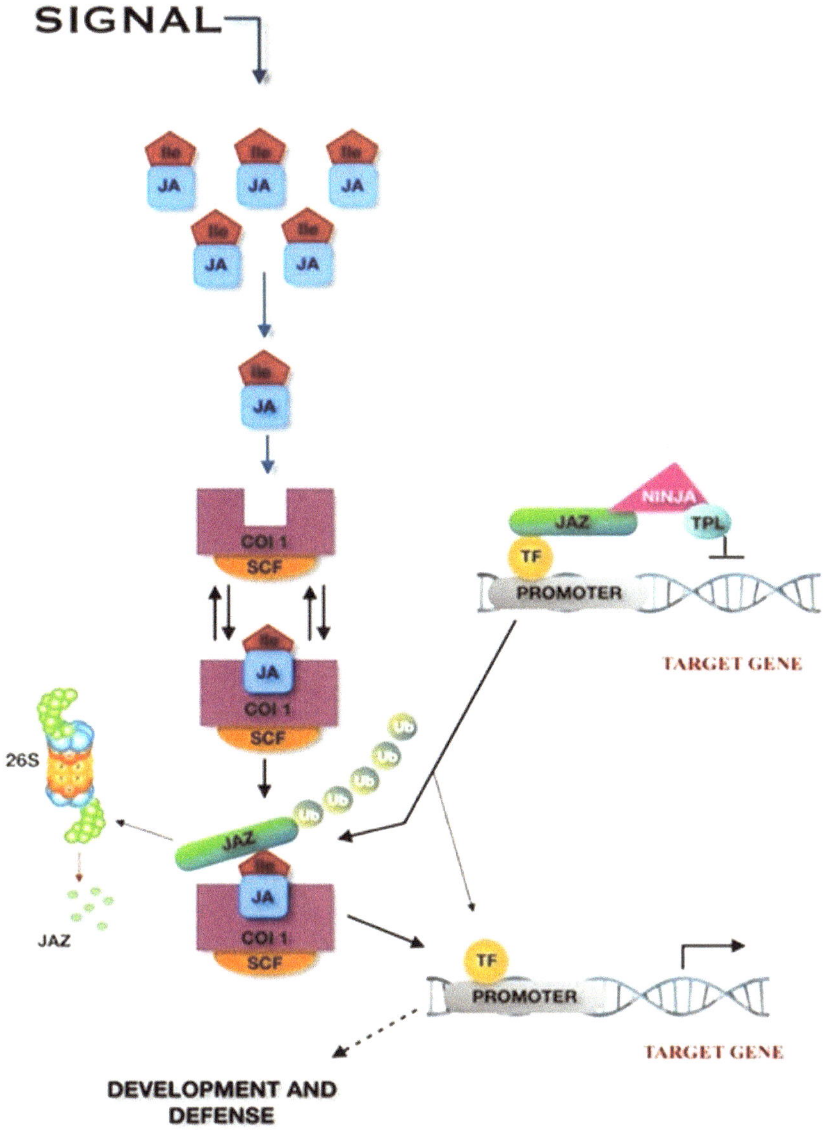

FIGURE 9.5 Signal transduction and perception of JA-Ile.

JA-Ile complex promptly synthesized and recognized by the COI1, an F-box protein, which takes on the various JA-ZIM domain (JAZ) proteins combinations and with the help of 26S proteasome, it got ubiquitinated followed by degradation of the domain. JAZ proteins degradation causes depression of TPL (TOPLESS) and the NINJA proteins (NOVEL INTERACTOR OF JAZ) inhibition of the responsive gene expression directed by TFs as a result freeing the TFs to modulate the defense responses and plant development. Skp-Cullin-F-box (SCF); TF.

9.7.3 JAZ PROTEINS—THE REPRESSORS OF JA SIGNALING

JAZ proteins are also said to be the repressors of JA signaling and help in ubiquitin-dependent degradation.[113,114] To date, 13 different JAZ repressors have been identified in *Arabidopsis* belonging to the plant-specific TIFY family in the existence of the dominant TIF[F/Y]XG core motif that is confined to the ZIM domain (larger amino acids conserved region, 28).[115,116] The JAZ proteins also contain an *N*-terminal (NT) domain (relatively conserved) and a C-terminal JA-associated (Jas) domain (highly conserved).[117] A short sequence of amino acids is adequate for JA-Ile complex or COR (coronatine) binding within the JAZ domain has been termed as JAZ degron. JAZ degron form a binary structure consisting of a loop and an amphipathic-helix which binds to the COI1 receptor and acts as a cover of the binding pocket of the JA-Ile complex.[118] These domains provide the specificity for protein-protein interactions which determines the differential materialization of JA complexes in the existence or nonexistence of the active stimulating hormone.[119] Five out of 13 JAZ proteins, namely, JAZ-5, -6, -7, -8, and -13, carry a supplementary EAR motif (a 7 amino acid motif) that unswervingly creates a synergy with TOPLESS (TPL) protein in absence of novel interactor of JAZ (NINJA) adaptor.[116] Based on the study by Chung et al.,[120] JAZ mutants lacking JA-hypersensitive phenotypes, show the functional superfluous between the JAZ proteins. However, the concerned literature advocates separated functions of *Arabidopsis* JAZ repressors in JA signaling.

9.7.4 JA-ILE RECEPTOR-COI1

A crucial step to understanding jasmonic acid signaling alarms the identification of the JA-Ile receptor.[104] Katsir et al.,[121] reported that COI1 is a critical constituent of a receptor to coronatine and JA-Ile complex because coronatine fixes directly to the COI1- JAZ complex capable to JA-Ile to contest with it for specific binding indicating that JA-Ile and coronatine get recognized by the matching receptor. It has also been observed based on biochem-assays, COI1 is capable of specifically binding to JA-immobilized sepharose, interacting directly with coronatine through Biotin tagged photo affinity probe. The JAZ repressor proteins have interaction with COI1 upon JA-Ile or coronatine binding and thus serve as the receptor for JA, which at first bind with the active molecules of JA providing a suitable platform for creating linkage with the repressor proteins of JAZ.[122] Crystal structure studies further revealed that the formation of the Horseshoe-shaped solenoid domain is done through LRRs (leucine-rich repeats) of COI1 that houses the JA-Ile binding pocket. On the contrary, the surface of the coronatine-bound COI1 pocket interacts with JAZ1 degron.[118]

9.7.5 *MYC2—TRANSCRIPTION FACTOR OF JASMONIC ACID SIGNALING*

MYC2 is a key regulator of JA signaling in Arabidopsis that encodes a TF-bHLH, that has the binding affinity with the G-box (CACGTG) and its related hexamers.[123,124] MYC2 also has a positive regulation on the two branches of JA genes, one of which is involved in plant responses to wounding caused by insects, and another branch is involved in the responses to infection caused by pathogens.[125,126] MYC2 is also involved in negatively regulating a group of intermediate TFs responsible for the expression of downstream pathogen-responsive genes. Interestingly, positive regulation of MYC2 on wound responses takes place early (within 6 h after JA application), while negative regulation of pathogen response takes a longer time (48 h after JA application).[127] The direct regulation of immediate JA-responsive genes together with jasmonate Zim genes (JAZs) and JA metabolic genes has also been credited to MYC2.[128] Moreover, JA-mediated inhibition of primary root growth also occurs under the influence of MYC2. It has been well observed that JA prevents primary root growth and modulates the activity of root meristem through repressing stem cell TFs, namely, Plethora 1 (PLT1) and Plethora 2 (PLT2) taking part in the auxin pathway synchronizing the root stem cell niche maintenance.[129] These studies, at the molecular level, exemplify root growth inhibition by regulating JA and its interaction with PGR-auxin.

9.8 PLANT PATHOGEN HIJACKING CORE COMPONENTS OF JA SIGNALING

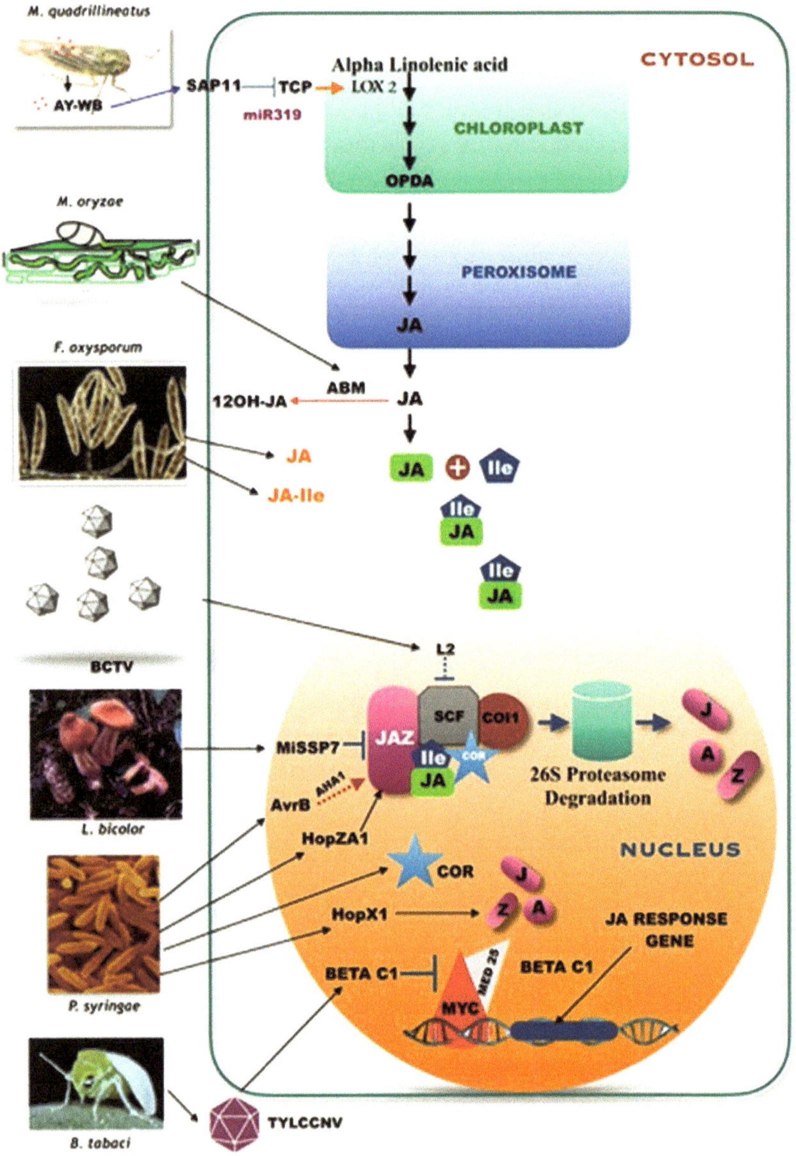

FIGURE 9.6 Different approaches have been retained by several microbial pathogens and insects to impound JA signaling.

The figure depicts the vir (virulence) constituents that attack the fundamental elements of the JA biosynthesis or signaling pathway. The Hyperactivation expression of LOX2 results in JA biosynthesis suppression via AY-WB phytoplasma released by an insect vector *Macrosteles quadrillineatus.*[130] *Magnaporthe oryzae* a fungus, invigorate JA hydroxylation to reduce signaling of JA via Abm effector.[131] *Fusarium oxysporum* results in the production of JA or JA-Ile complex and leads to turning on JA signaling.[132] The BCTV L2, a viral protein via CSN5 inhibits the activity of ubiquitin ligase complex-SCFCOI1.[133] A symbiotic associator *L. bicolor* conquers the deterioration of JAZ protein through MiSSP7 effector.[134] The Avr-B or COR effector is released by A hemibiotrophic bacterium *P. syringae* that increases the interaction between coreceptor proteins JAZ and COI1.[135,136] HopZ1a acetylates JAZ proteins and excites the JAZ degradation in a COI1-dependent manner.[137] Deterioration of JAZ protein is stimulated by HopX1 in a COI1-independent manner resulting in switching on the JA signaling.[138] TYLCCNV is employed by *B. tabaci* which is an insect vector to diminish the expression of MYC2-mediated gene via straight interaction between the MYC2 and C1 effector.[139]

9.9 ROLE OF JA IN REGULATING PLANT DEFENCE AGAINST INSECTS

The importance of JA in plant defense against various insect attacks has been revealed after several genetic studies. The mutants, that is, opr3 and aos or the coi1 receptor gene of JA biosynthesis genes in Arabidopsis are more susceptible to insects in comparison to wild-type plants.[65] Conversely, the higher order mutant jazD (a jazdecuple mutant defective in JAZ17, -9, -10, and -13) and jazQ (a jazquintuple mutant defective in JAZ1, JAZ3, JAZ4, JAZ9, and JAZ10) are strongly resistant against herbivores when compared.[140] Providentially, a single JAZ gene is known from *Marchantia polymorpha* and the role of MpJAZ in insect resistance is yet to be worked out however the essential role of MpCOI1 has already been investigated.[141,142] The downstream TFs in JA responses against herbivory are regulated by JAZ, including MYC-2/3/4/5, bHLH3/13/14/17, and the WD-repeat/bHLH/MYB complex.[143,144] According to Schweizer et al.,[145] MYC2, MYC3, MYC4, and MYC5 (bHLH-IIIe TFs) extravagantly regulates the responses of JA and plant resistance to herbivory, whereas mutants myc-2/3/4 and the mutants myc-2/3/4/5 are extremely susceptible to herbivory. The MYCs commonly induce proteinase inhibitors (PIs) that lead to a weak digestive system and function in insects.[125] Additionally, it is capable of controlling the biological synthesis of some essentially required secondary metabolites for plant defense against herbivory. Further, MYC2 is also behind the surge in the synthesis of plant terpenes (crucial in direct and indirect resistance against herbivory) by activating the expression of terpene synthesis (TPS) genes (TPS-10, TPS-11, and TPS-21).[145] A recent report has also been brought out for negative regulation of JA responses through JAMs which are considered to be the IIId subfamily of the bHLH family (basic helix-loop-helix) TFs such as bHLH-3, -13, -14, and -17.[146] JA-hypersensitive phenotypes were displayed by bhlh-17 (single mutant jam1) bhlh-3, bhlh-13, bhlh-14, and bhlh-17 (quadruple mutant jam1) exhibited increased level resistance to insect attack.[147] Despite lacking a conserved transcriptional repression domain, the IIId subfamily exhibits the ability of binding DNA and acts as transcriptional repressors to promoters of genes, activated by transcriptional activators, namely, MYC2 and WD-repeat/TT8/MYB75.[148] The physical barrier (trichomes) and chemical repellent (anthocyanins) are well-known hindrances in insect herbivory. The formation and synthesis of such defense mechanisms are firmly regulated by the WD-repeat/bHLH/MYB.[149] JAZs are known to have evident interference regulation of TT-8, GL-3, and EGL-3 (bHLH component), and GL-1 and MYB-75 (R2R3-MYB component) which are of WD-repeat/bHLH/MYB complex. At low levels, JAZs can hamper the synthesis and transcriptional activation of the WD-repeat/bHLH/MYB complex, the resultant effect is a suppression of trichome formation and anthocyanin accumulation in plant.[150,151]

During insect attack, COI1 binds with an E3 ubiquitin ligase SCF which is a bioactive JA, and thereafter JAZ is recruited for ubiquitination and degradation resulting in the inactivation of downstream TFs to trigger JA responses. The biosynthesis of PIs and secondary metabolites including terpenes and glucosinolates are upregulated by MYC2/3/4 to render plants unpalatable or show repellency to insects. The WD-repeat/bHLH/MYB complex enhances insect resistance by promoting the accumulation of trichome formation and anthocyanin. In addition to the WD-repeat/bHLH/MYB complex, FIL, which is also a JAZ target can augment the biosynthesis of anthocyanin. In contrast, bHLH-3, bHLH-13, bHLH-14 and bHLH-17, antagonize MYC-2, MYC-3, and MYC-4 along with

WD-repeat/bHLH/MYB complex are capable to repress the production of anthocyanin and functions in plant defense mechanism against insect as negative regulators.

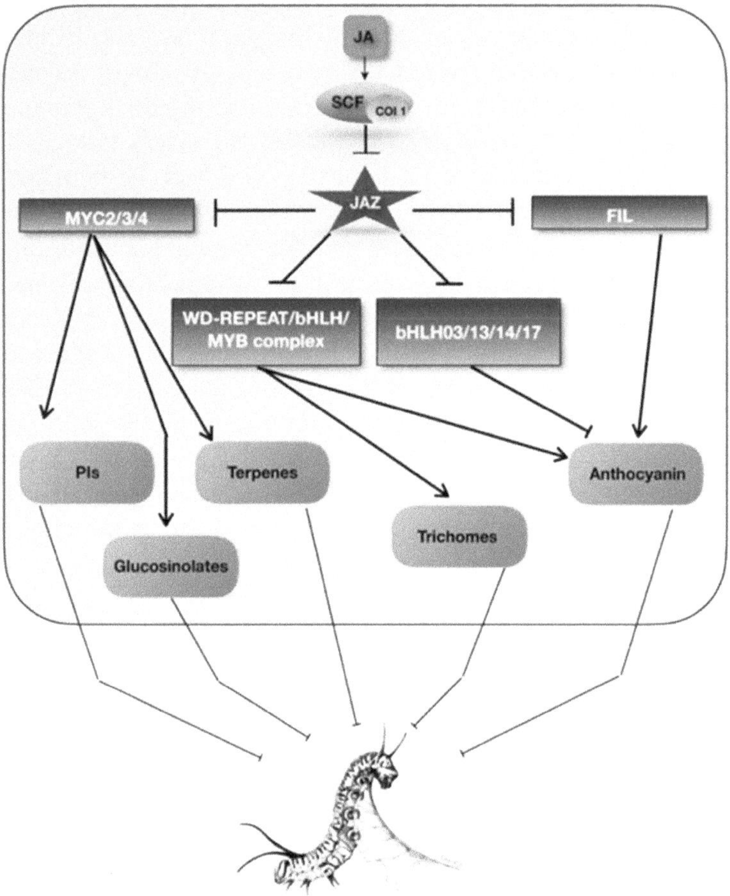

FIGURE 9.7 Anti-insect defense responses regulated by jasmonates. Anti-insect defense responses are regulated by jasmonates.

9.10 CROSSTALK

Plant hormones are essential and play an important part in regulating how plants interact with their intricate biotic and abiotic surroundings.[152] Each hormone begins a distinct biochemical route, which is then combined into a complex network of additive, antagonistic, and synergistic interactions. Crosstalk between pathways is referred to as hormone crosstalk. The hormone crosstalk is crucial for customizing plant responses to different insects and microorganisms in different internal and external situations by influencing the topology of the immunological network. In addition to providing immune system robustness, crosstalk also drives the specificity of induced defense responses against the quantity of biotic communicators. Recent research using both dry and wet lab techniques has significantly improved our comprehension of the extensive effects of hormone crosstalk on immune crosstalk systems' basic concepts and network operation. Hormone interaction is modulated at multiple levels of regulation, including protein stability, gene transcription, and hormone homeostasis, according to molecular studies. JAs are recognized plant hormones imparting plant defense regulation against biotic stresses caused by herbivores, and necrotrophic pathogens such as *B. cinerea.* On the contrary, SA regulates the defense mechanism of hemibiotrophic and biotrophic pathogens such as *P. syringae.*[153] The defense pathways mediated by JA and SA, generally antagonize each other and result in improved resistance against necrotrophs, which may have a clear correlation with greater sensitivity to biotrophs.[154,155] Studies on the role of plant hormones against microbial diseases suggest that microbes evolve some interference strategies to interrupt plant hormone signaling pathways leading to hormonal imbalance for their own benefits. Also, the coronatine (phytotoxin) production by a group of pathovars of *P. syringae* is another example of JA hormonal manipulation by pathogens.[154,118]

KEYWORDS

- **herbivory**
- **JA biosynthesis**
- **microbial and viral pathogens**
- **plant elicitors**
- **plant defense mechanism**
- **secondary metabolites**
- **signaling**

REFERENCES

1. Freeman, B. C.; Beattie, G. A. An Overview of Plant Defences Against Pathogens and Herbivores. *Plant Pathol. Microbiol. Pub.* **2008,** *94.* https://lib.dr.iastate.edu/plantpath_pubs/94
2. Doughari, J. H. An Overview of Plant Immunity. *J. Plant. Pathol. Microbiol.* **2015,** *6,* 11. DOI: 10.4172/2157-7471.1000322
3. Pablo, M. Prospects for Plant Productivity: From the Canopy to the Nucleus (Editorial). *J. Exp. Bot.* **2021,** *72* (11), 3931–3935. DOI: 10.1093/jxb/erab147
4. Erb, M. Plant Defences Against Herbivory: Closing the Fitness Gap. *Trends Plant Sci.* **2018,** *23,* 187–194.
5. Arvanitis, L.; Wiklund, C.; Münzbergová, Z.; Dahlgren, J. P.; Ehrlen, J. Novel Antagonistic Interactions Associated with Plant Polyploidization Influence Trait Selection and Habitat Preference. *Ecol. Lett.* **2010,** *13,* 330–337.
6. Ehrlén, J.; Münzbergová, Z. Timing of Flowering: Opposed Selection on Different Fitness Components and Trait Covariation. *Am. Nat* **2009,** *173,* 819–830.
7. Münzbergova, Z.; Skuhrovec, J. Data on Herbivore Performance and Plant Herbivore Damage Identify the Same Plant Traits as the Key Drivers of Plant–Herbivore Interaction. *Insects* **2020,** *11,* 865. DOI: 10.3390/insects11120865
8. Labandeira, C. C. The Four Phases of Plant-Arthropod Associations in Deep Time. *Geol. Acta* **2006,** *4* (4), 409–438.
9. Anonymous. 2022. https://www.newworldencyclopedia.org/entry/Herbivory
10. Pandit, M. A.; Kumar, J.; Gulati, S.; Bhandari, N.; Mehta, P.; Katyal, R.; Rawat, C. D.; Mishra, V.; Kaur, J. Major Biological Control Strategies for Plant Pathogens. *Pathogens* **2022,** *11* (2), 273. doi.org/10.3390/pathogens11020273
11. Bruehl, G. W. Plant Pathology, a Changing Profession in a Changing World. *Ann. Rev. Phytopathol.* **1991,** *29,* 1–12.
12. Doi, Y.; Teranaka, M.; Yora, K.; Asuyama, H. Mycoplasma or PLT Group- Like Microorganisms Found in the Phloem Elements of Plants Infected with Mulberry Dwarf, Potato Witches Broom, Aster Yellows, or Paulownia Witches Broom. *Ann. Phytopathol. Soc. Jpn.* **1967,** *33,* 259–266.
13. Kumar, V.; AlMomin, S. Plant Defense against Pathogens: The Role of Salicylic Acid. *Res. J. Biotechnol.* **2018,** *13* (12), 97–103.
14. Isah, T. Stress and Defense Responses in Plant Secondary Metabolites Production. *Biol. Res.* **2019,** *52,* 39. doi.org/10.1186/s40659-019-0246-3
15. Köhl, J.; Kolnaar, R.; Ravensberg, W. J.; Mode of Action of Microbial Biological Control Agents Against Plant Diseases: Relevance Beyond Efficacy. *Front. Plant Sci.* **2019,** *10,* 845. DOI: 10.3389/fpls.2019.00845
16. Ellis, J.; Catanzariti, A. M.; Dodds, P. The Problem of How Fungal and Oomycete Avirulence Proteins Enter Plant Cells. *Trends Plant Sci.* **2006,** *11,* 61–63.
17. Andersen, E. J.; Ali, S.; Byamukama, E.; Yen, Y.; Nepal, M. P. Disease Resistance Mechanisms in Plants. *Genes* **2018,** *9* (7), 339. https://doi.org/10.3390/genes9070339
18. Abd El Rahman, T. A.; El Oirdi, M.; Gonzalez-Lamothe, R.; Bouarab K., Necrotrophic Pathogens Use the Salicylic Acid Signaling Pathway to Promote Disease Development in Tomato. *MPMI* **2012,** *25* (12), 1584–1593. http://dx.doi.org/10.1094/MPMI-07-12-0187-R

19. Perfect, S. E.; Green J. R. Infection Structures of Biotrophic and Hemi-Biotrophic Fungal Plant Pathogens. *Mol. Plant Pathol.* **2008.** https://doi.org/10.1046/j.1364-3703.2001.00055.x
20. Spanu, P. D.; Panstruga, R. Editorial: Biotrophic Plant-Microbe Interactions. *Front. Plant Sci.* **2017,** *8*, 192. DOI: 10.3389/fpls.2017.00192.
21. Alberts, B.; Johnson, A.; Lewis, J. *Molecular Biology of the Cell*, 4th ed.; Garland Science. Introduction to Pathogens: New York, 2002. https://www.ncbi.nlm.nih.gov/books/NBK26917/
22. Abdulkhair, W. M.; Alghuthaymi, M. A. Plant Pathogens. In: *Plant Growth*; Rigobelo, E. C., Eds.; IntechOpen, 2016. DOI: 10.5772/65325.
23. Casadevall, A.; Pirofski, L. A. Host-Pathogen Interactions: Basic Concepts of Microbial Commensalism, Colonization, Infection, and Disease. *Infect. Immun.* **2000,** *68* (12), 6511–6518. https://doi.org/10.1128/IAI.68.12.6511-6518.2000
24. Yaktayo-Chang, J. P.; Tang, H. V.; Mendoza, J.; Christensen, S. A., Block, A. K. Plant Defence Chemicals Against Insect Pests. *Agronomy* **2020,** *10*, 1156. DOI: 10.3390/agronomy10081156.
25. Quiroz, A., Pettersson, J.; Pickett, J. A.; Wadhams, L. J.; Niemeyer, H. M. Semio-Chemicals Mediating Spacing Behaviour of Bird Cherry-Oat Aphid, *Rhopalosiphumpadi* Feeding on Cereals. *J. Chem. Ecol.* **1997,** *23* (11), 2599–2607.
26. Karban, R.; Baldwin, I. T. *Induced Responses to Herbivory*; Chicago University Press: Chicago, 1997; 319 pp.
27. Bonaventure, G. Perception of Insect Feeding by Plants. *Plant Biol.* **2012,** *14* (6), 872–880.
28. Bonaventure, G.; Van Doorn, A.; Baldwin, I. T. Herbivore-Associated Elicitors: FAC Signaling and Metabolism. *Trends Plant Sci.* **2011,** *16*, 294–299.
29. Bonaventure, G.; Van Doorn, A.; Baldwin, I. T.; Herbivore-associated elicitors: FAC signaling and metabolism. *Trends Plant Sci.* **2011,** *6* (6), 1360–1385.
30. Eichenseer, H. Salivary Glucose Oxidase: Multifunctional Roles for *Helicoverpa zea*? *Arch. Ins. Biochem. Physiol.* **1999,** *42*, 99–109.
31. Alborn, H. T.; Hansen, T. V.; Jones, T.H; Bennett, D. C.; Tumlinson, J. H.; Schmelz, E. A.; Teal P. E. Disulfo-Oxyfattyacids from the AMERICAN BIRD GRASSHOPPER Schistocerca Americana, Elicitors of Plant Volatiles. *Proc. Natl. Acad. Sci. USA.* **2007,** *104*, 12976–12981.
32. Bergey, D. R.; Cardenas, M. O.; de Moura, D. S.; Ryan, C. A. A Wound- and Systemin-Inducible Polygalacturonase in Tomato Leaves. *Proc. Natl. Acad. Sci. USA.* **1999,** *96*, 1756–1760.
33. Schmelz, E. A. Fragments of ATP Synthase Mediate Plant Perception of Insect Attack. *Proc. Natl. Acad. Sci. USA.* **2006,** *103*, 8894–8899.
34. Allmann, S.; Baldwin, I. T. Insects Betray Themselves in Nature to Predators by Rapid Isomerization of Green Leaf Volatiles. *Science* **2010,** *329*, 1075–1078.
35. Turlings, T. C. J.; Tumlinson, J. H.; Lewis, W. J. Exploitation of Herbivore-Induced Plant Odors by Host-Seeking Parasitic Wasps. *Science* 1990, *250*, 1251–1253.
36. Schwachtje, J.; Baldwin, I. T. Why Does Herbivore Attack Reconfigure Primary Metabolism? *Plant Physiol.* 2008, *146*, 845–851.
37. Kessler, D.; Diezel, C.; Baldwin, I. T. Changing Pollinators as a Means of Escaping Herbivores. *Curr. Biol.* 2010, *20*, 237–242.

38. Musser, R. O.; Hum-Musser, S. M.; Eichenseer, H.; Peiffer, M.; Erwin, G.; Murphy, J. B.; Felton, G. W. Herbivory: Caterpillar Saliva Beats Plant Defences. *Nature* 2002, *416*, 599–600.
39. Nishad, R.; Ahmed, T.; Rahman, V. J.; Kareem, A.; Modulation of Plant Defense System in Response to Microbial Interactions. *Front. Microbiol.* **2020,** *11*, 1298. doi.org/10.3389/fmicb.2020.01298
40. Fürstenberg-Hägg, J.; Zagrobelny, M.; Bak, S. Plant Defense Against Insect Herbivores. *Int. J. Mol. Sci.* **2013,** *14* (5), 10242–10297. DOI: 10.3390/ijms1405 10242
41. Peiffer, M.; Felton, G. W. The Host Plant as a Factor in the Synthesis and Secretion of Salvary Glucose Oxidase in the Larval *Helicoverpa zea. Arch. Insect. Biochem.* **2005,** *58*, 128–137.
42. Engelberth, J.; Koch, T.; Schueler, G.; Bach-Mann, N.; Rechtenbach, J.; Boland, W. Ion Channel-Forming Alamethicin Is a Potent Elicitor of Volatile Biosynthesis and Tendril Coiling. Crosstalk Between Jasmonate and Salicylate Signaling in Lima Bean. *Plant Physiol.* **2000,** *125*, 369–77.
43. Halitschke, R.; Schittko, U.; Pohnert, G.; Boland, W.; Baldwin, I. T. Molecular Interactions Between the Specialist Herbivore *Manduca sexta* (Lepidoptera, Sphingidae) and Its Natural Host *Nicotiana attenuata.* III. Fatty Acid–Amino Acid Conjugates in Herbivore Oral Secretions Are Necessary and Sufficient for Herbivore-Specific Plant Responses. *Plant Physiol.* **2001,** *125*, 711–17.
44. Schittko, U.; Preston, C. A.; Baldwin, I. T. Eating the Evidence? *Manducasexta* Cannot Disrupt Specific Jasmonate Induction in *Nicotiana attenuata* by Rapid Consumption. *Planta* **2000,** *210*, 343–346.
45. Winz, R. A.; Baldwin, I. T. Molecular Interactions Between the Specialist Herbivore *Manduca sexta* (Lepidoptera, Sphingidae) and Its Natural Host *Nicotiana attenuata.* IV. Insect Induced Ethylene Reduces Jasmonate Induced Nicotine Accumulation by Regulating Putrescine N-Methyl-Transferase Transcripts. *Plant Physiol.* **2001,** *125*, 2189–2202.
46. Paxton, J. D. Phytoalexins—A Working Redefinition. *Phytopathol. Z.* **1981,** *101*, 106–209.
47. Balint-Kurti, P. The Plant Hypersensitive Response: Concepts, Control and Consequences. *Mol. Plant Pathol.* **2019**. DOI: 10.1111/mpp.12821.
48. Métraux, J. P. Systemic Acquired Resistance. In: *Brenner's Encyc Genet*, 2nd ed.; 2013; pp 627–629. doi.org/10.1016/B978-0-12-374984-0.01509-6
49. Heil, M.; Bostock, R. M. Induced Systemic Resistance (ISR) Against Pathogens in the Context of Induced Plant Defences. *Ann. Bot.* **2002,** *89* (5), 503–512.
50. Wang, J.; Song, L.; Gong, X.; Xu, J.; Li, M. Functions of Jasmonic Acid in Plant Regulation and Response to Abiotic Stress. *Int. J. Mol. Sci.* **2020,** *21* (4), 1446. DOI: 10.3390/ijms21041446
51. Kende, H.; Zeevaart, J. The Five "Classical" Plant Hormones. *Plant Cell* **1997,** *9*, 1197–1210.
52. Pruski, K.; Astatkie, T.; Nowak, J. Jasmonate Effects on In Vitro Tuberization and Tuber Bulking in Two Potato Cultivars (*Solanum tuberosum* L.) Under Different Media and Photoperiod Conditions. *In Vit. Cell Dev. Biol. Plant* **2002,** *38*, 203–209.

53. Xu, J.; Li, Q.; Li, Y.; Yang, L.; Zhang, Y.; Cai, Y. Effect of Exogenous Gibberellin, Paclobutrazol, Abscisic Acid, and Ethrel Application on Bulblet Development in *Lycoris radiata. Front. Plant Sci.* **2020,** *11*, 615547. DOI: 10.3389/fpls.2020.615547
54. Hernández, H. A.; Ledezma-Rodríguez, M.; Avilez-Montalvo, R. N.; Juárez-Gómez, Y. L.; Skeete, A.; Avilez-Montalvo, J.; De-la-Peña, C.; Loyola-Vargas, V. M. Signaling Overview of Plant Somatic Embryogenesis. *Front. Plant Sci.* **2019,** *10*, 77. DOI: 10.3389/fpls.2019.00077
55. Ikeuchi, M.; Sugimoto, K.; Iwase, A.; Plant Callus: Mechanisms of Induction and Repression. *Plant Cell* **2013,** *25* (9), 3159–3173. DOI: 10.1105/tpc.113.116053
56. Kamińska, M. Role and Activity of Jasmonates in Plants Under In Vitro Conditions. *Plant Cell Tiss. Org. Cult.* **2021,** *146*, 425–447. doi.org/10.1007/s11240-021-02091-6
57. Ueda, J.; Kato, J. Isolation and Identification of a Senescence-promoting Substance from Wormwood (*Artemisia absinthium* L.). *Plant Physiol.* **1980,** *66*, 246–249.
58. Dathe, W.; Preiss, A.; Schade, W.; Sembdner, G.; Schreiber, K. Endogenous Plant Hormones of the Broad Bean, *Vicia faba* L. (-)-Jasmonic Acid, a Plant Growth Inhibitor in Pericarp. *Planta* **1981,** *153*, 530–535.
59. Farmer, E. E.; Ryan, C. A. Interplant Communication: Airborne Methyl Jasmonate Induces Synthesis of Proteinase Inhibitors in Plant Leaves. *Proc. Natl. Acad. Sci. USA* **1990,** *87*, 7713–7716.
60. Staswick, P. E. Novel Regulation of Vegetative Storage Protein Genes. *Plant Cell* **1990,** *2*, 1–6.
61. Boter, M.; Ruíz-Rivero, O.; Abdeen, A.; Prat, S. Conserved MYC Transcription Factors Play a Key Role in Jasmonate Signaling Both in Tomato and Arabidopsis. *Genes Dev.* **2004,** *18*, 1577–1591.
62. Falk, K. L.; Kastner, J.; Bodenhausen, N.; Schramm, K.; Paetz, C.; Vassao, D. G.; Reichelt, M.; von Knorre, D.; Bergelson, J.; Erb, M. The Role of Glucosinolates and the Jasmonic Acid Pathway in Resistance of *Arabidopsis thaliana* Against Molluscan Herbivores. *Mol. Ecol.* **2014,** *23*, 1188–1203.
63. Tsutsui, H.; Notaguchi, M. The Use of Grafting to Study Systemic Signaling in Plants. *Plant Cell Physiol.* **2017,** *58* (8), 1291–1301. DOI: 10.1093/pcp/pcx098.
64. Cheng, Z.; Sun, L.; Qi, T.; Zhang, B.; Peng, W.; Liu, Y.; Xie, D. The bHLH Transcription Factor MYC3 Interacts with the Jasmonate ZIM-Domain Proteins to Mediate Jasmonate Response in Arabidopsis. *Mol. Plant* **2011,** *4*, 279–288.
65. Wasternack, C.; Hause, B. Jasmonates: Biosynthesis, Perception, Signal Transduction and Action in Plant Stress Response, Growth and Development: An Update to the 2007 Review in Annals of Botany. *Ann. Bot.* **2013,** *111*, 1021–1058.
66. Vick, B.; Zimmerman, A. D. C. The Biosynthesis of Jasmonic Acid: A Physiological Role for Plant Lipoxygenase. *BiochemBiophys Res. Commun.* **1983,** *111* (2), 470–477.
67. Browse, J. Jasmonate Passes Muster: A Receptor and Targets for the Defense Hormone. *Annu. Rev. Plant Biol.* **2009,** *60*, 183–205.
68. Browse, J. The Power of Mutants for Investigating Jasmonate Biosynthesis and Signaling. *Phytochem* **2009,** *70* (13–14), 1539–1546.
69. Wasternack, C. Jasmonates: An Update on Biosynthesis, Signal Transduction and Action in Plant Stress Response, Growth and Development. *Ann. Bot.* **2007,** *100*, 681–697.
70. Schaller, A.; Stintzi, A. Enzymes in Jasmonate Biosynthesis—Structure, Function, Regulation. *Phytochem* **2009,** *70*, 1532–1538.

71. Weber, H.; Vick, B. A.; Farmer, E. E. Dinor-Oxo-Phytodienoic Acid: A New Hexadecanoid Signal in the Jasmonate Family. *Proc. Natl. Acad. Sci. USA.* **1997,** *94*, 10473–10478.
72. McConn, M.; Browse, J. The Critical Requirement for α-Linolenic Acid Is Pollen Development, Not Photosynthesis, in an Arabidopsis Mutant. *Plant Cell* **1996,** *8*, 403–416.
73. McConn, M.; Creelman, R. A.; Bell, E. Jasmonate Is Essential for Insect Defense in Arabidopsis. *Proc. Natl. Acad. Sci. USA.* **1997,** *94*, 5473–5477.
74. Li, C.; Liu, G.; Xu, C. The Tomato Suppressor of Prosystemin-Mediated Responses2 Gene Encodes a Fatty Acid Desaturase Required for the Biosynthesis of Jasmonic Acid and the Production of a Systemic Wound Signal for Defense Gene Expression. *Plant Cell* **2003,** *15*, 1646–1661.
75. Ishiguro, S.; Kawai-Oda, A.; Ueda, J.; Nishida, I.; Okada, K.; The Defective in Anther Dehiscience Gene Encodes a Novel Phospho- Lipase A1 Catalysing the Initial Step of Jasmonic Acid Biosynthesis, Which Synchronizes Pollen Maturation, Anther Dehiscence, and Flower Opening in Arabidopsis. *Plant Cell* **2001,** *13*, 2191–2209.
76. Hyun, Y.; Choi, S.; Hwang, H. J. Cooperation and Functional Diversification of Two Closely Related Galactolipase Genes for Jasmonate Biosynthesis. *Dev. Cell* **2008,** *14*, 183–192.
77. Bannenberg, G.; Martinez, M.; Hamberg, M. Diversity of the Enzymatic Activity in The Lipoxygenase Gene Family of *Arabidopsis thaliana. Lipids* **2009,** *44*, 85–95.
78. Glauser, G.; Dubugnon, L.; Mousavi, S. A. Velocity Estimates for Signal Propagation Leading to Systemic Jasmonic Acid Accumulation in Wounded Arabidopsis. *J. Biol. Chem.* **2009,** *284*, 34506–34513.
79. Caldelari, D.; Wang, G.; Farmer, E. E. Arabidopsis lox3 lox4 Double Mutants Are Male Sterile and Defective in Global Proliferative Arrest. *Plant Mol. Biol.* **2011,** *75*, 25–33.
80. Chauvin, A.; Caldelari, D.; Wolfender, J. L. Four 13-Lipoxygenases Contribute to Rapid Jasmonate Synthesis in Wounded *Arabidopsis thaliana* Leaves: A Role for Lipoxygenase 6 in Responses to Long-Distance Wound Signals. *New Phytol.* **2013,** *197*, 566–575.
81. Yan, J.; Li, H.; Li, S.; Yao, R.; Deng, H.; Xie, Q.; Xie, D. The Arabidopsis F-Box Protein Coronatine Insensitive1 Is Stabilized by SCFCOI1 and Degraded via the 26S Proteasome Pathway. *Plant Cell* **2013,** *25*, 486–498.
82. Acosta, I. F.; Laparra, H.; Romero, S. P. Tassel Seed1 Is a Lipoxygenase Affecting Jasmonic Acid Signaling in Sex Determination of Maize. *Science* **2009,** *323*, 262–265.
83. Bostock, R. M. Signal Crosstalk and Induced Resistance: Straddling the Line Between Cost and Benefit. *Annu. Rev. Phytopathol.* **2005,** *43*, 545–580.
84. Werck-Reichhart, D.; Bak, S.; Paquette, S. Cytochromes p450. *Arabidopsis Book* **2002,** *1*, e0028.
85. Howe, G. A.; Lee, G. I.; Itoh, A. Cytochrome P450-Dependent Metabolism of Oxylipins in Tomato: Cloning and Expression of Allene Oxide Synthase and Fatty Acid Hydroperoxide L. Yase. *Plant Physiol.* **2000,** *123*, 711–724.
86. Park, J. H.; Halitschke, R.; Kim, H. B. A Knock-Out Mutation in Allene Oxide Synthase Results in Male Sterility and Defective Wound Signal Transduction in Arabidopsis Due to a Block in Jasmonic Acid Biosynthesis. *Plant J.* **2002,** *31*, 1–12.

87. Li, H.; Yao, R.; Ma, S.; Hu, S.; Li, S.; Wang, Y.; Yan, C.; Xie, D.; Yan, J. Efficient ASK-Assisted System for Expression and Purification of Plant F-Box Proteins. *Plant J.* **2017,** *92*, 736–743.
88. Brash, A. R.; Baertschi, S. W.; Ingram, C. D. Isolation and Characterization of Natural Allene Oxides: Unstable Intermediates in the Metabolism of Lipid Hydroperoxides. *Proc. Natl Acad. Sci. USA* **1988,** *85*, 3382–3386.
89. Stenzel, I.; Hause, B.; Miersch, O. Jasmonate Biosynthesis and the Allene Oxide Cyclase Family of *Arabidopsis thaliana. Plant Mol. Biol.* **2003,** *51*, 895–911.
90. Ziegler, J.; Stenzel, I.; Hause, B. Molecular Cloning of Allene Oxide Cyclase: The Enzyme Establishing the Stereochemistry of Octadecanoids and Jasmonates. *J. Biol. Chem.* **2000,** *275*, 19132–19138.
91. Stenzel, I.; Otto, M.; Delker, C. Allene Oxide Cyclase (AOC) Gene Family Members of *Arabidopsis thaliana*: Tissue- and Organ-Specific Promoter Activities and In Vivo Heteromerization. *J. Exp. Bot.* **2012,** *63*, 6125–6138.
92. Stintzi, A.; Browse, J. The Arabidopsis Male-Sterile Mutant, opr3, Lacks the 12-Oxophytodienoic Acid Reductase Required for Jasmonate Synthesis. *Proc. Natl. Acad. Sci. USA.* **2000,** *97*, 10625–10630.
93. Koo, A. J.; Chung, H. S.; Kobayashi, Y. Identification of a Peroxisomal Acyl-Activating Enzyme Involved in the Biosynthesis of Jasmonic Acid in Arabidopsis. *J. Biol. Chem.* **2006,** *281*, 33511–33520.
94. Kienow, L.; Schneider, K.; Bartsch, M. Jasmonates Meet Fatty Acids: Functional Analysis of a New Acyl-Coenzyme A Synthetase Family from *Arabidopsis thaliana. J. Exp. Bot.* **2008,** *59*, 403–419.
95. Zhai, Q.; Yan, C.; Li, L.; Xie, D.; Li, C. Jasmonates. In: *Hormone metabolism and Signaling in Plants*; 2017; pp 243–272. doi.org/10.1016/B978-0-12-811562-6.00007-4
96. Li, C.; Schilmiller, A. L.; Liu, G. Role of Beta-Oxidation in Jasmonate Biosynthesis and Systemic Wound Signaling in Tomato. *Plant Cell* **2005,** *17*, 971–986.
97. Schilmiller, A. L.; Koo, A. J.; Howe, G. A. Functional Diversification of Acyl-Coenzyme A Oxidases in Jasmonic Acid Biosynthesis and Action. *Plant Physiol.* **2007,** *143*, 812–824.
98. Delker, C.; Zolman, B. K.; Miersch, O. Jasmonate Biosynthesis in *Arabidopsis thaliana* Requires Peroxisomal Beta-Oxidation Enzymes—Additional Proof by Properties of pex6 and aim1. *Phytochemistry* **2007,** *68*, 1642–1650.
99. Suza, W. P.; Staswick, P. E. The Role of JAR1 in Jasmonoyl-l-Isoleucine Production During Arabidopsis Wound Response. *Planta* **2008,** *227*, 1221–1232.
100. Staswick, P. E.; Tiryaki, I. The Oxylipin Signal Jasmonic Acid Is Activated by an Enzyme That Conjugates It to Isoleucine in Arabidopsis. *Plant Cell* **2004,** *16*, 2117–2127.
101. Swiatek, A.; Van Dongen, W.; Esmans, E. L. Metabolic Fate of Jasmonates in Tobacco Bright Yellow-2 Cells. *Plant Physiol.* **2004,** *135*, 161–172.
102. Gidda, S. K.; Miersch, O.; Levitin, A. Biochemical and Molecular Characterization of a Hydroxyjasmonate Sulfotransferase from *Arabidopsis thaliana. J. Biol. Chem.* **2003,** *278*, 17895–17900.
103. Koch, T.; Bandemer, K.; Boland, W. Biosynthesis of Cis-Jasmone: A Pathway for the Inactivation and the Disposal of the Plant Stress Hormone Jasmonic Acid to the Gas Phase? *HelvChim Acta* **1997,** *80*, 838–850.

104. Sun, J. Q.; Jiang, H. L.; Li, C. Y. Systemin/Jasmonate-Mediated Systemic Défense Signaling in Tomato. *Mol. Plant* **2011,** *4* (4), 607–615.
105. Feys, B. F.; Benedetti, C. E.; Penfold, C. N. Arabidopsis Mutants Selected for Resistance to the Phytotoxin Coronatine are Male-Sterile, Insensitive to Methyl Jasmonate, and Resistant to a Bacterial Pathogen. *Plant Cell* **1994,** *6*, 751–759.
106. Mewis, I.; Appel, H. M.; Hom, A. Major Signaling Pathways Modulate Arabidopsis Glucosinolate Accumulation and Response to Both Phloem-Feeding and Chewing Insects. *Plant Physiol.* **2005,** *138*, 1149–1162.
107. Song, S. S.; Qi, T. C.; Huang, H. The Jasmonate-ZIM Domain Proteins Interact with the R2R3-MYB Transcription Factors MYB21 and MYB24 to Affect Jasmonate-Regulated Stamen Development in Arabidopsis. *Plant Cell* **2011,** *23*, 1000–1013.
108. Xie, D. X.; Feys, B. F.; James, S.; Nieto-Rostro, M.; Turner, J. G. COI1: An Arabidopsis Gene Required for Jasmonate-Regulated Defence and Fertility. *Science* **1998,** *280*, 1091–1094.
109. Devoto, A.; Nieto-Rostro, M.; Xie, D. X. COI1 Links Jasmonate Signaling and Fertility to the SCF Ubiquitin-Ligase Complex in Arabidopsis. *Plant J.* **2002,** *32*, 457–466.
110. Ren, C.; Pan, J.; Peng, W. Point Mutations in Arabidopsis Cullin1 Reveal Its Essential Role in Jasmonate Response. *Plant J.* **2005,** *42*, 514–524.
111. Lee, S.; Sakuraba, Y.; Lee, T. Mutation of *Oryza Sativa* Coronatine Insensitive 1b (OsCOI1b) Delays Leaf Senescence. *J. Integr. Plant Biol.* **2015,** *57*, 562–576.
112. Liao, Y.; Wei, J.; Xu, Y. Cloning, Expression and Characterization of COI1 Gene (AsCOI1) from *Aquilaria sinensis* (Lour.) *Gilg Acta Pharm. Sin. B* **2015,** *5*, 473–481.
113. Thines, B.; Katsir, L.; Melotto, M.; Niu, Y.; Mandaokar, A.; Liu, G.; Nomura, K.; He, SY.; Howe, G. A.; Browse, J. JAZ Repressor Proteins Are Targets of the SCF (COI1) Complex During Jasmonate Signaling. *Nature* **2007,** *448*, 661–665.
114. Yan, Y.; Stolz, S.; Chételat, A.; Reymond, P.; Pagni, M.; Dubugnon, L.; Farmer, E. E., A Downstream Mediator in the Growth Repression Limb of the Jasmonate Pathway. *Plant Cell* **2007,** *19*, 2470–2483.
115. Shyu, C.; Figueroa, P.; DePew, C. L. JAZ8 Lacks a Canonical Degron and Has an EAR Motif That Mediates Transcriptional Repression of Jasmonate Responses in Arabidopsis. *Plant Cell* **2012,** *24*, 536–550.
116. Thireault, C.; Shyu, C.; Yoshida, Y. Repression of Jasmonate signaling by a Non TIFY JAZ Protein in Arabidopsis. *Plant J.* **2015,** *82*, 669–679.
117. Pauwels, L.; Goossens, A. The JAZ Proteins: A Crucial Interface in the Jasmonate Signaling Cascade. *Plant Cell* **2011,** *23*, 3089–3100.
118. Sheard, L. B.; Tan, X.; Mao, H. Jasmonate Perception by Inositol- Phosphate-Potentiated COI1–JAZ Co-receptor. *Nature* **2010,** *468*, 400–405.
119. Gimenez-Ibanez, S.; Boter, M.; Ortigosa, A.; García-Casado, G.; Chini, A.; Lewsey, M. G.; Ecker, J. R.; Ntoukakis, V.; Solano, R. JAZ2 Controls Stomata Dynamics During Bacterial Invasion. *New Phytol.* **2017,** *213*, 1378–1392.
120. Chung, H. S.; Cooke, T. F.; Depew, C. L. Alternative Splicing Expands the Repertoire of Dominant JAZ Repressors of Jasmonate Signaling. *Plant J.* **2010,** *63*, 613–622.
121. Katsir, L.; Schilmiller, A. L.; Staswick, P. E.; He, S. Y.; Howe, G. A. COI1 Is a Critical Component of a Receptor for Jasmonate and the Bacterial Virulence Factor Coronatine. *Proc. Natl. Acad. Sci. USA* **2008,** *105*, 7100–7105.

122. Yan, J.; Zhan, C.; G, M. The Arabidopsis Coronatine Insensitive1 Protein Is a Jasmonate Receptor. *The Plant Cell* **2009,** *21*, 2220–2236.
123. Dombrecht, B.; Xue, G. P.; Sprague, S. J. MYC2 Differentially Modulates Diverse Jasmonate-Dependent Functions in Arabidopsis. *Plant Cell* **2007,** *19*, 2225–2245.
124. Fernández-Calvo, P.; Chini, A.; Fernandez-Barbero, G. The Arabidopsis bHLH Transcription Factors MYC3 and MYC4 Are Targets of JAZ Repressors and Act Additively with MYC2 in the Activation of Jasmonate Responses. *Plant Cell* **2011,** *23*, 701–715.
125. Boter, M.; Ruiz-Rivero, O.; Abdeen, A. Conserved MYC Transcription Factors Play a Key Role in Jasmonate Signaling Both in Tomato and Arabidopsis. *Gene Dev.* **2004,** *18*, 1577–1591.
126. Lorenzo, O.; Chico, J. M.; Sánchez-Serrano, J. J.; Solano, R. Jasmonate-Insensitive1 Encodes a MYC Transcription Factor Essential to Discriminate Between Different Jasmonate-Regulated Defence Responses in Arabidopsis. *Plant Cell* **2004,** *16*, 1938–1950.
127. Zhai, Q.; Zhang, X.; Wu, F.; Feng, H.; Deng, L.; Xu, L.; Zhang, M.; Wang, Q.; Li, C. Transcriptional Mechanism of Jasmonate Receptor COI1-Mediated Delay of Flowering Time in Arabidopsis. *Plant Cell* **2015,** *27*, 2814–2828.
128. Grunewald, W. Vanholme, B.; Pauwels, L. Expression of the Arabidopsis Jasmonate Signaling Repressor JAZ1/TIFY10A Is Stimulated by Auxin. *EMBO Rep.* **2009,** *10*, 923–928.
129. Chen, Q.; Sun, J. Q.; Zhai, Q. Z. The Basic Helix-Loop-Helix Transcription Factor MYC2 Directly Represses PLETHORA Expression During Jasmonate-Mediated Modulation of the Root Stem Cell Niche in Arabidopsis. *Plant Cell* **2011,** *23*, 3335–3352.
130. Sugio, A.; Maclean, A. M.; Kingdom, H. N.; Grieve, V. M.; Manimekalai, R.; Hogenhout, S. A. Diverse Targets of Phytoplasma Effectors: From Plant Development to Defense Against Insects. *Annu. Rev. Phytopathol.* **2011,** *49*, 175–195.
131. Patkar, R. N.; Benke, P. I.; Qu, Z.; Chen, Y. Y. C.; Yang, F.; Swarup, S.; Naqvi, N. I. A Fungal Monooxygenase-Derived Jasmonate Attenuates Host Innate Immunity. *Nat. Chem. Biol.* **2015,** *11*, 733–740. DOI: 10.1038/nchembio.1885.
132. Cole, S. J.; Yoon, A. J.; Faull, K. F.; Diener, A. C. Host Perception of Jasmonates Promotes Infection by *Fusarium oxysporum* Formaespeciales That Produce Isoleucine- and Leucine-Conjugated Jasmonates. *Mol. Plant Pathol.* **2014,** *15*, 589–600.
133. Lozano-Dura´n, R.; Rosas-Dı´az, T.; Gusmaroli, G.; Luna, A. P.; Taconnat, L.; Deng, X. W.; Bejarano, E. R. Geminiviruses Subvert Ubiquitination by Altering CSN-Mediated Derubylation of SCF E3 Ligase Complexes and Inhibit Jasmonate Signaling in *Arabidopsis thaliana. Plant Cell* **2011,** *23*, 1014–1032.
134. Plett, J. M.; Daguerre, Y.; Wittulsky, S.; Martin, F. Effector MiSSP7 of the Mutualistic Fungus *Laccariabicolor* Stabilizes the Populus JAZ6 Protein and Represses Jasmonic Acid (JA) Responsive Genes. *PNAS* **2014,** *111* (22), 8299–8304. https://doi.org/10.1073/pnas.132267111
135. Zhang, L.; Yao, J.; Withers, J. Host Target Modification as a Strategy to Counter Pathogen Hijacking of the Jasmonate Hormone Rcceptor. *Proc. Natl. Acad. Sci. USA* **2015,** *112*, 14354–14359.

136. Zhou, C.; Wu, Y.; Yang, Y.; Du, M.; Zhang, X.; Guo, Y.; Li, C.; Zhou, J. M. An Arabidopsis Plasma Membrane Proton ATPase Modulates JA Signaling and Is Exploited by the *Pseudomonas syringae* Effector Protein AvrB for Stomatal Invasion. *Plant Cell* **2015,** 1–10. DOI 10.1105/tpc.15.00466.
137. Jiang, S.; Yao, J.; Ma, K. W.; Zhou, H.; Song, J.; He, S. Y.; Ma, W. Bacterial Effector Activates Jasmonate Signaling by Directly Targeting JAZ Transcriptional Repressors. *PLoS Pathog.* **2013,** *9*, e1003715.
138. Gimenez-Ibanez, S.; Boter, M.; Fernández-Barbero, G.; Chini, A.; Rathjen, J. P.; Solano, R. The Bacterial Effector HopX1 Targets JAZ Transcriptional Repressors to Activate Jasmonate Signaling and Promote Infection in Arabidopsis. *PLoS Biol.* **2014,** *12*, e1001792.
139. Li, R.; Weldegergis, B. T.; Li, J. Virulence Factors of Gemini Virus Interact with MYC2 to Subvert Plant Resistance and Promote Vector Performance. *Plant Cell* **2014,** *26*, 4991–5008.
140. Guo, Q.; Yoshida, Y.; Major, I. T.; Wang, K.; Sugimoto, K.; Kapali, G.; Havko, N. E.; Benning, C.; Howe, G. A. JAZ Repressors of Metabolic Defence Promote Growth and Reproductive Fitness in Arabidopsis. *Proc. Natl. Acad. Sci. USA* **2018,** *115*, E10768–E10777.
141. Monte, I.; Ishida, S.; Zamarreño, A. M. Ligand–Receptor Co-Evolution Shaped the Jasmonate Pathway in Land Plants. *Nat. Chem. Biol.* **2018,** *14*, 480–488.
142. Monte, I.; Franco-Zorrilla, J. M.; García-Casado, G.; Zamarreño, A. M.; García-Mina, J. M.; Nishihama, R.; Kohchi, T.; Solano, R. A Single JAZ Repressor Controls the Jasmonate Pathway in *Marchantia polymorpha. Mol. Plant* **2019,** *12*, 185–198.
143. Howe, G. A.; Major, I. T.; Koo, A. J. Modularity in Jasmonate Signaling for Multi-Stress Resilience. *Ann. Rev. Pl. Biol.* **2018,** *69*, 387–415.
144. Song, S.; Huang, H.; Wang, J.; Liu, B.; Qi, T.; Xie, D. MYC5 Is Involved in Jasmonate-Regulated Plant Growth, Leaf Senescence and Defence Responses. *Plant Cell Physiol.* **2017,** *58*, 1752–1763.
145. Schweizer, F.; Fernández-Calvo, P.; Zander, M.; Diez-Diaz, M.; Fonseca, S.; Glauser, G.; Lewsey, M. G.; Ecker, J. R.; Solano, R.; Reymond, P. Arabidopsis Basic Helix–Loop–Helix Transcription Factors MYC2, MYC3, and MYC4 Regulate Glucosinolate Biosynthesis, Insect Performance, and Feeding Behaviour. *Plant Cell* **2013,** *25*, 3117–3132.
146. Fonseca, S.; Fernández-Calvo, P.; Fernández, G. M. bHLH003, bHLH013 and bHLH017 are New Targets of JAZ Repressors Negatively Regulating JA Responses. *PLoS One* **2014,** *9*, e86182.
147. Nakata, M.; Mitsuda, N.; Herde, M.; Koo, A. J.; Moreno, J. E.; Suzuki, K.; Howe, G. A.; Ohme-Takagi, M. A bHLH-Type Transcription Factor, ABA- Inducible BHLH-Type Transcription Factor/JA-Associated MYC2-Like1, Acts as a Repressor to Negatively Regulate Jasmonate Signaling in Arabidopsis. *Plant Cell* **2013,** *25*, 1641–1656.
148. Qi, T.; Wang, J.; Huang, H.; Liu, B.; Gao, H.; Liu, Y.; Song, S.; Xie, D. Regulation of Jasmonate-Induced Leaf Senescence by Antagonism Between bHLH Subgroup IIIe and IIId Factors in Arabidopsis. *Plant Cell* **2015,** *27*, 1634–1649.
149. Ishida, T.; Kurata, T.; Okada, K.; Wada, T. A Genetic Regulatory Network in the Development of Trichomes and Root Hairs. *Ann. Rev. Pl. Biol.* **2008,** *59*, 365–386.

150. Qi, T.; Huang, H.; Wu, D.; Yan, J.; Qi, Y.; Song, S.; Xie, D. Arabidopsis DELLA and JAZ Proteins Bind the WD-Repeat/bHLH/MYB Complex to Modulate Gibberellin and Jasmonate Signaling Synergy. *Plant Cell* **2014,** *26*, 1118–1133.
151. Qi, T.; Song, S.; Ren, Q.; Wu, D.; Huang, H.; Chen, Y.; Fan, M.; Peng, W.; Ren, C.; Xie, D. The Jasmonate-ZIM-Domain Proteins Interact with the WD-Repeat/bHLH/MYB Complexes to Regulate Jasmonate-Mediated Anthocyanin Accumulation and Trichome Initiation in *Arabidopsis thaliana. Plant Cell* **2011,** *23*, 1795–1814.
152. Bhattacharya, A.; Effect of High-Temperature Stress on the Metabolism of Plant Growth Regulators; In: *Effect of High Temperature on Crop Productivity and Metabolism of Macro Molecules*; Academic Press, 2019; pp 485–591.
153. Allu, A. D.; Brotman, Y.; Xue, G-P.; Balazadeh, S. Transcription Factor ANAC032 Modulates JA/SA Signaling in Response to *Pseudomonas syringae* Infection. *EMBO Rep.* 2016, *17* (11), 1578–1589.
154. Gimenez-Ibanez, S.; Solano, R. Nuclear Jasmonate and Salicylate Signaling and Crosstalk in Defense Against Pathogens. *Front. Plant Sci.* **2013,** *4*, 72.
155. Gimenez-Ibanez, S.; Chini, A.; Solano, R. How Microbes Twist Jasmonate Signaling Around Their Little Fingers. *Plants* **2016,** *5* (1), 9. DOI: 10.3390/plants5010009.

Index

L

M

O

P